On the History
of Statistics
and Probability

STATISTICS

Textbooks and Monographs

A SERIES EDITED BY

D. B. OWEN, Coordinating Editor
Department of Statistics
Southern Methodist University
Dallas, Texas

PAUL D. MINTON
Virginia Commonwealth University
Richmond, Virginia

JOHN W. PRATT
Harvard University
Boston, Massachusetts

OTHER VOLUMES IN PREPARATION

William G. Cochran

H. O. Hartley

Jerzy Neyman

On the History of Statistics and Probability

Proceedings of a Symposium on the American Mathematical Heritage
to Celebrate the Bicentennial of the United States of America, held at
Southern Methodist University, May 27-29, 1974

Featuring articles by W. G. COCHRAN,
H. O. HARTLEY, and JERZY NEYMAN

edited by D. B. OWEN

Department of Statistics
Southern Methodist University
Dallas, Texas

MARCEL DEKKER, INC.   New York and Basel

MARCEL DEKKER, INC.
270 Madison Avenue, New York, New York 10016

LIBRARY OF CONGRESS CATALOG CARD NUMBER: 75-32473
ISBN: 0-8247-6390-4
Current printing (last digit):
10 9 8 7 6 5 4 3 2 1

PRINTED IN THE UNITED STATES OF AMERICA

In reading the papers in this volume, it becomes clear that the major growth in statistical science has been in the past 50 years, and although a goodly amount existed in the field of probability many years earlier, probability has also seen a rapid development in the last 50 years. It therefore seemed especially appropriate that a conference on the history of statistics and probability be held while many of the developers of these fields were alive. However, in organizing this symposium, it immediately became clear that it would be impossible to invite all of the people who should be included in this undertaking--people who were developers of the field and were widely recognized for their contributions. Hence, three principal speakers were chosen--Jerzy Neyman from the West Coast; H. O. Hartley from the Central States; and W. G. Cochran from the East Coast. Although all three were born overseas, their contributions to America were clearly definable and outstanding. Each principal speaker was asked to help choose some additional speakers, which they all did.

This process gave a more coherent program, but again, many people who should have been included were not and we indicate this fact by entitling this volume, *On the History of Statistics and Probability*, with the emphasis on "On" and making no pretense that it is complete.

Statisticians have paid little attention to the history of their subject, and only a few anecdotes exist on the early developments. This volume should provide a few more stories for use in some pioneering class in the history of statistics and probability to be given in the future.

We have among us some very exceptional individuals—not only highly competent technically, but also clearly great people to know as human beings. It is hoped that the truly great nature of these individuals will become evident through these papers in contrast to the papers they usually publish, which are highly technical and devoid of anecdotal content.

D. B. Owen

CHARLES E. ANTONIAK
    University of California, Berkeley

DAVID R. BRILLINGER
    University of California, Berkeley

WEI-CHING CHANG[1]
    University of Toronto

HERMAN CHERNOFF
    Stanford University
    Massachusetts Institute of Technology

WILLIAM G. COCHRAN
    Harvard University

DONALD A. DARLING
    University of California, Irvine

KJELL A. DOKSUM
    University of California, Berkeley[2]

J. L. DOOB
    University of Illinois, Urbana-Champaign

BRADLEY EFRON
    Stanford University

THOMAS S. FERGUSON
    University of California, Los Angeles

MORRIS H. HANSEN
    Westat, Incorporated

BOYD HARSHBARGER
    Virginia Polytechnic Institute and State University

---

[1] Present affiliation: *Travel Alberta.*
[2] Present affiliation: *University of Oslo.*

H. O. HARTLEY
  Texas A & M University

PETER W. M. JOHN
  University of Texas, Austin

OSCAR KEMPTHORNE
  Iowa State University

JACK C. KIEFER
  Cornell University

WILLIAM G. MADOW[1]
  Stanford University

NATHAN MANTEL
  George Washington University

DAVID S. MOORE
  Purdue University

JERZY NEYMAN
  University of California, Berkeley

PREM S. PURI[2]
  Purdue University

FRED D. RIGBY
  Texas Tech University

ELIZABETH L. SCOTT
  University of California, Berkeley

ROBERT H. TRAXLER[3]
  University of California, Berkeley

---

[1]Present address: *New Hampshire Avenue, Washington, D. C.*

[2]Present affiliation: *University of California, Berkeley.*

[3]Present affiliation: *New Mexico State University.*

Jasper Adams, *Stephen F. Austin State University*
Roberto Alanis, *Instituto Tecnologico y de Estudios Superiores de Monterrey*
Henry Ander, *Texas A & M University*
Charles E. Antoniak, *University of California, Berkeley*
K. K. Arora, *E. R. Squibb*
Ersen Arseven, *Texas A & M University*
Eddy Baird, *Southern Methodist University*
J. P. Basu, *University of Texas at Dallas*
James Beckett, *Southern Methodist University*
U. Narayan Bhat, *Southern Methodist University*
Richard P. Bland, *Southern Methodist University*
John Boddie, *Southern Methodist University*
Tom Bohannan, *Tarleton State College*
Robert Bodwell, *Southern Methodist University*
Don Book, *Texas A & M University*
Thomas Boullion, *Texas Tech University*
K. O. Bowman, *Union Carbide Corporation, Nuclear Division*
Bernie C. Boyle, *E-Systems Incorporated*
Rich Burdick, *Texas A & M University*
T. L. Bratcher, *University of Southwestern Louisiana*
Richard H. Browne, *University of Texas Health Science Center at Dallas*
Manuel Cardenas, *Texas A & M University*
Ranajit Chakraborty, *University of Texas Center for Demographics & Population Genetics*
Satish Chandra, *Tuskegee Institute*
Wei-Ching Chang, *University of Toronto*
Herman Chernoff, *Stanford University*
Danny Childers, *Texas A & M University*
W. D. Clark, *Stephen F. Austin State University*
William A. Coberly, *NASA/Johnson Space Center*
Tim Coburn, *Oklahoma State University*
William G. Cochran, *Harvard University*
Richard Cooper, *Trinity University*
Alfred E. Crofts, Jr., *Pan American University*
Donald A. Darling, *University of California, Irvine*

James M. Davenport, *Texas Tech University*
John H. Davis, *Sun Oil Company*
Cynthia Ann Dolan, *University of Texas, Austin*
J. L. Doob, *University of Illinois, Urbana-Champaign*
Wanzer Drane, *Southern Methodist University*
Benjamin Duran, *Texas Tech University*
Danny Dyer, *University of Texas, Arlington*
Roger Erickstad, *Le Tourneau College*
Jim Evans, *University of Texas, Austin*
John H. Farmer, *National Center for Toxicological Research*
Thomas S. Ferguson, *University of California, Los Angeles*
Morris D. Finkner, *New Mexico State University*
William H. Frawley, *Southland Corporation*
R. J. Freund, *Texas A & M University*
Kay Fromme, *Stephen F. Austin State University*
C. E. Gates, *Texas A & M University*
Jim Gentle, *Texas A & M University*
Larry Gianchetta, *Texas A & M University*
Jean D. Gibbons, *University of Alabama*
Marilyn Gilchrist, *Mountain View, Dallas County Community College*
Stan Gordon, *Anderson-Clayton Foods*
H. L. Gray, *Southern Methodist University*
Richard F. Gunst, *Southern Methodist University*
Shanti S. Gupta, *Purdue University*
Allan L. Gutjahr, *Soccoro, New Mexico*
W. T. Guy, *University of Texas, Austin*
Roy Haas, *Southern Methodist University*
Carl E. Hall, *University of Texas, El Paso*
Gaineford J. Hall, Jr., *University of Texas, Austin*
Morris H. Hansen, *Westat, Incorporated*
Ronald B. Harrist, *University of Texas, Houston*
Boyd Harshbarger, *Virginia Polytechnic Institute and State University*
H. O. Hartley, *Texas A & M University*
Mort Hawkins, *University of Texas, Houston*
Erwin Hearne, III, *Southern Methodist University*
Tom Herring, *Texas Tech University*
James Hess, *Southern Methodist University*
Harold Hietala, *Southern Methodist University*
Glen Houston, *Southern Methodist University*
Yueh-ling Hsiao, *Southern Methodist University*
H. H. Hunt, *Texas A & M University*
Bill Iwig, *Texas A & M University*
Omer Jenkins, *Texas A & M University*
Peter W. M. John, *University of Texas, Austin*
Thomas Johnson, *Southern Methodist University*
Marvin J Karson, *University of Alabama*
Anand Katiyar, *McNeese State University*
Gary Kelley, *Southern Methodist University*
Judy Kelley, *Southern Methodist University*
Oscar Kempthorne, *Iowa State University*
Robert L. Kieschnick, *Austin, Texas*

Patty Knowles, *Fort Worth ISD*
Walter Knowles, *Alcon Laboratories*
Samuel R. Knox, *Millsaps College*
Ralph Kodell, *Texas A & M University*
Ignacy Kotlarski, *Oklahoma State University*
A. M. Kshirsagar, *Texas A & M University*
Margaret Land, *Oklahoma State University*
Rachael LaRoe, *Mary Hardin-Baylor College*
Bob Leslie, *U.S. Service Automobile Association*
William L. Lester, *Tuskegee Institute*
Nathan Lewis, *Southern Methodist University*
Truman Lewis, *Texas Tech University*
Loretta Li, *Bishop College*
William D. Love, *Baylor College of Dentistry*
Dan Lurie, *Medical University of South Carolina*
Daniel B. McCallum, *University of Arkansas*
Ennis D. McCune, *Stephen F. Austin State University*
John A. McGuire, *Auburn University*
Donald McIntire, *Southern Methodist University*
Ronald McPherson, *Southern Methodist University*
Archer McWhorter, Jr., *University of Houston*
Edward Mansfield, *Southern Methodist University*
Nathan Mantel, *George Washington University*
D. P. Martin, *University of Texas at Arlington*
Robert L. Mason, *Medical University of South Carolina*
J. C. Mathew, *Iowa State University*
James Matis, *Texas A & M University*
John Michael, *Southern Methodist University*
David Minaldi, *Anderson-Clayton Foods*
Billy J. Moore, *City of Dallas Health Department*
David S. Moore, *Purdue University*
Marion Moore, *University of Texas at Arlington*
Michael C. Moore, *Danray Incorporated*
Frederick W. Morgan, *Southern Methodist University*
C. B. Murray, *University of Houston*
C. J. Nelson, *National Center for Toxicological Research*
Jerzy Neyman, *University of California, Berkeley*
Patrick L. Odell, *University of Texas, Dallas*
Michael O'Hagan, *Texas Instruments Incorporated*
Melchor Ortiz, *Texas A & M University*
D. B. Owen, *Southern Methodist University*
Elinor S. Pape, *University of Texas, Arlington*
Frances Pascale, *Albertus Magnus College*
Mary Ann Pate, *University of Wisconsin and Police Foundation*
Jagdish K. Patel, *Southern Methodist University*
Shirlene Pearson, *Southern Methodist University*
Charles Perry, *Seguin Lutheran College*
Ralph Price, *Southern Methodist University*
Howard Prier, *E-Systems, Incorporated*
Prem S. Puri, *Purdue University*
Robert Raymond, *University of Minnesota, Morris*

Campbell B. Read, *Southern Methodist University*
Joan S. Reisch, *Southern Methodist University and University of Texas Health Science Center at Dallas*
Fred D. Rigby, *Texas Tech University*
Larry Ringer, *Texas A & M University*
Paula Roberson, *Southern Methodist University*
Carl T. Russell, *University of Texas, Austin*
Johnny Russell, *Texas A & M University*
William R. Schucany, *Southern Methodist University*
Elizabeth L. Scott, *University of California, Berkeley*
Robert L. Seilken, Jr., *Texas A & M University*
Edward Seymour, *Southern Methodist University*
Davis W. Smith, *Louisiana State University, Baton Rouge*
Mark Smith, *Texas Instruments Incorporated*
Julianne Souchek, *University of Texas, Austin*
Randy Spoeri, *Texas A & M University*
Dan Stone, *Danray Corporation*
David Stones, *University of Texas, Dallas*
George P. Steck, *Sandia Laboratories*
John L. Stover, *University of Texas, Austin*
Jerrell T. Stracener, *Aerospace Corporation*
David Stevenson, *Chevrolet Marketing Systems Division, General Motors*
Monty J. Strauss, *Texas Tech University*
Shiaw Y. Su, *Sun Oil Company*
Philip A. Sugar, *Sun Oil Company*
Dalton Tarwater, *Texas Tech University*
An-ling Terng, *Southern Methodist University*
Eugene Tidmore, *Baylor University*
Robert H. Traxler, *University of California, Berkeley*
Danny W. Turner, *Baylor University*
Kala Uprety, *Southern Methodist University*
Larry Vadala, *University of Texas, Arlington*
John W. Van Ness, *University of Texas, Dallas*
William Waide, *Oklahoma State University*
Elbert Walker, *New Mexico State University*
William Wallace, *General Dynamics*
K. T. Wallenius, *Office of Naval Research*
John T. Webster, *Southern Methodist University*
Dan Weiser, *Mobil Research and Development Corporation*
R. O. Wells, *Rice University*
Alan Wheeler, *Southern Methodist University*
John W. White, *Southern Methodist University*
John T. White, *Texas Tech University*
Wayne A. Woodward, *Southern Methodist University*
Thomas Woteki, *University of Texas, San Antonio*
Lynn Zeigler, *University of Texas, Dallas*

On May 27-29, 1974, the second in a series of conferences on the
American Mathematical Heritage was held at SMU's Bob Hope Theater as
an early event in the Bicentennial Celebration of the United States.
This second conference dealt with the History of Statistics and
Probability and was attended by over 180 participants including
visitors from Mexico and Canada.

The principal sponsors of the conference were the Office of Naval
Research and the Mobil Foundation, Inc. The Texas Bicentennial Program, Inc., arranged for directors of future conferences in the Mathematical Heritage Series to attend.

The entire conference was video-taped in black and white. At the
time of this writing, no plans have been set to use these video-tapes.
There were four round table discussion panels with three of them before
an audience. The fourth panel was held in a closed room with the
audience watching on closed circuit television. Even though this
volume does not contain a copy of these discussions, it seems appropriate to record the participants.

Panels Held in the Theater

Panel A-1

Chairman:	Herman Chernoff
Panelists:	W. G. Cochran
	M. H. Hansen
	O. Kempthorne
	P. W. M. John
	R. Traxler
Topic:	Submitted Questions

Panel A-2

Chairman: W. R. Schucany
Panelists: H. O. Hartley
            Jerzy Neyman
            Robert Traxler
Topic: Solution of Statistical Distribution Problems by
            Monte Carlo

Panel A-3

Chairman: Alan Wheeler
Panelists: P. W. M. John
            Nathan Mantel
            H. O. Hartley
Topic: Optimization in Statistics or Impact of Mathematical
            Programming

Panel Held in Closed Room

Panel B

Chairman: Charles Antoniak
Panelists: Oscar Kempthorne
            E. L. Scott
            W. G. Cochran
            J. Neyman
            H. O. Hartley
Topic: Submitted Questions

The papers given in this volume appear in the same order as on
the program of the symposium, except that additional papers by Traxler,
Doksum, and Brillinger are included.

1.

EARLY DEVELOPMENT
OF TECHNIQUES IN
COMPARATIVE EXPERIMENTATION

WILLIAM G. COCHRAN was born July 15, 1909, in Rutherglen, Scotland. He is now a U.S. citizen. He has received M.A. degrees from Glasgow University in 1931 and Cambridge University in 1938, an honorary A.M. degree from Harvard University in 1957, and honorary L.L.D. degrees from Glasgow University in 1970 and Johns Hopkins University in 1975. He is a Fellow of the American Academy of Arts and Sciences (elected in 1971), and a member of the National Academy of Sciences (elected in 1974). He was Statistician, Rothamsted Experimental Station, England (1934-1939), Professor of Mathematical Statistics, Iowa State College (1939-1946), Associate Director, Institute of Statistics, University of North Carolina (1946-1948), Professor of Biostatistics, School of Hygiene and Public Health, Johns Hopkins University (1948-1957), and since 1957 has been Professor of Statistics, Harvard University. He is a Fellow of the American Statistical Association (President, 1953; Journal Editor, 1945-1949), of the Institute of Mathematical Statistics (President, 1946), of the American Association for the Advancement of Science (Vice-President, 1966), and an Honorary Fellow of the Royal Statistical Society. He is a member of the Biometric Society (President, 1954-1955) and of the International Statistical Institute (President, 1967-1971). He held a Guggenheim Fellowship from 1964 to 1965, was the S. S. Wilks Memorial Medal winner for the American Statistical Association in 1967, and received the Outstanding Statistician award of the Chicago Chapter of the American Statistical Association in 1974. He is author and coauthor of five books and over 100 research papers.

EARLY DEVELOPMENT
OF TECHNIQUES IN
COMPARATIVE EXPERIMENTATION

William G. Cochran

Department of Statistics
Harvard University
Cambridge, Massachusetts

The historical development of statistical ideas and techniques as regards their practical application has been relatively little studied. It seems likely that in time we will come across centuries-old examples of individual comparative experiments or discussions of the principles to be followed in comparative experimentation. Thus, Stigler (1973) has noted that in the eleventh century the Arabic doctor Avicenna laid down seven rules for medical experimentation on human subjects, including a recommendation for replication and the use of controls and a warning of the dangers of confounding variables. In 1627, Francis Bacon published an account of the effects of steeping wheat seeds for 12 hours in nine different concoctions (e.g., water mixed with cow dung, urine, three different wines), with unsteeped seed as controls, on the speed of germination and the heartiness of growth. He noted that these comparisons (made in single replication) were important to profitability, as most of the steepings were cheap. The results suggested the inadvisability of wasting wine, since the claret treatment proved inferior to the control, whereas wheat steeped in the other two wines did not germinate at all. The winner was seed steeped in urine.

As far as I know, however, the study of the practical conduct of comparative experiments as we know it today was begun and pursued during the eighteenth and nineteenth centuries in agricultural field experimentation. I shall start with the great English agronomist, Arthur Young, who made comparative field experimentation his main occupation for a number of years.

ARTHUR YOUNG'S CONTRIBUTIONS

In 1763, at the age of 22, Arthur Young inherited a farm in England from his father. He decided to devote himself to experiments designed to discover the most profitable methods of farming. In succeeding years he carried out on his farm a large number of field experiments on the principal crops in his region, publishing their results and his conclusions in a three-volume book *A Course of Experimental Agriculture* (1771).

What were Young's ideas on the conduct of field experiments? In many ways, they were surprisingly modern. First, he stressed that experiments must be *comparative*. When comparing a new method with a standard method, *both* must be included in the experiment, even if the farmer already knows a great deal about the general performance of the standard method in past years or on different fields. Young's reason was that the performance of any method on specific fields in a specific year was sufficiently affected by soil fertility, drainage, and climate that only a comparison on the same fields and weather could be trusted. An example of his method of conducting field experiments is his early comparison of sowing wheat in rows by the drill (the new husbandry) compared with broadcasting the seed (the old husbandry). A single comparison or trial was conducted on large plots--an acre or a half-acre in a field split into halves--one drilled, one broadcast. Of the two halves, Young (1771) writes: "the soil exactly the same; the time of culture, and in a word every circumstance equal in both." Not surprisingly, he said nothing about how he decided which half to drill, which to broadcast.

Second, Young had the idea that because of this variability caused
by soil fertility, drainage, insect pests, and other factors, the results of a single trial could not be trusted. The drill-broadcast
comparison involved seven trials (replications) in seven fields, in
which he used his overall impression of the seven sets of results in
drawing conclusions. He went further, noting that his comparative
results might change in a different year with differing weather. For
this reason, he repeated his experiments on different plots in five
successive years, drawing his overall conclusions from a summary table
of the annual mean profit per acre for drilling and broadcasting. It
was not until the present century that this amount of replication became standard practice.

To revert to Young's insistence on the comparative nature of
experiments, he considered bringing in the yields for past crops by
the old method (broadcasting) in his overall summary of the results,
but he rejected this idea. He wrote "nothing of this sort should be
done, even on the same soil, unless the experiments were absolutely
comparative; it may be a matter of amusement or curiosity, but of no
utility, no authority." This issue persists today. In reviewing the
present state of knowledge about the relative merits of two therapies
for hospitalized patients, we may find a few well-controlled experiments and a larger number of doctors' observations on their experiences
with one or the other therapy. Young would seem to suggest that to
consider the latter group is a waste of time.

Young stressed careful measurement. All expenses were recorded
on each half in pounds, shillings, pence, hapennies, and farthings
and at harvest a sample of wheat from each half was sent to market on
the same day to determine the selling price.

Young had no quantitative technique for statistical analysis. But
he had remarkable awareness of two problems that still plague us today
in experimentation and statistical studies generally. One is the
problem of bias in the investigator. Each of Young's books begins with
what we would now call a review of literature, and I was surprised at how

many past field experiments (mostly in single replication) he found
to review in the 1760s. He wrote (1771):

> The many volumes upon agriculture which I have turned over,
> guarded me against a too common delusion, and ever fatal in an
> inquiry after truth; then adopting a favorite notion, and form-
> ing experiments with an eye to confirm it. There is scarcely a
> modern book on agriculture, but carries marks of this unhappy
> vanity in the author, which must render its authority doubtful
> to every sensible reader. The design of perusing such works
> was to find practical and experimental directions in doubtful
> points; and my disappointment gave me a disgust at favorite
> hypotheses.

He went on to give examples in which the authors slanted both the
data they presented and their summary remarks to support their favorite.

I imagine that any statistician who does much consulting on the
planning of experiments has had the same experience as I have, of
experimenters, including competent ones, who begin the consultation
by saying "I want to do an experiement to show that ... ." He knows
the answer. In experiments with human subjects, special protective
devices such as "double-blindness" have had to be adopted in which,
if feasible, neither the subject nor the person measuring the response
knows which treatment is being measured. It has been noted that this
difficulty may be present in what is now called social experimentation--
the attempt to use randomized comparative experiments in the evaluation of the effects of social programs. Some sociologists take the
view that new social programs, undertaken with laudable aims, will
sometimes, perhaps often, be found to have few beneficial effects.
The administrator in charge of the program will not find it easy to
accept and publicize a negative finding about the value of the program
that he has been directing.

Furthermore, Young was well aware of the problem of drawing
conclusions that extend beyond the experimental results to the broader
set of conditions to which we are interested in applying the results.
He stressed the pitfalls in what we now call inferences from sample
to population. He warned that his conclusions could not be trusted
to apply on a different farm, with different soil and farm management
practices. On a more subtle issue, he also noted that his results

would not apply as a guide to long-term agricultural policy. The reason was that drilling made the field easier to weed than did broadcasting. This might then enable wheat--the principal cash crop-- to be grown more frequently on a field, instead of the older method of leaving the field fallow in alternate years, or then growing a coarser but less profitable crop in order to smother out the weeds.

On this question of inferences from the experimental results, a statistician would insist that, ideally, experiments should be conducted under a representative or random sample of the broad set of conditions (the population, as we call it) to which the results are to be applied. But for reasons of expense and feasibility, this is a requirement seldom satisfied in practice even today. It can be done fairly well in very simple experiments--as when I am comparing the closeness of shaving myself (as the population) by an electric shaver or a specific type of blade, or the relative time it takes me to drive home from the office by routes A and B. But in more complex experiments, it is quite hard, in teaching, to find examples in which the requirement was deliberately satisfied. Often, for example, the effects of different procedures on the behavior of young adults are compared with the behavior of the students who happen to be around the psychology department of University Y at a specific time. As Student (1926) put it, "in some cases it is only by courtesy that experiments can be considered to be a random sample of any population." In clinical trials on ill patients, where it is obviously desirable that superior therapies be adopted in medical practice, we may not be able to do much better than describe the relevant characteristics of the patients in the experiment as a guide to the kind of population to which we judge the experimental results to be applicable, with supplementary analyses that may guide speculation about the relevance of the results to populations of a different type.

THE NINETEENTH CENTURY

If I may digress from my main subject for a moment, this issue of statistical inference from the results was faced again ca. 1830, when a proposal was made to form a statistics section of the British

Association for the Advancement of Science. The Association quite
properly appointed a committee, under the distinguished chairmanship
of Thomas Malthus, and asked it to report on the question: Is
statistics a branch of science? The committee readily agreed that
insofar as statistics dealt with the collection and orderly tabulation
of data--that was science. But on the question "Is the statistical
interpretation of the results scientifically respectable?," a violent
split arose into pros and cons. The cons won, and they won again a
few years later in 1834 when the Statistical Society of London, later
to become the Royal Statistical Society, was formed. Their victory
was symbolized in the emblem chosen by the Society. This was a fat,
neatly bound sheaf of healthy wheat--presumably representing the
abundant data collected, and well-tabulated. On the binding ribbon
was the Society's motto--the Latin words, *Aliis exterendum,* which
literally means "Let *others* thrash it out." It seems strange that
statisticians in England should have begun their organization by
timidly proclaiming to the world what they would *not* do. As a Scot
who has lived and worked among the English, this does not sound like
the English to me.

I do not know the full reason for this attitude. A contributing
factor may have been that most of the senior statisticians of the time
were heads of official statistical agencies, whose tasks were confined
to collection and tabulation of data. Interpreters of data bearing on
social, economic, or political matters often disagreed violently as to
the conclusions flowing from the data, a practice that they have
continued to this day. By 1840, the Society was already beginning
to strain against the limitation, and its meetings have always been
noted for their vigorous discussions of all aspects of statistics as
we regard it.

Returning to experimentation, an illustration of the ideas that
were current in the nineteenth century was James Johnston's book
Experimental Agriculture (1849), which was devoted to advice on the
practical conduct of field experiments regarded as a scientific problem.
Among points made by Johnston were the following:

1. He stressed the importance of doing experiments and doing them well. A badly made experiment was not merely time and money wasted, but led to the adoption of incorrect results into standard books, to loss of money in practice by the erroneous advice it gave, and to the neglect of further researches. (I have heard this point made recently with regard to medical experiments on seriously ill patients, where there is often a question for the doctor if it is ethical to conduct an experiment, but from the broader view-point a question of whether it is ethical *not* to conduct experiments.)

2. He had observed that plots near one another tended to give similar yields. Hence, he recommended that repetitions of the same treatment be scattered, with the consequence that *different* treatments were placed on *neighboring* plots. A common rule in helping to achieve this was to place repetitions of the same treatment in relation to one another by the Knight's move in chess.

3. He hinted, though without elaboration, at the value of factorial experimentation, by insisting that with two fertilizers a and b, the four treatments--none, a, b, and ab-- should all be compared.

4. He realized the vital need for a quantitative theory of variation, writing "I have elsewhere drawn attention to the importance of this question--'when are results to be considered as identical?' ... As yet we do not possess any such system of mean results, though few things would at present do more to clear up our ideas as to the precise influence of this or that substance on the growth of plants."

As Johnston did not know, considerable progress on the mathematical side toward a quantitative theory of variation had already been made by the early nineteenth century by workers such as De Moivre, Gauss, and Laplace. Results available by about 1820 were the concept of the standard deviation or standard error as a measure of the amount of variation in a population. Then there were the theory of least squares, the distribution of the means of samples from a normal distribution, and some forms of the Central Limit Theorem, with applications in particular to errors of observation in astronomy. Formulas were available for calculating what was called the *probable error* of the average of a sample of independent normal observations. This was a quantity such that the actual error in the mean was equally likely to exceed or fall short of this quantity. I have not come across specific attempts to apply these methods in drawing conclusions from field trials during the nineteenth century, but I would be surprised if such attempts do not exist.

Many will recall, for instance, Charles Darwin's discussion in 1876 of the heights attained by corn plants in his comparison of crossed and self-fertilized corn (*Zea mays*) in pot experiments, which Fisher reproduced in his book, *Design of Experiments* (1935). This experiment was similar to Young's comparison of drilling and broad-casting wheat in format, except that Darwin had 15 replications as against Young's 7. Darwin sent his data to Francis Galton for statistical advice. Galton's discussion makes it clear that he was aware of the probable error and of the result that the averages of independent samples from a normal distribution are themselves normally distributed. But, as I interpret his writing, two gaps caused him to abandon his attempts to apply these mathematical results to Darwin's data. The first--the one noted by Fisher--was that even assuming the 15 differences in height (crossed minus selfed) to be normally dis-tributed, the calculation of the probable error required a good esti-mate of the standard deviation of the differences, which Galton felt 15 observations were insufficient to provide. The second problem was that 15 observations were also not nearly enough to tell him the actual law of distribution followed by the individual differences in height. As Galton put it: "the real difficulty lies in our ignorance of the precise law followed by the series."

Toward the end of the nineteenth century and into the twentieth, there were two main lines of development in the layout of field experiments. Field experimentation in agriculture has the convenient property that the size and shape of the experimental unit--the plot-- is to a large extent controllable by the experimenter. Investigations of different sizes and shapes of plots led to a realization that for a given total area of land, replicated small plots (e.g., 100 square meters, or 1/40 acre) were a good choice (e.g., Wagner, 1898). These were much smaller than the quarter- or half-acre plots used by Young.

The introduction of many new varieties of farm crops following Mendel's work produced two reasons for even smaller plots with higher

numbers of replications. Because the difference in the cost of using
variety A rather than variety B in practical farming was small, lying
only in the cost of seed, relatively small differences in mean yield
became of economic importance, so that experiments of greater discriminating power were needed. Second, with new varieties sometimes only
a limited amount of seed was available for experimentation.

This period also saw numerous extensions of Johnston's recommendation that unlike treatments be placed on plots near one another.
Highly ingenious systematic plans were produced, the objective of
which was to obtain the maximum precision in the comparison of different
treatments. Three examples of such plans are shown below.

SONNE	KNIGHT'S MOVE
7 treatments	5 treatments
4 replications	5 replications

```
    A  F  E  G        A  B  C  D  E

    B  G  D  F        C  D  E  A  B

    C  A  C  E        E  A  B  C  D

    D  B  B  D        B  C  D  E  A

    E  C  A  C        D  E  A  B  C

    F  D  G  B

    G  E  F  A
```

HALF-DRILL STRIP

2 treatments, number of replications flexible

ABBA | ABBA | ABBA | ABBA | ABBA | . . . .

The Sonne plan automatically eliminated systematic differences
in fertility between columns. Furthermore, if the rows were numbered
from 1 to 7, the mean position of each letter was the same (4), so
that any *linear* trend in fertility from row to row was also automatically eliminated from comparisons among the treatment means. The
Knight's move is a particular 5 × 5 Latin square with the

additional property that no treatment appears twice on a diagonal--
helpful in the event of fertility gradients parallel to the diagonals.

The half-drill strip, invented by E. S. Beavan, was very convenient
for comparing two varieties of a cereal. In the drill, the seed boxes
on the left half were sown with variety A, on the right with B. A
plot (one half-drill wide) extended the length of one side of the
chosen area. Having sown plots AB, the horse turned at the end, going
back to sow BA and complete the sandwich ABBA. Any linear trend within
a sandwich was eliminated from the mean yield, even if the trend differed from sandwich to sandwich. Systematic plans of these types became the recommended methods for use in field experiments.

More detailed information about the methods used in experimentation
from Francis Bacon on can be found in Young's (1771) review of literature and in two reviews, one by Fussell (1935) and one by Crowther
(1937), which I found most helpful.

THE TWENTIETH CENTURY--INITIAL WORK TOWARD
QUANTITATIVE INTERPRETATION OF RESULTS

The first major step in attempting to quantify results of experimentation was Student's 1908 paper, "The probable error of a mean."
'Student' was the pen name of William D. Gosset. On leaving Oxford,
Gosset went to work in 1899 as a Brewer in Dublin for Messrs. Guinness,
who liked their workers to use pen names if publishing papers. It
apparently is not known how or when Student became interested in
statistics, but many of the statistical problems that he encountered
at Guinness were small-sample problems, for which the available large-
sample results were at best only a dubious approximation. As we have
noted, the probable error 0.67σ of a normally distributed value x--
the result that Galton was trying to use--requires knowledge or a good
estimate of σ. In his 1908 paper, Student set out to find the distribution of the amount of error $(\bar{x} - \mu)$ in the sample mean, when divided
by s, where s was the estimate of σ from a sample of any known size.
From this distribution of $(\bar{x} - \mu)/s$, the probable error of a mean $\bar{x}$
could be calculated for any size of sample. He was well aware of the
point that worried Galton--a small sample was insufficient to determine

the form of the distribution of x. But he chose the normal distribution for simplicity, giving his opinion: "it appears probable that
the deviation from normality must be very severe to lead to serious
error." I believe that subsequent work has justified this judgment,
if we have independence and no wild outliers.

  Although his mathematical analysis of the problem was incomplete,
he got the right answer. The steps were as follows. He worked out
the first four moments of the distribution of s^2, and noted that they
were the same as those for a Pearson's Type III curve, which he
guessed, correctly, was the distribution followed by s^2. He then
showed that s^2 and $\bar{x}$ are uncorrelated, and assumed (again correctly)
that they were independent. From these results, finding the distribution of $(\bar{x} - \mu)/s$ was an easy integral. He constructed a table of
the cumulative function for sample sizes from 4 to 10. (In modern
notation, his table in a column headed n is that for $t/\sqrt{n-1}$ with
(n - 1) degrees of freedom.) He also checked that for n = 10, his
t-table agreed quite well with the corresponding normal table.

  This remarkable paper had two further sections. Student did what
we would now call two Monte Carlo studies, using the heights and the
left middle finger measurements of 3,000 criminals, arranged in 750
random samples of size 4. Both empirical t-distributions based on
the 750 calculated values of t agreed well with his table by Karl
Pearson's χ^2 test of goodness of fit. A further section gave three
illustrations of the use of the table in practice, one to a paired
experiment like Young's, but which compared the hours of sleep produced
by two soporific drugs, one to pot experiments on wheat, and one to
a field experiment on barley.

  The t-distribution did not spread like wildfire. In his foreword
to Student's *Collected Papers*, (1942), McMullen wrote "For a long time
after its discovery and publication the use of this test hardly spread
outside Guinness's brewery". Even in September 1922, 14 years later,
we find Student writing to Fisher: "I am sending you a copy of Student's
Tables as you are the only man that's ever likely to use them!" Young
research workers who feel that the world is very slow to appreciate

their results might be heartened by this example. The world is
indeed a little slow at times to realize how brilliant we are.

Soon after the publication of Student's 1908 paper, two papers
at last appeared the objective of which was to explain and use
probability results in planning and analyzing agricultural experiments.
The first was the result of the collaboration by Wood, an agronomist,
and Stratton, an astronomer, and was entitled "The Interpretation of
Experimental Results" (Wood and Stratton, 1910). This paper first
illustrated the kinds of frequency distributions of the measurements
found on small plots--normal, skew, and even trimodal. Then, after
remarking that "no two branches of study could be more widely separated
than Agriculture and Astronomy," they noted that both branches suffered
from variability caused by the weather. They explained the astronomer's
method of estimating the accuracy of his averages by the use of the
probable error. Its calculation and meaning were illustrated by data
on percent dry matter in mangels and further examples. It was used
to perform something like both a one-tailed and a two-tailed test of
significance of the difference between two treatment means. Estimates
were made of the numbers of replications needed both in feeding experiments on animals and yield experiments on crops in order that an
observed difference of a given size between two treatment means would
be significant at the 1 in 30 level. (Thus, although they did not find
the number of replications necessary to control the power of the tested
difference at a given level--this had to wait for the work of Neyman,
et al. (1935)--they were getting close.) For these calculations they
needed, of course, estimates of the probable error for a single animal
or plot, obtained from a survey of replicated experiments.
The second paper on the conduct of field experiments, "The
experimental error of field trials", was written by two agronomists,
Mercer and Hall (1911). They investigated the effects of size and
shape of plot, type of experimental plan, and number of replications
on the standard deviation of the mean, using data for wheat and mangels
all treated alike. They recommended five replications of 40th acre
plots as giving 2% for the standard deviation of a mean. As with the
writers in the nineteenth century, a systematic distribution of the

treatments was recommended, with replications of the same treatments
scattered and unlike treatments placed near one another. An appendix
by Student proposed an efficient systematic plan, with sandwiches
ABBA in two directions, for comparing two treatments. Neither paper
used Student's t-table, although Student gave two references to it.

Both papers used the results of what were called *uniformity
trials,* in which a large number of small plots, all treated similarly,
were harvested. These results provided data from which the standard
errors obtained from different sizes and shapes of plots, numbers of
replications, and different experimental plans could be estimated.
Agriculture is, I think, unique in the amount of effort devoted to
the study of the variability with which field experiments had to cope.
A catalog prepared by Cochran (1937) contained references to 191
uniformity trials on field crops and 31 trials on tree crop experiments.

To summarize from this incomplete review, it took roughly a full
century to accomplish two major steps. These steps were (1) to begin
applying the probability theory already available in astronomy so as
to provide objective quantitative methods for the interpretation of
the results from experiments on variable material, and (2) to establish
detailed methods for the efficient practical conduct of field experiments. Further major changes were soon to come, however.

ENTER FISHER

Fisher joined the Rothamsted Experimental Station in 1919. As
I have noted elsewhere (Cochran, 1973), his first solution in 1922
to a problem in designing an experiment dealt with an experiment in
which the measurement was a binomial variable that was nonlinear in
the quantity to be estimated. This quantity was the density of tiny
organisms in a liquid, where the test could detect whether a sample
contained no organisms or at least one organism, but could not count
the number of organisms. Nonlinear problems, which occur also in
quantal bioassay, have come to be extensively studied in more complex
situations from 1960 onward, but this work stands apart from Fisher's
writings on field experimentation.

His initial steps in the development of the analysis of variance, as applied to a three-way classification, appeared in a paper by Fisher and Mackenzie (1923), which dealt with the analysis of a factorial experiment on potatoes. In the same year, in his paper "On testing varieties of cereals," Student (1923) reproduced the algebraic instructions for a two-way analysis of variance, which he had received from Fisher in correspondence. Student applied them to an experiment on barley with eight varieties in 20 replications. Both experiments had been laid out systematically. Furthermore, the potato experiment was actually of the split-plot type, for which Fisher's analysis was not fully correct. Evidently, he was feeling his way at that time.

Student's paper had two minor points of interest. He knew that with only *two* varieties in 20 replications, he could find the variance of their differences, from the 20 individual differences within replications. How was he to extend this to *eight* varieties? With eight varieties, 28 different pairs could be formed, and he realized that a natural, but unappealing, method was to repeat the preceding calculation 28 times, finding the average of the 28 variances of differences. Fisher's analysis of variance table gave the identical result much more quickly.

Second, Student introduced the term "variance"--the square of the standard deviation--as useful in studying the contributions of different sources to the variability of an observation. In a later footnote he pointed out that he was not the author of this term, as Fisher had used it since 1918.

Fisher's earliest writings on general strategy in field experiments were contained in the first edition of his book *Statistical Methods for Research Workers* (1925) and in a paper of 10-1/2 small pages (1926), entitled "The arrangement of field experiments." This short paper presented nearly all his principal ideas on the planning of experiments.

Fisher began with an explanation of the concept of a test of significance. Manure was applied to an acre of land, whereas a neighboring acre was left unmanured but was otherwise handled similarly. The yield on the manured plot was 10% higher. What

reason was there to think that the 10% increase was the result of the
manure and not soil heterogeneity? Fisher noted that if the experimenter could say that in 20 years of past experience with the *same*
treatment on the two acres the difference had never before reached 10%:

> ... the evidence would have reached a point which may be called
> the verge of significance; for it is convenient to draw the line
> at about the level at which we can say 'Either there is something
> in the treatment or a coincidence has occurred such as does not
> occur more than once in twenty trials.' This level, which we
> may call the 5 per cent point, would be indicated, though very
> roughly, by the greatest chance deviation observed in twenty
> successive trials.

(Fisher noted that it would take about 500 years of previous experience
to determine the 5% significance level reasonably accurately by what
we call the Monte Carlo method.) I have given this argument in some
detail, because students sometimes ask "How did the 5% significance
level or Type I error come to be used as a standard?" I am not sure,
but this is the first comment known to me on the choice of 5%. Fisher
went on:

> If one in twenty does not seem high enough odds, we may, if we
> prefer it, draw the line at one in fifty (the 2 per cent point)
> or one in a hundred (the 1 per cent point). Personally, the
> writer prefers to set a low standard of significance at the 5
> per cent point, and ignore entirely all results which fail to
> reach this level.

Fisher sounds fairly casual about the choice of 5% for the significance
level, as the words 'convenient' and 'prefers' have indicated.

He then remarked that if the experimenter had the actual yields
for, say, the 10 past years under uniform treatment, and if he could
trust the theory of errors, he could calculate the 5% significance
level of the difference by using an estimated standard error and
Student's t-table, which Fisher had included in the first edition of
Statistical Methods a year earlier.

Fisher next pointed out that methods had been devised for
obtaining from the results of the experiment itself an estimate of
the standard error of the difference between two treatment means--
thereby providing the technique that Johnston was calling for nearly

80 years earlier. Since this estimate had to come from the differences between plots treated alike, the solution had to lie in
replication. However, the recommended systematic plans in current
use, which scattered the replicates and juxtaposed unlike treatments,
deliberately violated a basic assumption made in the mathematical
theory behind the probable error. If the experimenter's judgment
was correct, differences between the scattered replicates would
consistently overestimate the real errors of the differences between
the means of unlike treatments. If his judgment was wrong, the
differences between replicates would underestimate the relevant errors.
The only way to be certain of a valid estimate of error was to ensure
that the relative positions of plots treated alike did not differ in
any systematic or relevant way from those of plots treated differently.

There was one easy solution. Having decided on the numbers of
treatments and replicates, the experimenter could assign treatments
to plots entirely at random, e.g., by drawing numbered balls from a
well-mixed bag. Fisher seemed to regard the truth of this claim as
obvious, giving no mathematical discussion.

Fisher realized that to an experienced experimenter, replacement
of a carefully chosen systematic design by a randomized design might
seem to involve an undesirable loss of precision as the price of this
valid estimate of error. Not so, he said. By the randomized blocks
design, with compact blocks of neighboring plots, the goal of having
unlike treatments near one another could be combined with independent
randomization *within each block*. He stressed, however, that it was
now necessary to change the method of calculating the standard error.
This plan automatically eliminated consistent differences between
block fertility levels from the actual errors of the differences between treatment means; therefore, they must also be eliminated from
the estimate of error by the analysis of variance method which had
been developed by Fisher. The Latin square plan went a step further,
with each treatment occurring once in each row and each column of
the square. Thus, fertility gradients along both the length and the
breadth of a field could be eliminated from the actual and the

estimated errors. He counted and classified the Latin squares up to
the 6 × 6, offering to send the user squares drawn at random from all
possible squares of the desired size.

The final section of this 1926 paper, entitled "Complex
experimentation" contained Fisher's advocacy of what we now call
factorial design, in which he indulged in a brief general philosophic
statement.

> No aphorism is more frequently repeated in connection with field
> trials, than that we must ask Nature few questions, or, ideally,
> one question, at a time. The writer is convinced that this view
> is wholly mistaken. Nature, he suggests, will best respond to a
> logical and carefully thought out questionnaire; indeed, if we
> ask her a single question, she will often refuse to answer until
> some other topic has been discussed.

He then gave the field plan of a 3 × 2 × 2 factorial experiment
on oats in 8 randomized blocks of 12 plots, the treatments being 3
amounts of nitrogen fertilizer, either in the form of ammonium sulfate
or ammonium chloride, applied early or late in the season, in all
12 combinations.

He stressed the following advantages of this approach.

1. The first was its efficiency. Every plot yield or observation provided some information about the effects of each of
   the three factors, whereas in a single-factor experiment a
   plot yield provided information only on the single factor
   under investigation. The 96 plots in this experiment supplied
   32 replications on the average differences between sulfate
   and chloride, between early and late application, and between
   different amounts of nitrogen. To obtain the same number of
   replications, single-factor experiments would have required
   224 plots (more than twice as many).

2. Many questions about the interrelationships between the effects
   of the factors could be investigated; for instance, was the
   effect of nitrogen the same in early and late application?

3. The advantage that Fisher regarded as the most important was
   that the average effect of any factor was averaged over a
   number of variations in the conditions as regards the other
   factors, giving these results, as he put it, a very much
   wider inductive basis than given by single-question methods.

4. The final paragraph of this paper noted that it would sometimes be advantageous to sacrifice information deliberately
   on certain interrelations (interactions) believed to be

unimportant, confounding them with block differences in order
to reduce the size of block and thus increase the precision
of the more important comparisons. This subject, confounding,
came to be extensively worked out later.

These four techniques--blocking, randomization, factorial design,
and the analysis of variance--explained more fully in *The Design of
Experiments* (Fisher, 1935), came to be cornerstones in comparative
experimentation on variable material.

After 1930, work on the relevant mathematical properties of
randomization sets began to appear. Papers by Eden and Yates (1931),
Tedin (1931), Bartlett (1935), and Welch (1937) investigated numerically or algebraically how well the discrete distribution of F over
the randomization set approximated the tabular distribution for
randomized blocks and Latin squares under the null hypothesis.

Although generalization from this limited number of studies is
risky, these studies indicate that the 5% level in the F table corresponds to a randomization test level somewhere between 3% and 7%.
Under a less restrictive model in which treatments affect individual
plots differentially, Neyman et al. (1935) showed that the error in
randomized blocks remain unbiased over the set for the F test of the
null hypothesis of no average treatment effects, but the error is
subject to some bias in the Latin square, which leans very heavily
on strict additivity.

Consider any standard design (e.g., randomized blocks) with
fixed mean, block and treatment effects μ, β_j, τ_k, and with
given experimental errors e_{ij} of any structure on each plot. Under
an additive model, the observed yield y_{ij} on the ith plot in the jth
block will be

$$y_{ij} = \mu + \tau + \beta_j + e_{ij}$$

where τ is the treatment effect that the randomization happens to
assign to that plot. Calculate any comparison $\Sigma L_i \bar{y}_i$ among the
treatment means for each possible outcome of the randomization. Its
variance over the randomization set can be proved to be identically
equal to the average over the set of its estimated variances, found

by using the error mean square and the rules in the analysis of variance. It is in this sense that randomization guarantees a valid estimate of error, irrespective of the nature of the experimental errors on the individual plots. The assumption of strict additivity is, of course, crucial here. For example, with several levels of nitrogen, it is often found that the variance of $N_{\ell in}$ from block to block exceeds that of N_{quad}, so that errors must be calculated separately for the two comparisons. This is a counterexample with a relation between τ and e. But randomization still makes the separate errors valid.

The proponents of systematic designs, including Student, were by no means immediately converted to randomization. Debating papers on both sides of the argument appeared at intervals for about the next 12 years. The protagonists occasionally indulged in derogatory remarks about one another's scientific acumen, a practice that I have heard Fisher describe privately as "just some Billingsgate"-- the name of the London fish market noted for the salty language of its vendors. Claims made by those who favored systematic plans were

1. It was better, scientifically, to have smaller *real* errors when comparing different treatment means, even at the expense of some overestimation in the *estimated* standard errors attached to these means. (An argument on this point can be quite stimulating.)

2. Some specific outcomes of a randomization looked obviously unwise or risky, e.g., getting AB, AB, AB, AB, AB, AB, in a paired experiment laid out in six blocks in a single strip. Sometimes, as in this case, such undesirable plans could be avoided, still meeting Fisher's conditions, by introducing additional blocking into the plan as in a crossover design. However, in more complex cases, attempts to develop general methods for finding randomization sets free from undesirable elements have had only limited success.

3. By the adoption of a more realistic mathematical model applicable to a systematic plan, a method of analysis that provided a sufficiently valid estimate of the standard error might be found for systematic designs. I gather that Neyman (1935) looked into this type of approach more generally in a paper that was published in 1929, which I have not seen. In competent statistical hands this approach might have worked quite satisfactorily. It is sometimes used as a rescue operation when an unfortunate outcome of a randomization is

unnoticed until the responses were measured, e.g., Outhwaite
and Rutherford (1955). As a simple example, Student (1938)
suggested that with Beavan's sandwich design, the sandwich
be regarded as the unit, the error being calculated from
the variation in $(\bar{A} - \bar{B})$ between sandwiches. If this did
not provide enough degrees of freedom, he suggested using
the variation between half-sandwiches, with one degree of
freedom subtracted for the average of the linear effects
which this plan removed.

The debate on randomization had mostly died out by 1938. As
comparative experimentation spread rapidly beyond agriculture, one
circumstance favored the adoption of Fisherian randomized plans. In
agriculture, systematic plans had emerged from over 60 years of
extensive study of the nature of variations in soil fertility. In
other areas of research, investigators often had no comparable knowledge or data about the nature of the variations they faced beyond a
few trends easily handled by blocking. They were less likely to feel:
"I can do better than randomized blocks."

I detect a hint of this attitude in Student's (1931a) discussion
of the Lanarkshire milk experiment that had been carried out in 1930.
This experiment was one of the earlier ones in which something like
randomization was used. Moreover, it was in the difficult area now
called social experimentation. The experiment involved 20,000 school-
children, of whom 5,000 received 3/4 pint daily of raw milk, 5,000
received 3/4 pint daily of pasteurized milk, and 10,000 received no
milk supplement in school. In a given school, only two treatments
were compared, either None vs Raw or None vs Pasteurized. The
assignment of children to treatment in a class was either by ballot
(which I take it implies randomization) or alphabetically.

Student approved this invocation of the goddess of chance.
Unfortunately, the plan that was used might be called improved
randomization. If the original allocation gave an undue proportion
of well- or ill-nourished children to one treatment, the teachers
could make substitutions. It looked as if what enough teachers did
instead was to give the milk to those children who needed it most.
At the start of the experiment the "no milk" children averaged 3
months growth in weight and 4 months growth in height superior to

the "milk" children. Student noted that even with statistical adjust-
ments, these biases made conclusions about the effects of milk doubtful.

Comparing the effects of pasteurized and raw milk, Fisher and S.
Bartlett (1931) concluded from these data that pasteurized milk was
inferior to raw milk with regard to the increases in both weight and
height. Student's view was that since the assignment of raw or pasteurized milk to schools was not random, he would be very chary of
drawing any conclusion on this question from these data. It would have
been tempting to ask Fisher and Student: "Whose side are you on?"

Student's recommended plan was that if a large-scale experiment was wanted, randomized blocks of two treatments (e.g., raw
and pasteurized), to be used in a class, children in the same block
being of the same age and sex and of similar height, weight,
and physical condition. He also suggested that randomized blocks of
50 pairs of identical twins would probably give more reliable information as well as be much less expensive. In another paper (1931b) in
the same year, he recommended fully systematic plans for yield trials
in agriculture as superior to Fisher's randomized blocks and Latin
squares. He seemed to be saying: "In an unfamiliar area, better
stick to randomization."

ACKNOWLEDGMENT

This work was supported by Contract No. N00014-67A-0298-0017 with
the Office of Naval Research, Navy Department.

REFERENCES

Bacon, F. (1627), *Sylva Sylvarum, or a Naturall History*, pp. 109-110.
    William Rawley, London.

Bartlett, M. S. (1935), The effect of non-normality on the t
    distribution. *Proc. Cambridge Phil. Soc. 31*, 223-231.

Cochran, W. G. (1937), Catalogue cf uniformity trial data. *J. Roy.
    Statist. Soc. B 4*, 233-253.

Cochran, W. G. (1973), Experiments for non-linear functions. *J. Am.
    Statist. Assoc. 68*, 771-781.

Crowther, E. M. (1936), The technique of modern field experiments. *J. Roy. Ag. Soc. Engl. 97,* 1-28.

Eden, T., and Yates, F. (1931), On the validity of Fisher's z test when applied to an actual example of non-normal data. *J. Ag. Sci. 23,* 6-17.

Fisher, R. A. (1925), *Statistical Methods for Research Workers,* Oliver and Boyd, Edinburg.

Fisher, R. A. (1926), The arrangement of field experiments. *J. Ministry Ag. Sept. 33,* 503-513.

Fisher, R. A. (1935), *The Design of Experiments.* Oliver and Boyd, Edinburgh.

Fisher, R. A., and Mackenzie, W. A. (1923), The manurial response of different potato varieties. *J. Ag. Sci. 13,* 311-320.

Fisher, R. A., and Bartlett, S. (1931), Pasteurized and raw milk. *Nature 127,* 591-592.

Fussell, G. E. (1935), The technique of early field experiments. *J. Roy. Ag. Soc. Engl. 96,* 78-88.

Johnston, J. F. W. (1849), *Experimental Agriculture, Being the Results of Past and Suggestions for Future Experiments in Scientific and Practical Agriculture.* W. Blackwood and Sons, Edinburgh.

McMullen, L. (1942), Preface to *Student's Collected Papers.* Cambridge Univ. Press, Cambridge.

Mercer, W. B., and Hall, A. D. (1911), The experimental error of field trials. *J. Ag. Sci. 4,* 109-132.

Neyman, J., Iwaszkiewicz, K., and Kolodziejczyk, S. (1935), Statistical problems in agricultural experimentation. *J. Roy. Statist. Soc. Suppl. 2,* 108-180.

Outhwaite, A. D., and Rutherford, A. (1955), Covariance analysis as an alternative to stratification in the control of gradients. *Biometrics 11,* 431-440.

Stigler, S. M. (1973), Gergonne's 1815 paper on the design and analysis of polynomial regression experiments, Tech. Rept. 344, Univ. of Wisconsin, Madison.

Student (1908), The probable error of a mean. *Biometrika 6,* 1-24.

Student (1923), On testing varieties of cereals. *Biometrika 15,* 271-294.

Student (1931a), The Lanarkshire milk experiment. *Biometrika 23,* 398-407.

Student (1931b), Yield trials. *Balliere's Encyclopedia of Scientific Agriculture,* 1342-1360. Balliere and Co., London.

Student (1938), Comparison between balanced and random arrangements of field plots. *Biometrika 29,* 363-379.

Tedin, O. (1931), The influence of systematic plot arrangements upon the estimate of error in field experiments. *J. Ag. Sci. 21*, 191-208.

Wagner, P. (1898), *Suggestions for Conducting Reliable Practical Experiments* [English transl.]. Chemical Works, London.

Welch, B. L. (1937), On the z-test in randomized blocks and Latin squares. *Biometrika 29*, 21-52.

Wood, T. B., and Stratton, F. J. M. (1910), The interpretation of experimental results. *J. Ag. Sci. 3*, 417-440.

Young, A. (1771), *A Course of Experimental Agriculture*. Exshaw et al., Dublin.

THE ANALYSIS OF
VARIANCE AND FACTORIAL DESIGN

OSCAR KEMPTHORNE was born January 31, 1919, in St. Tudy, Cornwall,
England. He received his B.A. in 1940, M.A. in 1943, and Sc.D.
in 1960, all from Cambridge University. He worked at the
Rothamsted Experimental Station from July, 1941 to December,
1946, and then joined the faculty of Iowa State College where
he has remained. He became naturalized in 1955. Since 1964
he has been a Distinguished Professor of Science and Humanities.
He has been a Visiting Professor at Oklahoma State University
in 1962, at the University of Buenos Aires, Argentina, in 1963,
and at Stanford University in 1964. He has been a Fellow of
the American Statistical Association, The Institute of Mathematical Statistics, and the American Association for the Advancement of Science, since 1952. He is a Fellow of the Royal
Statistical Society and a member of the International Statistical Institute, The American Society of Naturalists, and the
Biometric Society. He was on the Editorial Board of the *Annals
of Mathematical Statistics* and has been a Director of the
American Statistical Association. He was also President of the
Biometric Society, ENAR. He gave the Fisher Memorial Lecture
for the Joint Statistical Meeting in 1965. He is an Associate
Editor of *Biometrics*. He is author and coauthor of six books
and over 100 research papers.

THE ANALYSIS OF
VARIANCE AND FACTORIAL DESIGN

Oscar Kempthorne

Statistical Laboratory
Iowa State University
Ames, Iowa

PREFATORY REMARKS

I wish first to express some views and reactions to the theme of
our symposium "The History of Statistics and Probability." I react
immediately to the word "history." I recall the history I learned
as a child in Great Britain in which the American Revolutionary War
was presented only as a minor episode in the history of Great Britain,
and I recall various histories that I have read, which each consists
of the interpretation of one individual to a vast panorama of human
activity, this interpretation being almost entirely an interaction
between the ideas and values of the historian and the actual chron-
icle of what really happened. This seems to illustrate an unavoidable
problem, that history is in the mind of the beholder, and that an
unbiased history is impossible to achieve. Every history has a strong
personal or subjective element.

The subjectivity of an historical approach must be realized by
the listener and the reader. The history that will be presented
here is very subjective and very abbreviated. It is very abbreviated
because statistics as a human mental activity goes back to antiquity.

The enumeration of real populations with regard to immediately
apparent attributes was surely practiced many thousands or perhaps
even hundreds of thousands of years ago. The statistics were
collected for political and social purposes. The big break occurred
with the collection of statistics to develop models of the real
world. These statistics would not only give the status at the time
of collecting the statistics, but would also be used to predict the
future or "pieces of the future" that one could contemplate as
"formable" or "potential," and held the possibility that some
"pieces of the future" could be avoided and others could be sought
by plan.

So when I think about the history of statistics, my mind ranges
over all the attempts of humanity to construct models of the universe.
That I am encompassing a vast terrain is obvious, and I realize that
I may well be judged foolish. To illustrate this vastness, I give
one example-the collection of astronomical observations by Tycho
Brahe and the analyses of these observations by Kepler. On the one
hand, we have a tremendous data collection activity, and on the other
hand, a most remarkable data analysis activity. It may be said
"That is not statistics: that is physics." I would merely reply
that this example contains two of the most critical aspects of
statistical thinking, *as it exists today*.

What are the main aspects of statistics? I give a very succinct
listing. (Please recall in assessing this list my remark above about
the nature of history).

1. Collection of data by processes that may be deemed to have
    interpersonal validity

2. Development of models which are consonant with all the data
    available which will involve, in general, approximation and
    approximative models

3. Development of ideas of probability for these purposes:

    (a) to separate the mass of data into two parts, a part
         that is deterministic and may be treated by the
         processes of classical mathematics, e.g., of the
         nineteenth century, and a part that exhibits no
         discernible regular structure; this second part is
         encompassed by probability calculus, which was well

      developed a century ago with modern formulations of
      increasing generality and abstractness by von Mises,
      Kolmogorov, and others

  (b)   to develop models that incorporate the occurrence of
       randomness

  (c)   to make predictions about the future

4.   Development of survey methodology by which opinions about a
     population could be formed with some degree of logic and
     objectivity by examination of only part of the population

5.   Development of the experimental method in which populations
     are generated at the choice of the investigator

6.   Attempted development of a logic of decision

7.   Attempted development of logic of induction

I deem it essential to mention all of these aspects because a viewing of our program here will not give the overall broad picture. On the basis of the program, it would appear that this symposium attempts to address items 3-6 of my listing. The student and reader should be warned that extremely critical portions of the history are omitted. Furthermore, there should be a warning that there are very deep controversies in the logic and philosophy of statistics. These came to light in the past two decades. I will give only two examples. First, with regard to probability, there is little disagreement about the mathematics of probability, although the subject has changed remarkably, and one simply cannot read the modern literature without a strong basis in measure theory. But we do not appear to be any closer to a consensus on the meaning of probability than we were a century ago. Second, those at the symposium and readers of the presented material will note the great importance given in survey and experiment methods to the use of random sampling and experiment randomization. We must mention, however, that there is a group of keen minds that regards (or appears to regard) these ideas as irrelevant and highly misleading.

So let us all be aware that this symposium presents part of the history from personal viewpoints of the participants. We are entitled to trust that what is given has more claim to objectivity and truth than a mere presentation in good faith by serious workers.

THE DESIGN AND ANALYSIS OF EXPERIMENTS

It is easy and natural to make a historical judgement about what ideas lie in this corpus of knowledge by examining the several books with some or all of the words in their titles. A historian should try, I believe, to take a larger view.

The design of experiments is often presented as a collection of ideas by which questions of science and technology may be answered or by which one of a class of terminal decisions can be chosen wisely. Each experiment is performed alone, and is treated as a single separate fact. The caricature of an experiment is that it is a study that leads to one star, or two stars, or three stars. But the nature, except in rather special terminal decision contexts, is that a single experiment is only one step, a very important one for the particular experimenter, in the long, sustained effort of humanity to build a validated model of the real world. It is an easy task to describe the single experiment as a single, isolated experience, and that is what is done in almost all our books. To see how all experimentation fits into the grand problem and to construct a plausible model for the overall learning process seems an almost insuperable task. So in this area as in most, the great effort has been on small, single, isolatable subproblems.

THE ANALYSIS OF VARIANCE

I imagine there would be little argument against the proposition that the analysis of variance is the most widely used tool of modern (post-1900) statistics. We have only to look at the statistical methods that are deemed essential by workers in the substantive fields of biology, psychology, sociology, education, agriculture, and so on. Statisticians do not bear down on these fields, and tell the workers "You and your students must be slightly educated in the analysis of variance". The demand for knowledge of this topic comes from the substantive fields.

But what is the analysis of variance? I have seen expositions

to the effect that the analysis of variance is merely an appendage
(very useful, to be sure) to the probability and statistical theory
associated with linear models and the multivariate Gaussian or normal
distribution. It arises by the application of likelihood ratio-
testing ideas. This is so far from my views as a statistician and
historian as to be ludicrous. But let me warn the listener and
reader that I may be in error.

The problem lies partly, I believe, in the simple technical fact
of our present-day language that everybody is agreed that variance
is a property (simple in simple cases) of a random variable that is
taught in a first course on statistics. Ergo, the analysis of
variance deals with analysis of this sort of variance. But my thesis
will be that this is only a small part of the actual story.

It would be interesting to have a completely definitive presentation of the history of the ideas of analysis of variance. It is not
uncommon to state that the idea was given in its modern form by Fisher
(1918), when he partitioned the total variance of a human attribute
into portions attributed to heredity, environment, and other factors,
which led to an equation

$$\sigma^2 = \sigma_1^2 + \sigma_2^2 + \cdots + \sigma_k^2$$

where σ^2 is the total variance, and the several σ_i^2 are variances
associated with certain forces. For such a use, the title "analysis
of variance" is totally appropriate.

However, this presents, in my opinion, a very limited view. The
views that I shall present arise from the history of this century.
The originator of the basic ideas seems clearly to be R. A. Fisher.
I find it remarkable how prescient is Fisher's (1920) examination of
the yield of dressed grain from Broadbalk. (This is paper 3 of
Fisher's "Contributions to mathematical statistics.") In this paper,
the word "variance" is used very cryptically to indicate portions of
the mean square deviation of observations which can be isolated by
some sort of linear model.

THE ANALYSIS OF VARIANCE AS A GENERAL PYTHAGOREAN THEOREM

I suppose there are few classical theorems of mathematics that
are more important to elementary and advanced mathematics (if as a
statistician, I may express an opinion on the latter) than the original Pythagorean Theorem. Let me present it in a slightly unusual
form.

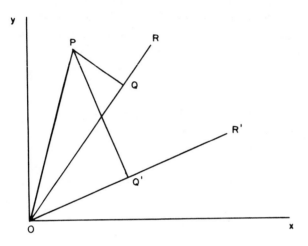

We have the ordinary plane, with origin O and axes Ox, Oy; we have a
point P and a line through the origin OR. We drop a perpendicular
from P to the line OR, the foot of this perpendicular being Q. Then

$$OP^2 = OQ^2 + QP^2$$

If, however, we had, instead, the line OR' , the foot of the
perpendicular would be Q' and we would have

$$OP^2 = OQ'^2 + Q'P^2$$

So for each line OR, or OR', or whatever, we have a decomposition of
OP^2 into two nonnegative parts.

When we turn to three dimensions, the plot is more complicated. If we have the point P and a single line OR, we have nothing new because we can reduce the above in the plane given by OP and OR, and we have the previous two-dimensional plot. But we can introduce two lines OR_1 and OR_2, and we can then sketch the following:

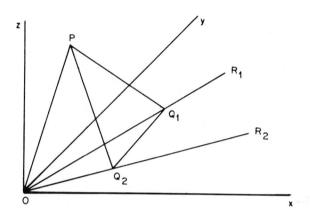

Here

$$OQ_2 \perp Q_2P$$
$$OQ_1 \perp Q_1P$$
$$Q_1Q_2 \perp OQ_1$$

and

$$OP^2 = OQ_1^2 + Q_1Q_2^2 + Q_2P^2$$

If we take different lines OR_1', OR_2' we get a different partition of OP^2.

What is the n-dimensional picture? I am totally unable to draw it. I have to go to algebra--to be Cartesian.

I shall give one view of the general nature of the analysis of variance for an arbitrary finite number of dimensions. The extension to a countable infinity of dimensions is a matter of somewhat intricate plumbing. The extension to an infinity of dimensions of arbitrary order is, I understand, rather tricky--but perhaps not.

Consider an n vector $y' = (y_1, y_2, \ldots, y_n)$. Consider k sets of $n \times 1$ vectors. Denote each set by the columns of the matrix $X_i (n \times p_i)$. At this point, we can take one of two routes. We can consider approximating the vector y by the least squares method, using a sequence of models

$$y = X_1\beta_1 + d_1$$
$$y = X_1\beta_1 + X_2\beta_2 + d_{12}$$
$$y = X_1\beta_1 + X_2\beta_2 + X_3\beta_3 + d_{123}$$
$$\cdot \quad \cdot \quad \cdot \quad \cdot \quad \cdot \quad \cdot \quad \cdot \quad \cdot \quad \cdot \quad \cdot \quad \cdot$$
$$y = X_1\beta_1 + X_2\beta_2 + \cdots + X_k\beta_k + d_{12 \cdots k}$$

Here the vector y and the matrices $\{X_i\}$ are given. The d terms are deviations or disturbances. I shall outline this procedure. However, before doing so, I should mention that the matter can be presented in terms of column spaces and orthogonal projections on to these spaces. I prefer the former because it can be presented with very little mathematical vocabulary, whereas the latter requires a course in finite dimensional vector spaces. Perhaps the point I am trying to make here can be exemplified by the fact that it is quite uncontrovertible that there are tens of thousands of scientific workers who have some understanding of the analysis of variance but who have never been exposed to vector spaces, projections, and the like. Perhaps, I can also, add to my justification for adopting the prosaic presentation by the simple remark that the analysis of variance is a data analysis procedure and that when one wishes to determine an analysis of variance one has to compute projections--when one does this, one is drawn into the mode of presentation that I shall follow. As a mathematician, of course, one can merely say "Let the projection of y be Py."

The story in terms of fitting a sequence of models is rather simple. Let me emphasize that I am not considering stochastic models which involve random variables. I am talking about approximative models. Then the simple facts are as follows

1. Consider finding Xb such that

    $$(y - Xb)'(y - Xb) = \min_{\beta \in R_p} (y - X\beta)'(y - X\beta)$$

    $$= \min_1, \text{ say}$$

2. The normal equations are

    $$X'Xb = X'y$$

3. These equations are consistent for all $y \in R_n$

4. The solution for Xb is unique and provides the global minimum

5. Now consider fitting by least squares the model

    $$y = X\beta + Z\gamma + d$$

    The minimum sum of squares is

    $$(y - Xb - Zc)'(y - Xb - Zc) = \min_2, \text{ say}$$

    where b, c satisfy the normal equations

    $$\begin{pmatrix} X'X & X'Z \\ Z'X & Z'Z \end{pmatrix} \begin{pmatrix} b \\ c \end{pmatrix} = \begin{pmatrix} X'y \\ Z'y \end{pmatrix}$$

6. Obviously, the minimum sum of squares in this case is less than or equal to the minimum sum of squares in the former case, because the variables γ in the second problem can be thought of as occurring in the first problem, but as being forced to be zero.

7. So $\min_1 - \min_2 \geq 0$. Clearly, $\min_1$ and $\min_2$ are quadratic functions of the vector y. So $\min_1 - \min_2$ is a quadratic function of the vector y, which is greater than or equal to zero.

8. Suppose, now, that we have sequence of models, such that a model in the sequence is a restriction of the succeeding one obtained by forcing a parameter (scalar or vector) to be null Let the minimum sum of squares for model i in the sequence be M_i.

Then with the sequence of models $1, 2, \ldots, k$ we have

$$M_1 \geq M_2 \geq M_3 \geq \cdots \geq M_k$$

Let $T = \sum_i y_i^2$, and $R_i = T - M_i$, where R_i is the sum of squares removed by fitting the ith model. Then $R_k \geq R_{k-1} \geq \cdots \geq R_2 \geq R_1$ and we have

$$T = R_1 + (R_2 - R_1) + (R_3 - R_2) + \cdots + (R_{k-1}) + (T - R_k)$$

This is the primitive analysis of variance. Note that in the presentation there has been no mention of random variables, and no mention except *in the name* of the word "variance." This idea of the analysis of variance was present, very clearly I believe, in Fisher's mind.

At the same time, as I have indicated, Fisher also had in mind the notion of partitioning the total variance of a random variable into components that could be attributed to identifiable causes. It was natural, as we shall see later, to work out mathematical statistics of the process by assuming that the disturbances in the sequence of models were independent Gaussian random variables with zero mean and constant variance. Early work was done by Irwin (1931), and the main development was done by Cochran (1934) leading to the very famous Cochran's Theorem. Additional clarification was given by Madow (1938).

It has become quite clear in the ensuing years that the process is one of algebra and is, in fact, a general Pythagorean Theorem. The algebraic ideas were used by Cochran but, seemingly, only as a means to a mathematical statistical end.

Let me now sketch the matter from the viewpoint of algebra. I use items (1)-(4) of the previous listing starting from the number 9:

9.   Hence, there exists a matrix B such that

      $X'XB = X'$

10.   XB is unique, symmetric, and idempotent

      Denote it by P_X.

11. $P_X X = X$

12. $P_X(I - P_X) = \phi$

13. $y = P_X y + (I - P_X)y = u + v$

14. $u'v = 0$

15. $y'y = u'u + v'v$

Now let $Z_i = (X_1, X_2, \ldots, X_i)$. Let $B_{12\cdots i}$ be such that

$$Z_i' Z_i B_{12\cdots i} = Z_i'$$

Let

$$P_{12\cdots i} = Z_i B_{12\cdots i}$$

Then after some rather easy manipulations and proofs, we can write

$$Y = P_1 y + (P_{12} - P_1)y + (P_{123} - P_{12})y$$

$$+ \cdots + (P_{12 \cdots k} - P_{12 \cdots k-1})y + (I - P_{12 \cdots k})y$$

$$= u_1 + u_{21} + u_{321} + \cdots + u_{k\,\overline{k-1}\,\cdots\,21} + u$$

For brevity, write

$$S_1 = P_1, \quad S_{21} = P_{12} - P_1, \quad S_{321} = P_{123} - P_{12}, \quad S = I - P_{12 \cdots k}$$

and so on. Then

$$Y = S_1 y + S_{21} y + S_{321} y + \cdots + S_{k\,\overline{k-1}\,\cdots\,21} y + Sy$$

Then from the construction of the matrices, or operators if you wish, any two of the components on the right-hand side are orthogonal, and we have

$$\sum_i^n y_i^2 = y'y = y'S_1 y + y'S_{21} y + \cdots + y'S_{k\,\overline{k-1}\,\cdots\,21} y + y'Sy$$

We have decomposed the vector y into orthogonal parts, just as in the case of the simple Pythagorean Theorem. We have made an "analysis variance" of the set of numbers $y_1, y_2, \ldots, y_n$, which constitute the vector y.

It is of historical interest, I think, that Fisher (1939, p. 239) disposed of the matter very simply with the diagram measuring about 1×1 inch:

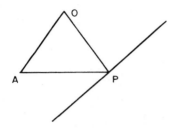

We see that given the vector y and the matrices $X_1, X_2, \cdots, X_k$, there are k! analyses of variance, one for each order in which we may place the X matrices. The situation is very simple if each X_i matrix has dimensions n × 1, and there is orthogonality so that $X_i'X_j$ equals 0 for all i ≠ j.

The topic is related, clearly, to the area of orthogonal functions and orthogonal polynomials. But this mathematical theory is rather simple in terms of logic, even though it may be difficult in terms of mathematical derivations, because when dealing with mathematical functions, over an interval, say, one has a y vector to be "explained," which has uncountably infinite dimensionalty. So one "knowns" a huge amount. From one viewpoint, the mathematical theory is a very special limiting case in which great difficulties arising from finiteness of knowledge do not occur.

Why is the analysis of variance, regarded purely as an arithmetic or algebraic process of decomposing a vector into geometrically orthogonal pieces and of decomposing the squared length of that vector according to a generalized Pythagorean Theorem, so important to statistics? The answer is rather simple. In our attempts to build a useful and valid picture of the real world, we choose an area for investigation, we define a population of individual entities, we observe a variable y for each individual we wish to explain, and for each individual we observe variables, say, $x_1, x_2, \cdots, x_m$, which we hope will explain the variable y. If at all possible, we wish to use linear explanation, so we are led to the consideration of models such as

$$y = X_1 \beta_1 + X_2 \beta_2 + \cdots + X_k \beta_k + d_k$$

in which each column of each X matrix is the set of observations on an explanatory variable, each vector β_j is a vector of constants to be determined, and d_k is a vector of deviations or disturbances.

The difficulties in this assignment are great. A consideration of these difficulties leads us almost inexorably into the history of the analysis of variance of the past 60 years. To document at length this view would entail a huge amount of writing. I can merely give a sketch by enumerating some difficulties.

1. Do we have an univariate situation in which the single variable y is caused by the explanatory variables? Or is there simultaneous causation (like the Maxwell equations, due originally, it seems, to Heaviside), so that there are several simultaneously dependent variables?

2. Even if we have a single dependent variable y, in what way is y to be explained? We can apply any monotone transform to y giving y* (equals, for instance, log y or 1/y, or whatever), and then consider the explanation of y*, and hence, by reversing the transformation, the explanation of y.

3. We can apply the same idea to each explanatory variable, e.g., $\ln x_i$.

4. We can consider new explanatory variables that are mathematical functions of the given explanatory variables, such as x_1/x_2 or $x_1/(x_1 + x_2 + x_3)$, and so on.

5. We must formulate an attitude and an approach to the disturbance vector d in our model equation.

As regards item 1 , the question goes back at least one century with discussion of nonsense correlations and the like. In my opinion, the best work of this century is that of Sewall Wright (1921) and his ideas of path coefficients. It is interesting that this work was highly regarded for two decades or so only in the theory of genetics, and was displaced somewhat by other approaches. But in the past decade, the ideas have been taken up very keenly in the area of social statistics. On items 2 and 3 , there was considerable attention to this type of problem in bioassay (see, e.g., Finney (1947)). More recently, there has been a more general approach by Kruskal (1964), which has attracted many thinkers. On item 4,

there has been little activity although the problem is now part of
the folklore of statistics.

The matter that has been explored in the part four decades is
item 5. It is obvious from the geometry of the analysis of variance
and the facts that the matrices $S_1, S_{21}, \ldots, S_{k \overline{k-1} \cdots 21}$ and S are
symmetric idempotent, and that a product of any pair is null, that a
single orthogonal transformation diagonalizes every one of them.

These purely algebraic results then mesh very nicely with the
situation in which the elements of the vector y are independent
Gaussian with the same variance, because geometrically orthogonal
functions of such random variables are independent. It follows then
that distributions of quadratic forms $y'S_\alpha y/\sigma^2$ are independent central
or noncentral χ^2 distributions and the panoply of Fisherian statistical
tests may be developed.

It is of interest that this entire process goes through very
nicely only with homoscedastic independent Gaussian random variables.
A loose, general statement is that if any of the three requirements,
homoscedasticity, independence, or Gaussianity is not met, the
situation becomes very difficult, if not impossible. Also it is of
interest to note that when these conditions are met distribution
theory is available only for very simple statistics. Consider, for
example, the problem of finding the distribution of the sum of the
absolute values of residuals. I mention these examples because we
never know the correct model, or even the correct linear model,
assuming linearity of model. So application of the ideas requires
a working logic for deciding if a model is appropriate.

It seems rather clear that such a logic must be based on two
ideas. We may use analogies between our present data situation and
our historical experience of situations that are "like" this present
situation. Just how one can develop an Aristotelian logic for such
a process seems rather moot--indeed, we may wonder if any such logic
can ever be developed. In the past decade or so, the view has been
expressed, it appears, that this problem may be solved in a logical
way by the use of Bayesian arguments. The history of this idea goes

back for more than a century; it has been widely accepted, then
rejected totally, then brought back to prominence again. The leading exponents of this century have been Jeffreys, de Finetti, and
L. J. Savage. However, these workers have had different stances
with, for instance, Savage proposing a theory of how a single individual should make a terminal decision. To what extent such a theory
may be regarded as contributing to the development of a model of the
world in which we live seems strongly questionable. It seems eminently
clear that there is a dichotomy between the world of facts and the
world of values, as it appears, Wittgenstein and many others emphasized.
I do not imply, thereby, that these two worlds must be integrated
eventually and in human actions, but it does seem that there is a
world of facts that is largely independent of the world of values.

The other idea as to choice of model is that the model and the
data must be consonant in some sense. So there must be some procedures by which the goodness of fit of a proposed model may be examined.
The problem arises in purely mathematical work, of course, the
elementary example that comes to my mind being remainder formulas
for expansions. Goodness of fit in the statistical sphere must be
based, it seems, on probability ideas. A model that states that the
probability of the given data is zero is surely unacceptable. In
most cases of probability models, however, almost any result has
some probability and approach to goodness of fit of a model for
explaining a variable y in terms of some explanatory variables may
be made by examining residuals of observed data, i.e., $y - \hat{y}$, where
$\hat{y}$ is the fitted vector. Almost all statistical tests in linear
models and analysis of variance situations consist of examining
residuals and functions of residuals. I find it interesting that
the paper of Fisher (1920) referred to above has a section entitled
"Correlation of Residuals." Also, of course, the original Karl
Pearson χ^2 test criterion is a function of residuals, as are many of
the statistical tests developed during the past 50 years. There
has been a trememdous amount of activity in this direction. One
example that pleases me very highly is the one degree of freedom

for nonadditivity of Tukey (1949), which is aimed at analysis of variance situations and has a very simple structure, related completely to the algebraic decomposition of a vector y described earlier.

The big revolution in this area in the past two decades has been the development of the computer. Without a computer, the computation of residuals is impossible, but with a computer, one can obtain residuals, one can plot them in various ways, and one can get practically useful solutions to distributional problems that will defeat our most expert mathematical minds by Monte Carlo methods. I am very glad that a significant portion of this historical symposium is being devoted to uses of the modern computer.

I have talked about the univariate analysis of variance. The extension of the arithmetic or algebra to the multivariate analysis of variance is really quite trivial. Suppose that we have "independent" vectors $y_1, y_2, \ldots y_r$. Then, just as we have a sum of squares $y_i' S_\alpha y_i$, for the αth source in the univariate case, we now have r sums of squares $y_i' S_\alpha y_i$, and we also have $r(r - 1)/2$ sums of products $y_i' S_\alpha y_j$. The whole analysis of variance then becomes a partition of the matrix of total sums of squares and sums of products, as a sum of several matrices. With such a partition, one can consider several interesting operations, such as reducing the matrices to diagonal form with orthogonal transformations, examining the roots of each matrix and of (A - B), where A and B are two of the matrices in the partition, and so on.

When one adjoins random variable distributions as the source of disturbances one encounters a huge jump in mathematical difficulty. There has been a huge effort in the past three decades, following basic work by Hotelling and Fisher, by S. S. Wilks, S. N. Roy, C. R. Rao, T. W. Anderson, and others.

CLASSIFICATORY AND CONTINUOUS EXPLANATORY VARIABLES

I have stated that the analysis of variance is based on a linear model

$$y = X\beta + \text{deviation}$$

or

$$y = X_1\beta_1 + X_2\beta_2 + \cdots + X_k\beta_k + \text{deviation}$$

Each column of the matrices $X_1, X_2, X_3, \ldots$ contains the values of a hypothesized explanatory variable or factor for the individuals observed. No special account is taken in the earlier presentation of any particular characteristics that the X matrices may possess.

It is now useful to make a dichotomy. The explanatory variables may take values over some continuum, the corresponding factors of which are then called quantitative or continuous. In contrast, the explanatory variables may take the values 1 or 0, "1" occurring if the observation falls in a particular subclass of possibilities and "0" occurring if it does not fall in a particular subclass. Models with X matrices having such a property may usefully be called classificatory models. At times in the history of statistics, the term "analysis of variance" has been restricted to encompass only such cases. The early paper by Irwin (1931), for instance, deals with this case, although not in the terms given earlier. I now take up this area, which I will call Qualitative Factorial Situations. This will then lead naturally to factorial design.

QUALITATIVE FACTORIAL SITUATIONS

Consider the growth of a biological organism in the presence and absence of each of, say, five potentially nutritious stimuli. There are then $2^5 = 32$ combinations of stimuli. We may easily envisage more factors and more possibilities or levels for each factor, as, for example, a chemical reaction taking place with one of four catalysts. In general, then, we envisage a lattice of combinations of factors. This configuration obviously occurs, in experimental science, but it also occurs in observational science; for instance, we may envisage the possibility of explaining a quantifiable political attribute of individuals in terms of sex, male or female, political party with a few defined, say four possibilities, educational level, grouped into, say four classes. We envisage a population of individuals each of

whom has a particular sex, party, and educational level. Each
individual occurs at one of the points of a 2 × 4 × 4 lattice. Not
all such situations are "small." Let me give a realistic "large"
situation that may easily occur in agricultural science. Suppose we
have 100 varieties of corn, grown at each of five locations, in each
of three years with each of two nutritional regimes. Then, considering just a single attribute such as lysine content of the corn, we
shall obtain 100 × 5 × 3 × 2 = 2,500 numbers. Suppose, furthermore,
that there is no problem of error (or "wandering") of the data. Then
our problem is to make sense of this large set of numbers. There is
no problem of induction of inference. The task is simply to get into
one's mind the nature of the 2,500 numbers. One might take the view
that one should simply file the numbers with some sort of easy access
system, and then, if one needed one of them, go to the system. But,
generally speaking, such a suggestion is quite useless because
humanity is collecting such data sets by the million. Ways must be
found to condense such data sets to small sets of numbers that contain
the gist of the total data set. The most useful general-purpose
algorithm for looking at and condensing such data sets is the analysis
of variance and associated techniques. I mention this to counter the
idea that analysis of variance deals with random variables. I
deliberately choose a situation in which there are no random variables
and state that the analysis of variance is helpful in appreciating
the nature of the set of 2,500 numbers. The procedure is related
somewhat to the simple Chebychev idea that one can form a *partial
approximate* idea of the nature of a set of N numbers by computing
the mean and the mean square deviation.

FACTORIAL EXPERIMENT PROBLEMS

The general background is that we wish to determine the effects
of a number of experimental factors when applied to a collection of
experimental units, e.g., microbes, mice, men, plots of land, pieces
of metal, batches of oil, and so on. It may be possible in the so-
called exact sciences to obtain experimental units that are essentially

identical, e.g., tubes of a chemical solution prepared very carefully to be nearly identical. But the general case is that experimental units that are identical to each other are impossible to obtain. This situation may be illustrated by the following table.

We may envisage the existence of a number for each cell: y_{ij}, i indexing the unit and j the treatment. But we observe at best only one number in each row. The question is then: How can we form an idea of the totality of numbers $\{y_{ij}\}$? The Fisherian answer, introduced in the early 1920s, is to use randomization, which consists of drawing a sample from the table by use of a random-number generator. This leads to what one may call a fractional replication of the two-way table that is forced by the nature of the circumstances. The notion was introduced in a pure sampling context in the 1950s under the name "deep sampling."

   The next step in experimentation was the introduction of a grouping of the experimental units into blocks, with intense search for good incomplete plans by Yates (1935, 1936), Bose (1939), and others. A reference list would be huge. Also came the idea of two-way arrangement of units, and of split-plotting, which uses a double hierarchical arrangement of units, and so on.

Each of these schemes for observation uses what may be termed random fractional replication of the total treatment-unit array. Associated with each scheme is an analysis of variance which has aspects of validity and utility.

Procedures of blocking have some formal resemblance to ordinary schemes of fractional replication of fractional systems that I discuss in the following section.

FRACTIONAL REPLICATION

Consider a qualitative factorial situation in which we have factors A with a levels, B with b levels, C with c levels, and so on. Then the total number of fractional combinations is abc..., which may be a very large number. One can easily meet situations in applied science with 6, 8, or 10, or even 20 factors. If we have 10 factors each at 2 levels, there are, obviously, 1,024 combinations. The experimenter would be most unwise, generally speaking, to perform an experiment on all 1,024 combinations. He should make a partial examination of the total set. But how may he reasonably pick out a subset? Ideas were put forward initially by Yates (1935) and by Tippett (Fisher, 1935). The Yates problem was rather neat: If one has objects, a, b, c, ... to weigh and one can obtain the weight of any combination of objects with constant error variance, what combinations of the objects should be weighed? The Fisher-Tippett problem was that there were 6 factors each at 5 levels, with 3,125 combinations in all, and one wished to make a guess as to which factor was producing deviant observations.

The next step was made again by Fisher (1942) who developed the equivalent of fractional replication of treatment-block combinations for factors with a prime number of levels. In Fisher's work, the problem was called the development of confounding systems for factorial designs. The fractional replication of factorial systems was given its first definitive approach by Finney (1945).

The later 1940s saw a concurrence of ideas of factorial systems, hyper-Latin squares, and confounding for the case of a set of factors with the same prime power number of levels. All this development was based totally on Galois field theory with important work by Bose, Rao, and others.

During the past 30 years there has been a tremendous development of ideas of fractional replication of factorial systems, although the nonprime and unsymmetrical cases are not really at all easy to treat, because basic Galois field theory is not available. It is interesting that this area is related to communication theory, as discussed by Bose and others.

RANDOM SAMPLING OF A MULTIDIMENSIONAL LATTICE

Another stream of thought has tackled the problem of forming a partial approximative idea of a multidimensional rectangular lattice of numbers by sampling. The idea is that we have a population of numbers $y(x_1, x_2, \ldots, x_k)$, $x_i = 1, 2, \ldots, L_i$, or a total population if $N = \Pi L_i$ numbers. One way of proceeding is to take a random sample of ℓ_i of the levels of the ith factor from the totality of L_i levels. Then observe the numbers $y(x_1, x_2, \ldots, x_k)$ for the combinations of selected levels. A data analysis is then to perform an analysis of variance of the observations. The analysis of variance may be used to obtain estimates of the corresponding parameters for the full lattice of numbers. The basic properties of such sampling with regard to expectations of means and expectations of mean squares in the analysis of variance have been obtained [e.g., Cornfield and Tukey (1956) and Wilk and Kempthorne (1956)]. If all levels, of a factor are chosen in the sample, the factor is called "fixed"; if a random sample only is taken, the factor is called "random." This relates to the general development of fixed, mixed, and random models, which may be tackled in this way, or in other ways related to the multivariate distribution of Scheffé (1956) and other workers. The problems here merge clearly into the general components of variance problem first discussed by Fisher (1918).

CONTINUOUS FACTOR PROBLEMS

A fairly general problem is that we have a situation in which a
dependent variable y is considered to depend on continuous variables
$x_1, x_2, \ldots, x_k$, and that we may write

$$y(x) = \eta(x) + \text{error}$$

where $x' = (x_1, x_2, \ldots, x_k)$ and the vector x ranges over some region
in k-dimensional Euclidean space. In the early days, it was common
to lump this situation with that of qualitative factors. The general
problem is that $\eta(x)$ is unknown; we do not even know the form of $\eta(x)$.
In the case of a single factor, it could be

$$\eta(x) = \beta_0 + \beta_1 x$$

or

$$\eta(x) = \beta_0 + \beta_1 x + \beta_2 x^2$$

or

$$\eta(x) = e^{\beta_0 + \beta_1 x}$$

or

$$\eta(x) = \beta_0 + \beta_1 \tanh^{-1}\left(\frac{x}{\beta_2}\right)$$

and so on.

Clearly, the problem is horrendous. It is surely a nightmare to
the logician or to the pure mathematician. The functions are usually
thought of as being piecewise continuous, or else there seems to be
no possibility of solution. Consider the experimental problem of
determining a function that takes one value for rational values of x
and another value for irrational values of x. But it is of no help
to humanity or to natural human desires to throw up one's hands in
horror at the problem. Much progress has been made in the practical
problem, e.g., of the pure physicist or chemist, by collecting observations at various x values, and then subjecting the resultant data
to the ingenuity of the human mind.

It is only in the past 30 years that mathematically formulated approaches to the general problem area have been proposed and developed. It is fair history, I believe, to state that the first serious and sustained approach to a subproblem of the general problem was made by Box and Wilson (1951). They tackled the problem: Assume that $\eta(x)$, where x is a vector, is unimodal and continuous, and develop a strategy for finding the particular vector x at which $\eta(x)$ is a maximum. They used the ideas of steepest ascent. It would take me away from general history to discuss the merits and demerits of this proposal, which has been applied widely. Other algorithms have been proposed, e.g., Partan, by Shah et al. (1964). This area merges rather quickly into the general area of numerical optimization, and a rather large number of algorithms have been advanced. It is worth noting that almost any algorithm for numerical optimization on a computer can be adapted to be an algorithm for process optimization and vice versa.

The procedures mentioned in the previous paragraph are only superficially involved with the problem of experimental or measurement error. Explicit attention to the stochastic element has been given in what must be regarded, I think, as highly sophisticated mathematical work on the probability structure of long-term schemes for attaining an optimum in the presence of noise. The problem was named "Stochastic Approximation," and the early basic work was done by Robbins and Monro (1951) and by Kiefer and Wolfowitz (1952). A huge stream of mathematical-statistical work ensued.

Another stream of work connected with factorial situations goes under the name of "Optimal Design." A good review is given by Fedorov (1972). The basic work assumes that the form of (x) is *known* to be

$$\theta_1 f_1(x) + \theta_2 f_2(x) + \cdots + \theta_k f_k(x)$$

where $f_1(x)$, $f_2(x)$, ..., $f_k(x)$ are known. The estimation by least squares or best linear unbiased estimation of the vector $\theta' = (\theta_1, \theta_2, \ldots, \theta_k)$ is quite trivial. The problem is to make a choice of points x at which observations should be made. The problem is difficult because there are several measures of goodness of estimation

of the vector parameter, no one of which is totally compelling.
The basic work was done by de La Garza (1954), Guest (1958), Hoel
(1958), Kiefer (1959). Extensive developments have been made by
Kiefer and others. Work on other lines in the same area has been
done by Box and several coworkers. The prime difficulty in this
area from the viewpoint of the user is that the functional form of
$\eta(x)$ is usually unknown. Error control is important, but the problem
of bias in estimation cannot be omitted from a full consideration.

CONCLUDING REMARKS

The history of ideas of the first half of the twentieth century
is well described in the statistical texts listed at the end of the
references.

I think it is objectively verifiable that the past 50 years have
witnessed tremendous growth and development of an area of human
thought that barely existed a century ago except as a matter of
counting existent populations of social interest. One needs only to
compare the disciplines represented in university and college faculties
and in courses offered, for the years 1920 and 1974.

That this has happened is not to be wondered at, in my opinion,
because statistics with probability theory is clearly aimed at the
overall objective of understanding and developing modes of accommodation
with the real world. The subjects have really only just begun their
growth, remarkable though that growth has been.

REFERENCES

Bose, R. C. (1939). On the construction of balanced incomplete
    block designs. *Ann. Eugenics (London) 9,* 353-399.

Box, G. E., and Wilson, K. B. (1951). On the experimental attain-
    ment of optimum conditions. *J. Roy. Statist. Soc. B 13,* 1-45.

Cochran, W. G. (1934). The distribution of quadratic forms in a
    normal system, with applications to the analysis of covariance.
    Proc. Cambridge Phil. Soc. 30, 178-191.

Cornfield, J., and Tukey, J. W. (1956). Average values of mean
    squares in factorials. *Ann. Math. Statist. 27,* 907-949.

de La Garza, A. (1954). Spacing of information in polynomial regression. *Ann. Math. Statist. 25,* 123-130.

Fedorov, V.V. (1972). *Theory of Optimal Experiments.* Academic Press, New York.

Finney, D. J. (1945). The fractional replication of factorial experiments. *Ann. Eugenics (London) 12,* 291-301.

Finney, D. J. (1947). *Probit Analysis.* Cambridge Univ. Press, Cambridge.

Fisher, R. A. (1918). The correlation between relatives on the supposition of Mendelian inheritance. *Trans. Roy. Soc. Edinburgh 52,* 399-433.

Fisher, R. A. (1920). Studies in crop variation. I. An examination of the yield of dressed grain from Broadbalk. *J. Agr. Sci. 11,* 107-135.

Fisher, R. A. (1935). *The Design of Experiments.* Oliver & Boyd, Edinburgh.

Fisher, R. A. (1939). The sampling distribution of some statistics obtained from non-linear equation. *Ann. Eugenics (London) 9,* 238-249.

Fisher, R. A. (1942). The theory of confounding in factorial experiments in relation to the theory of groups. *Ann. Eugenics (London) 11,* 341-353.

Guest, P. G. (1958). The spacing of observations in polynomial regression. *Ann. Math. Statist. 29,* 294-298.

Hoel, P. G. (1958). Efficiency problems in polynomial estimation. *Ann. Math. Statist. 29,* 1134-1146.

Irwin, J. O. (1931). Mathematical theorems involved in the analysis of variance. *J. Roy. Statist. Soc. 94,* 284-300.

Kiefer, J. (1959). Optimum experimental designs. *J. Roy. Statist. Soc. B 21,* 272-319.

Kiefer, J., and Wolfowitz, J. (1952). Stochastic estimation of the maximum of a regression function. *Ann. Math. Statist. 23,* 462-466.

Kruskal, J. B. (1964). Multidimensional scaling by optimizing goodness of fit to a nonmetric hypothesis. *Psychometrika 29,* 1-27.

Madow, W. G. (1938). Contributions to the theory of multivariate statistical analysis. *Trans. Am. Math. Soc. 44,* 454-495.

Robbins, H., and Monro, S. (1951). A stochastic approximation method. *Ann. Math. Statist. 22,* 400-407.

Scheffé, H. (1956). Alternate models for the analysis of variance. *Ann. Math. Statist. 27,* 251-271.

Shah, B. V., Buehler, R. J. and Kempthorne, O. (1964).   Some
    algorithms for minimizing a function of several variables.
    J.S.I.A.M. 12, 74-92.

Tukey, J. W. (1949).   One degree of freedom for non-additivity.
    Biometrics 5, 232-242.

Wilk, M.B., and Kempthorne, O. (1956).   Some aspects of the analysis
    of factorial experiments in a completely randomized design. *Ann.
    Math. Statist. 27*, 950-985.

Wright, S. (1921).   Correlation and Causation.   *J. Agr. Res. 20*,
    557-585.

Yates, F. (1935).   Complex experiments. *J. Roy. Statist. Soc.
    (Suppl.) 2*, 181-247.

Yates, F. (1936).   Incomplete randomized blocks.   *Ann. Eugenics
    (London) 7*, 121-140.

STATISTICAL TEXTS FOR SUGGESTED READING

Cochran, W. G., and Cox, G. M. (1950).   *Experimental Designs*.   Wiley,
    New York.

Cox, D. R. (1958).   *The Planning of Experiments*.   Wiley, New York.

Kempthorne, O. (1952).   *The Design and Analysis of Experiments*.
    Wiley, New York.

Rao, C. R. (1952).   *Advanced Statistical Methods in Biometric
    Research*.   Wiley, New York.

Scheffé, H. (1959).   *The Analysis of Variance*.   Wiley, New York.

Snedecor, G. W. (1937).   *Statistical Methods*.   Iowa State College
    Press, Ames, Iowa.

3.

INDUSTRIAL EXPERIMENTATION
(1955-1965)

PETER W. M. JOHN was born August 20, 1923, in Porthcawl, Wales. He
received his B.A. degree in 1944, M.A. degree in 1948, and the
Postgraduate Diploma in Statistics in 1949, all from Oxford
University. He was a Royal Air Force officer during World War
II. He came to the United States in 1949 to be a temporary
instructor at the University of Oklahoma and has stayed ever
since. He became a naturalized U.S. citizen in 1955, the year
in which he received his P.D. degree in Mathematics at the
University of Oklahoma. From 1957 to 1961 he was a Research
Statistician for Chevron Research Corporation (Standard Oil of
California). He taught at the Berkeley and Davis Campuses of
the University of California before coming in 1967 to the
University of Texas at Austin, where he is a Professor of
Mathematics. His main research interests are in the design of
experiments and applied multivariate analysis. His book,
Statistical Design and Analysis of Experiments, was published
by Macmillan in 1971.

INDUSTRIAL EXPERIMENTATION
(1955-1965)

Peter W. M. John

Department of Mathematics
University of Texas
Austin, Texas

It has been 25 years since I attended my first and only formal course
of lectures on the design of experiments. It is trivial to say that
the subject has grown since that time. At the same time, I was also
reading a British government handbook on statistics by K. A. Brownlee,
which was, to my knowledge, the first book available that dealt with
industrial applications. In both the book and the course, we considered, as I recall, the basic ideas of the factorial experiment in the
context of that point in time. In those days, we had not yet heard
much about mixed, fixed, and random models; the basic technique for
the analysis seemed to me, in my innocence, to be to start at the foot
of the analysis of variance table and to work one's way up, testing
each interaction against the residual mean square, and pooling the nonsignificant sums of squares as we went with the aim of getting as many
degrees of freedom as possible into the error term. We learned a little about balanced incomplete block designs and about the complete 2^n
factorial design with confounding.

It was not until eight years later that I had occasion to think
about the design of experiments again. I had just become a research

statistician for a large oil company, which explains why I interpret
the title "Industrial Experimentation" very narrowly as confined to
applications in chemical engineering. Almost immediately, it became
apparent that my old lecture notes would not contain the answers to
the questions that were going to arise. There were fundamental
differences between the experiments that my new colleagues in
industry were performing and the experiments that I had learned about
eight years earlier within 100 miles of Rothamsted.

What were, or indeed are, these fundamental differences? There
are some who argue that industrial experiments have smaller variances
than do agricultural experiments because of the more sophisticated
nature of the equipment involved. That may well be the case when we
are talking about the repeatability of experiments carried out under
carefully controlled conditions in a laboratory, but it is not
necessarily true when one is faced with a random sample of taxi cabs
from greater Los Angeles. However, we do more often have reliable
prior estimates of variances that can be used in the analysis. More
important differences are the cost of experiments and the speed with
which they can be performed and analyzed. Some experimental runs on
an elaborate piece of equipment in a refinery may cost several thousand
dollars apiece. The cost of a few more plots of oats or barley may be
minimal. The agronomist has to plant his entire experiment at the same
time and must gather his data several months later. He had better be
able to answer his questions from that batch of data, because if he
needs more he must, like the Dallas Cowboys, "wait until next year."
The engineer can carry out an experimental run (i.e., obtain a datum
point) in a few hours, and have its results back from the analytical
laboratory perhaps the next day. Therefore, he will tend to think of
carrying out a small experiment, perhaps a fractional factorial design.
Having analyzed that, he will incorporate the results in the planning
of the second experiment and so on, reaching his target by a sequence
of small intermediate experiments. This is most apparent in response
surface methodology in which the scientist starts his investigation
with a set of operating conditions at which a plant is performing at
less than the maximum rate of production and makes his way by the

method of steepest ascent to the set of operating conditions at which
the yield of the plant is maximized.

TOPICS OF INTEREST

I intend now to say a few words about some of the ideas and topics that
were new and interesting to us in those days. Some of the so-called
new things were new to me only because I had just not heard about them
before, but others were topics at the frontier of research in experimental design sparked by industrial problems. Any list of topics has
to be to a certain extent personal, topics that were interesting and
exciting to me; that is right and proper within the terms of reference
of this symposium. Because of limitations of time and space, it must
also be selective; some things have had to be left out, which I should
like to have included.

ONE-WAY LAYOUT IN THE ANALYSIS OF VARIANCE

I begin with the one-way layout in the analysis of variance. I thought
that there, at least, was a subject that I had mastered. It was really
very simple. One computed the sums of squares for treatments and error,
performed an F test, and proceeded from there. If the value of F was
not significant, one declared that there were no differences between the
treatments and that was the end. If the F value was large enough to
justify one or two stars, one computed Fisher's least significant
difference and went to work comparing treatment means.

I soon came to the conclusion that the question to which the F test
is supposed to provide the answer was not very often relevant. My
colleagues were often thoroughly convinced before they started that the
treatments differed in their responses. Not even the most spendthrift
of them would have run an experiment to test the hypothesis of equal
means. Failure to obtain a significant value of F was only an indication that either r or α was too small. (This was not, of course,
entirely a waste of time if it convinced the experimenter that his prior
estimate of error, which was based on local myth, was really too low.)
The interesting questions to them were "which treatments differ and by
how much" and "which is the best treatment."

I am now inclined to think that the question of deciding which
treatments differed caused much more turmoil than it deserved. As I
mentioned before, we used to apply Fisher's least significant differ-
ence to all the differences $\bar{y}_h - \bar{y}_i$ with an error risk of α for each
comparison. The idea that we should perhaps be thinking of an error
rate α per experiment in the sense that, if the null hypothesis were
true, we should, with probability $1 - \alpha$, find all the differences
between means to be nonsignificant, had been raised by Keuls just
prior to World War II. After the war, several statisticians in America
worked on the problem with major contributions from Tukey whose modified t statistic was based upon the Studentized range, from Scheffé
with his S statistic, which is essentially a hunting license for looking at all possible contrasts between the treatments, and Duncan's
multiple range test. It was not as simple as I had thought.

MATRICES

In 1957 there were very few books available on the design of experiments. Kempthorne's book had appeared in 1952. The book by Cochran
and Cox preceded it in 1950. In 1954 a group of statisticians in
England, who were working in the several laboratories of Imperial
Chemicals Industries, combined their efforts in a book edited by O. L.
Davies, and in the same year Wiley published the book by Bennett and
Franklin on statistical analysis in chemical engineering. Mathematically, the most advanced of these four books is Kempthorne's, and one
of its most striking features was the use of matrices. Matrices are
used, but even then only lightly, in one chapter of Davies' (1954)
book; I do not recall that they are mentioned in the other two. I am
not going to claim that Kempthorne's was the first book I ever saw on
statistics that made use of matrices, because I did have in my library
Cramèr's (1946) book in the Princeton series; I think Kempthorne's was
the second. Kendall did not make use of matrices in his two volumes,
which appeared at the time of World War II (Kendall, 1943).

It may be somewhat surprising to our younger colleagues that the
use of matrix algebra is relatively new in statistics. Matrices appeared in the general syllabus for the B.A. in mathematics at Oxford in

the 1930s. They appeared in this country in the undergraduate curriculum for most postwar mathematics students only with the general
adoption of the trail-blazing textbook by Birkoff and MacLane (1941).

The early and mid-1950s thus saw the emergence of a growing
trickle of statistical practitioners who knew some matrix algebra,
and who were therefore equipped to think of the analysis of variance
and the design of experiments in the context of the general linear
hypothesis and the regression model. We also had access for the first
time to electronic computers and so we discovered multiple regression.

It became possible to fit hyperplanes to predict a response Y
from several predictor variables X_1, X_2, ... both quickly and conve-
niently. What had been impracticable because of the man-hours involved
in inverting 8 × 8 matrices on desk calculators suddenly became very
easy. I say 8 × 8 matrices because our computer was very small and
immature by modern standards, and that was about the number of predictors in the average data set, for a large problem.

Some of our earlier difficulties are no longer problems with the
improved computers. Our old machine had a certain amount of difficulty
in trying to invert matrices that were nearly singular. Such matrices
could occur in two ways. The first cause was high correlations between
the predictor variables; we did not realize at first that in fitting
a polynomial in X, where X takes the values 1, 2, 3, ..., the variables
X and X^2 are highly correlated and lead to a matrix that our computer
could not invert; that is a problem that can be taken care of easily
if the X values are equally spaced by using orthogonal polynomials, but
it is a bit of a nuisance if there are gaps. Another source of nearly
singular matrices lay in the regression program that came with the machine. It called for inverting the correlation matrix. Computing
the off-diagonal terms of this matrix involves dividing by the square
roots of sums of squares. Sometimes we would get enough round-off
error that if, for example, an engineer naively used for a subset of
his predictors a group of X variables that were constrained to add up
to 100%, the resulting matrix would be nonsingular with a determinant
that was nearly zero. The computer would huff and puff in its attempt
to invert it, until the operator mercifully called a halt to the

proceedings. That is what happened when we were lucky. If we were unlucky, the machine inverted the correlation matrix as presented and gave us a prediction equation, the coefficients of which had huge variances and covariances. This could be particularly unfortunate because some of the engineers liked to look only at the prediction equation and not to be bothered with such troublesome details as the covariance matrix of the estimated coefficients.

PREDICTOR VARIABLES IN REGRESSION

The most interesting question that arose in those days is still with us. How do you find the best subset of the predictor variables? There seemed to be two different groups of practitioners. Those favoring the "kitchen-sink" technique would set up a model that contained all the X variables at their disposal, and then drop variables one at a time by looking at the partial correlation coefficients or the increase in the error sum of squares. They continued until no more variables could be dropped without a substantial jump in the error sum of squares. The alternative was stepwise regression, in which we brought in, or added, the predictor variables one at a time; at each stage, we chose as the new addition that predictor which gave us the largest reduction in the sum of squares for error; we went on until some arbitrary stop rule told us that we had gone far enough.

Lately, more sophisticated procedures have been developed. They are discussed in a series of papers that have appeared in *Technometrics* written by Mallows (1973), Gorman and his colleagues at American Oil Company (1966), and Hocking and his students (1967); in the same journal, Hoerl and Kennard (1970) recently proposed a modification of the standard least-squares procedure called ridge analysis, which has caused a certain amount of controversy.

It is appropriate to mention in passing that the journal *Technometrics* is, itself, a product of the growing interest in industrial experimentation. Its godfathers were the American Statistical Association and the American Society for Quality Control. The journal is now in its sixteenth year and seems to be a healthy child.

FRACTIONAL FACTORIALS

Another area in which the emphasis on the regression approach led to advances was that of fractional factorials, especially with factors at two levels. The 2^n factorial design and the corresponding computational algorithm were introduced by Yates (1937) in a monumental paper which was published by the Imperial Bureau of Soil Science as a pamphlet. For n factors at two levels each we need 2^n points and this number increases very quickly. For seven factors we need 128 runs, and for 15 factors 32,748 are required. We might well ask how we can design an experiment to serve our purposes with only a fraction of this number of runs.

Fractional factorials were introduced by Finney (1945) in an agricultural setting, and his papers appeared in the *Annals of Eugenics*. Earlier writers spoke of the factors as taking levels denoted by zero for the low level and one for the high level. In the framework of the regression model, we associated the ith factor with a variable x_i, which takes only the values ± 1. We are thus experimenting at the vertices of an n-dimensional hypercube with coordinates ± 1. The expected response $E(Y)$ is the sum of first-degree terms $\beta_i x_i$ for the main effects, double products $\beta_{hi} x_h x_i$ for the two-factor interactions, and so on. We can take a 2^3 basic design with eight points in three factors A, B, C and make it into a 2^{7-4} fraction by adding the four extra factors in the following way. We introduce the fourth factor D by setting $x_4 = x_1 x_2$, or putting D = AB; then we put E = AC, F = BC, and G = ABC. From this eight-point design we can obtain unbiased estimates of all seven main effects, as long as we are prepared to assume that all the interactions can be ignored. Alternatively, we may say that the main effects are estimable if we suppress the interactions and confine ourselves to a first-order model. The extension to 15 factors in 16 runs is obvious.

These are two examples of saturated fractions. The 16-point design provides 15 degrees of freedom, every one of which is used to estimate some effect. There are no degrees of freedom left over for error.

The designs were not entirely new. They had occurred as blocks
in confounding arrangements suggested by Fisher (1942), when he showed
how a 2^n design could be divided into blocks in such a way that each
and every interaction with three or more factors, and only those, is
confounded with blocks. At the same time, in a completely different
context, Plackett and Burman (1946) had found a family of saturated
orthogonal main effects plans for 4t - 1 factors in 4t runs, which
were based upon Hadamard matrices. When 4t is a power of 2 the designs
that we developed earlier are isomorphic to the Plackett and Burman
designs, but the latter system also provides designs for 12, 20, 24,
... runs.

There were three new aspects in the postwar development of this
field. The first was the idea of the screening experiment, which has
its ancestry in the Plackett and Burman designs and perhaps, going
back further to Tippett's experiment with a Greco-Latin square. In
its simplest form, this is an initial experiment in which we take a
quick look at all the factors which might affect the response under
investigation. The underlying idea is that only a few of these factors
will have any effect. The estimates of the main effects of these
factors will stand out from the mass of mediocrity of the others, and
we shall choose these prominent factors for further more detailed
investigation. Saturated orthogonal main effects plans for factors
at two levels are clearly appropriate here, but unless n = 4t - 1 we
cannot find them. So we spent some time hunting for designs that were
nearly saturated, or designs in which part of the orthogonality property was sacrificed. Some experimenters even investigated supersatu-
rated designs in which the number of factors exceeds the number of
degrees of freedom.

So far we have considered models that had ignored the interactions.
By and large we are still prepared to suppress the interactions involving three or more factors, but the two-factor interactions are often
of interest. The second aspect to be mentioned is the classification
of fractions according to their ability to provide estimates. The

classification that has persisted is that of Box and Hunter (1961).   It
is a revision of an earlier classification proposed by Box and Wilson
(1951).   Fractions were divided into three types: resolution III in
which the main effects are estimable when all interactions are ignored
(we have just seen examples of them); resolution IV in which main
effects are estimable whether or not we ignore two-factor interactions,
but the latter are not all estimable; and resolution V in which we can
estimate all main effects and all two-factor interactions.

The Plackett and Burman designs are of little use as resolution
IV or V designs. The sequence of designs introduced by Finney have
the restriction that they must have a number of points which is a
power of 2; the smallest resolution V design in this series for eight
factors is the quarter replicate with 64 points. That is rather a
large number of points to estimate a mean, eight main effects, and 28
interactions, and so the hunt was on to find fractions that have
resolution IV or V and which have a number of points that are intermediate between the powers of 2. In the particular eight-factor
example that I have mentioned, there is a useful design that was
discovered independently by Addelman (1961) and myself (1962), which
needs only 48 points.

Some of the most important work in this area was carried out by
Webb (1968) and Margolin (1969) for resolution IV designs. Webb showed
that every resolution IV design for n factors must contain at least
2n points; such designs with exactly 2n points are called minimal.
Webb conjectured, and Margolin proved his conjecture, that every
minimal resolution IV design is a foldover design, which means a design in which the points occur in complementary pairs.

The third aspect to which I have time to give only a passing
reference is the graphic method by Cuthbert Daniel for deciding which
effects, if any, are significant in a fraction with 2^{n-k} points. This
is the method of the half-normal plot, which was discussed in detail
by Daniel in one of the early issues of *Technometrics* (1959).

RESPONSE SURFACE METHODOLOGY

The most important application of regression models was in the area
of response surface methodology. This is the area in which George
Box (1957) has made his major contribution as both a statistician
and a chemical engineer, first in the laboratories of Imperial Chemicals Industries, then in the Statistical Techniques Research Group at
Princeton, and since that time at the University of Wisconsin.

The ideas involved are most easily appreciated in the simple
case of a single response Y, which depends on two factors X_1 and X_2.
We can envision X_1 as denoting the temperature of the reactor and
X_2 as the pressure. We assume that we have "knobs" on the reactor
that enable us to control X_1 and X_2. Suppose that we are interested
in some response Y, such as the yield or the percentage conversion,
and that Y is measured along the third axis coming up at us out of
the paper. The situation is as if we had a bird's-eye view of X_1,
X_2 space with hills and valleys corresponding to the various values
of Y. The two-dimensional picture is that of a map with contours of
constant yield. We are not really interested in the topography of
the whole Euclidean plane, but somewhere, perhaps, is a mountain with
a summit, and that is at the maximum attainable value of Y.

It would be nice if we were able to run our plant at that point,
if we could find the values of X_1 and X_2 which optimize Y. But life
is not that simple. Suppose that X_2, instead of being the pressure,
is the strength of a catalyst. As time goes by, the catalyst loses
strength. Can we alter the temperature in such a way that as X_2
decreases we lose yield--we have to--but we lose it as slowly as
possible? We have, therefore, two questions. How do we find the top
of the mountain? After that, how do we map the region around the peak?

Suppose there are n factors; it is a lot simpler if n is 2, or
3, or 4. We begin our investigation at the point at which the plant
is running now. This may be the point that was recommended by the
engineers on the basis of their pilot plant experiments as being their
best guess at the optimum, but the whims of scaling up from the pilot
plant to the full-scale reactor may have made it decidedly suboptimal.

In the simplest terms, we take a main-effects fraction to find
the direction cosines of our path of steepest ascent. We explore
in this direction, striding along like a man with tunnel vision until
we stop climbing. Then we carry out a further fractional 2^n and find
a new direction in which to continue our climb. We continue this
process, zigging and zagging a bit because of our tunnel vision, until
we reach a point from which our probing steps to the North, South,
East, and West are all downhill, and we realize that we must be near
the peak. At this stage, we run a larger experiment with at least
three levels of each factor, and fit a second-degree model--the general
quadric. The center of the quadric will be our estimate of the optimal
point, and the first principal axis of the quadric, if it is an ellip-
soid, will give the answer to the second question--the direction of
the most gentle path from the top, the path along which the yield
decreases most slowly.

This has, indeed, been a fertile field for research and application. We need good fractional factorials to help us climb the mountain. When we get to the top, we shall need to be able to add a
second set of points to the last 2^n fraction to give us a good second-
order design. We found ourselves involved with questions about orthogonal designs, by which we meant designs in which the estimates of the
quadratic coefficients are uncorrelated, with rotatable designs and
with problems of orthogonal blocking, to mention only a few of the
exciting topics.

Before I leave this topic, I should like to say a few words in
praise of a by-product of response surface methodology. One of the
troubles with practical response surface optimization is that it
involves turning experimenters loose on an operating plant, where
they are not, in the normal way of doing business, always welcome.
Plant foremen, quite understandably, are apprehensive that researchers
will, at the very least, cause a temporary, if not permanent, loss of
production. Research engineers are dangerous enough, but the thought
of statisticians roaming around with a free hand is positively alarm-
ing. I commend to you, however, and to any engineers the method of
evolutionary operation, which George Box introduced to the literature
in 1957.

This is a method of running a plant so as to produce both product and information. By repeated small changes in two or three factors, which corresponds to running replicated 2^2 or 2^3 designs, in a framework that can be handled by the plant operators themselves, the plant is slowly nudged into more efficient operating conditions as a part of a standard operating procedure. The most important thing about the procedure is that it can, and should, be run entirely by the plant personnel. They can follow what is going on; they can become involved in the process. The changes are gentle; nothing drastic is done to foul up the plant, and there is no way to go but up.

THE GORDON RESEARCH CONFERENCES

No account of the advances in industrial experimentation in the 1950s and 1960s would be complete without a reference to the important part played by the Gordon Research Conferences. These conferences on topics in the general area of chemistry are held each summer in prep schools in New Hampshire. There are now nearly a hundred of these conferences scheduled every summer. For the past 20 years there has been a conference on Statistics in Chemistry and Chemical Engineering. The conferences last for a week with sessions morning and evening and lots of time for discussion. The conferees come from industry and the universities. Some of the industrial members are statisticians, and the formal presentations and discussion periods take the audience to the frontiers of applied statistics. In those earlier days there were also (and there still are) some nonstatisticians, senior scientists from industrial installations who had come to find out what the statisticians were doing, and how these ideas might be helpful to them in their own laboratories. They took the ideas back home and they spread the word. They also made helpful contributions to the discussions, especially in giving us insights into the problems that were arising in their own work.

It was at the Gordon Conferences that we met each summer to exchange ideas and to recharge our batteries. In some areas, we went back to our laboratories with new insights and new enthusiasms; in

others, we had to be content with the consolation of learning that
our friends were faced with the same frustrations we had, and to take
comfort in the knowledge that we had companions in adversity.

The decade from the mid-1950s to the mid-1960s was a very
exciting time in the development of industrial experimentation. I
am glad that I was there to take part in it. We have almost completed
another decade since that time, and perhaps there are some in the
audience who think I should conclude with some kind of prophesy about
the next 10 years.

I have been involved in multiple regression long enough to be
very wary of extrapolation, and other forms of soothsaying, and so I
shall, instead, give you a status report. I attended the Gordon
Conference again last summer after an absence of several years. I
am glad to report that we were not still discussing the same topics
that occupied us 10 or 20 years ago. You might get some indication
of the trends of the next decade from the list of topics that we did
discuss; we talked about shape studies, optimum experiments (in the
sense of Kiefer and Wolfowitz and of Henry Wynn), multivariate analy-
sis--both discrimination and classification, the analysis of vast
quantities of data, such as the Los Angeles smog study provides,
computer programs, and last, but not least, a topic that is dear to
the hearts of our hosts here today--jackknife statistics.

REFERENCES

Addelman, S. (1961). Irregular fractions of 2^n factorial experiments.
    Technometrics 3, 479-496.

Bennett, C. A., and Franklin, N. J. (1954). *Statistical Analysis in
    Chemistry and the Chemical Industry.* Wiley, New York.

Birkhoff, G., and MacLane, S. (1941). *A Survey of Modern Algebra.*
    Macmillan, New York.

Box, G. E. P. (1957). Evolutionary operation: A method for increasing
    industrial productivity. *Appl. Statist. 6*, 3-23.

Box, G. E. P., and Hunter, J. S. (1961). The 2^{k-p} fractional factorial
    designs. *Technometrics 3*, (I)311-352; (II)449-458.

Box, G. E. P., and Wilson, K. J. (1951). On the experimental attain-
    ment of optimal conditions. *J. Roy. Statist. Soc., B 13*, 1-45.

Brownlee, K. A. (1948). *Industrial Experimentation* (3rd ed.). H. M. Stationary Office, London.

Cochran, W. G., and Cox, G. M. (1950). *Experimental Designs* (1st ed.). Wiley, New York.

Cramèr, H. (1946). *Mathematical Methods of Statistics.* Princeton Univ. Press, Princeton, N. J.

Daniel, C. (1959). Use of half-normal plots in interpreting factorial two-level experiments. *Technometrics 1,* 311-342.

Davies, O. L. (1954). *Design and Analysis of Industrial Experiments.* Oliver & Boyd, London.

Duncan, D. B. (1955). Multiple range and multiple F-tests. *Biometrics 11,* 1-42.

Finney, D. J. (1945). The fractional replication of factorial arrangements. *Ann. Eugenics (London) 12,* 291-301.

Fisher, R. A. (1942). The theory of confounding in factorial experiments in relation to the theory of groups. *Ann. Eugenics (London) 11,* 341-353.

Gorman, J., and Toman, R. J. (1966). Selection of variables for fitting equations to data. *Technometrics 8,* 27-51.

Hocking, R., and Leslie, R. N. (1967). Selection of the best subset in regression analysis. *Technometrics 9,* 531-540.

Hoerl, A. E., and Kennard, R. W. (1970). Ridge regression: Biased estimation for non-orthogonal problems. *Technometrics 12,* 66-67.

John, P. W. M. (1962). Three-quarter replicates of 2^n designs. *Biometrics 18,* 172-184.

Kempthorne, O. (1952). *The Design and Analysis of Experiments.* Wiley, New York.

Kendall, M. G. (1943). *The Advanced Theory of Statistics,* Vol. I. Charles Griffin, London.

Kendall, M. G. (1946). *The Advanced Theory of Statistics,* Vol. II. Charles Griffin, London.

Kiefer, J., and Wolfowitz, J. (1959). Optimum designs in regression problems. *Ann. Math. Statist. 30,* 271-294.

Mallows, C. (1973). Some comments on C_p. *Technometrics 15,* 661-675.

Margolin, B. (1969). Orthogonal main effects plans permitting estimation of all two-factor interactions in $2^n 3^m$ factorial series of designs. *Technometrics 11,* 747-762.

Plackett, R. L., and Burman, J. P. (1946). The design of optimum multifactorial experiments. *Biometrika 33,* 305-325.

Scheffé, H. (1953). A method of judging all contrasts in the analysis of variance. *Biometrika 40,* 87-104.

Tippett, L. H. C. (1934). Applications of statistical methods to the control of quality in industrial production. Manchester Mathematical Society.

Tukey, J. W. (1953). The problem of multiple comparisons, dittoed manuscript, 396 pp. Princeton University, Princeton, N. J.

Webb, S. R. (1968). Non-orthogonal designs of even resolution. *Technometrics 10*, 291-300.

Wynn, H. P. (1972). Results in the theory and construction of D-optimal experimental designs. *J. Roy. Statist. Soc. B 34*, 133-147.

Yates, R. (1937). *The Design and Analysis of Factorial Experiments*. Imperial Bureau of Social Science, Harpenden, England.

4.

*SOME IMPORTANT EVENTS
IN THE HISTORICAL DEVELOPMENT
OF SAMPLE SURVEYS*

MORRIS H. HANSEN and *WILLIAM G. MADOW* are internationally recognized authorities on sample survey design and the conduct of surveys. The two-volume book that was coauthored with William Hurwitz serves as a standard reference work throughout the world. Both men have long associations with the Bureau of the Census, where Hansen was Associate Director and responsible for the Research and Development Program which contributed many innovations that influenced survey methods in both industry and government.

Hansen was born December 15, 1910, in Thermopolis, Wyoming. He holds a B.S. degree given in 1934 and an honorary L.L.D. degree given in 1959, both from the University of Wyoming. He received an M.A. degree from the American University in 1940. He is currently Senior Vice-President of Westat, Inc. He was President of the Institute of Mathematical Statistics in 1953 and of the American Statistical Association in 1960.

Madow was born February 22, 1911, in New York City. He earned a B.A. degree in 1932, an M.A. degree in 1933, and a Ph.D. degree in 1938, all from Columbia University. From 1939 to 1945 he worked for the U.S. government, chiefly at the Bureau of the Census. From 1949 to 1957 he was Professor of Mathematical Statistics and Chairman of the Statistical Research Laboratory, University of Illinois in Urbana. Since 1957 he has been staff scientist at the Stanford Research Institute and Consulting Professor in Statistics, Stanford University. He is a Fellow of the Institute of Mathematical Statistics and also of the American Statistical Association.

4.

SOME IMPORTANT EVENTS
IN THE HISTORICAL DEVELOPMENT
OF SAMPLE SURVEYS*

Morris H. Hansen

Westat Inc.
Rockville, Maryland

William G. Madow

Stanford Research Institute
Menlo Park, California

The events that have made the history of sample-survey theory and practice what it is are, like all historically important events, partly due to the needs of the times, and partly due to the availability of people who create or fill those needs. It is hard to imagine what statistics would be today had there been no R. A. Fisher or no J. Neyman, and it is impossible to guess what statistics might have become had it not been for the death of A. Wald in December 1950. There was a tremendous growth in the opportunities for mathematical statisticians during and after the Great Depression and World War II, often the result of clearly visible needs. One of the areas in which the needs were greatest was sampling.

In this paper, we make no effort to give a comprehensive review of the historical development of sample surveys. Rather, we endeavor to put into broad perspective some of the developments and contributions

---

*Dedicated to the memory of William N. Hurwitz.

to sample-survey concepts, theory, and practice that we feel have
been of particular importance in the evolution of the design and
execution of sample-surveys. We have no thought that others would
emphasize precisely the same developments. Relatively few priorities
in contributions are identified, partly because of considerable overlap or joint contributions in a particular area. Moreover, in general,
we attempt to emphasize the relatively early historical developments
that tended to establish the practices and principles that are widely
applied, even though these may have been extended, modified, and
improved in subsequent developments.

As there is no generally accepted definition, we begin by defining more specifically what we mean by a sample survey. Perhaps what
has distinguished sample surveys rather sharply from other areas of
statistical measurement and analysis is the design and execution of
surveys that provide estimates of characteristics of specific finite
populations. In such cases one might, at least conceptually, obtain
and summarize information on all members of the population. A sample
survey can, in shorter time schedules and at lower cost, provide
estimates of what might have been obtained from complete coverage
using the same observational or measurement procedures. It does this
by identifying a sample for which the relevant information is obtained,
and by providing estimates from the sample observations of the characteristics of the aggregate population. Estimates may be made for one,
for a few, or for many characteristics. A well-designed and controlled
sample survey permits the achievement of higher quality and more intensive training and control of interviewers, in addition to more effective measurement procedures and controls than can be accomplished in
a complete census.

Although we give primary consideration to sample surveys taken to
provide information for finite populations, we also recognize that
sample-survey methodology is important and widely applied in the more
general areas of analysis and inference about causal systems.

The problems of observational or measurement errors need special
comment for, because of them, the results of a complete census may

not, and ordinarily will not, be unique. The theory of finite
population sampling does not explain measurement error, but certain
features of sample-survey design and evaluative studies have made it
practicable to estimate the sizes and sources of some important types
of measurement errors. We are indeed fortunate if errors of observation are absent or trivial in a particular study (as might be the
case in estimating the characteristics recorded in a large file of
records from a sample of those records). In practice, errors of
measurement or observation are sometimes large, and often the most
important source of survey errors, both in sample surveys and censuses.
Consequently, to the extent that is feasible, sampling methodology
should include means for estimating measurement error and for interpreting the effects of measurement error on the outcomes of surveys.

Sample surveys require expenditures of funds, time, and other
scarce resources. Ordinarily there are numerous alternative resources,
procedures, and types of supplementary information that may be used
in sample selection or observation, or in making estimates from
samples. These can be used in various ways, guided by common sense,
experience, and theory. The effects of choosing among alternative
designs can make substantial or even striking differences in costs,
in the accuracy of the observations or measurements, and in the magnitudes of the errors of sample estimates. Consequently, the design
and execution of sample surveys become problems of management as well
as of systems analysis and design. The goals of sample surveys must
be carefully identified; designs and operating procedures must be
chosen to achieve those goals at near-minimum costs, subject to time
and other restraints.

In this setting, a sample survey is an economic production process,
intended to make the most efficient use of limited resources in order
to achieve specified goals. The design and execution of a survey
involve the efficient utilization of capital investment, determinations
of cost-effectiveness, and concepts of systems design and control as
well as those of statistical theory and the substantive area involved.

No attempt is made to describe early intuitive or common-sense
developments or applications of sampling except as they came to be

the means of identifying the need for statistical theory to guide
choices in design, estimation, and interpretation [see, however,
Jensen (1926), Stephan (1948), and Seng (1951)]. Major progress
results when theory is provided to support, guide, and modify common
sense and intuition.

Applications occur in many subject areas--social and economic,
agriculture, health, forestry, and a wide range of other areas of
human concern. Effective applications usually involve joint roles
for specialists in the subject matter and in sample-survey methodology,
although sometimes these are handled by the same person. Ordinarily,
the sample survey is an interdisciplinary activity involving the
blending of statistics and the subject matter into an effective total
survey design or system. The statistical tools used in sample surveys
include the theory of probability and many of the standard statistical
tools, adapted or extended as necessary or convenient.

SOME HIGHLIGHTS OF FINITE POPULATION SAMPLING DEVELOPMENT

Some Early Developments

Sampling as an intuitive tool has been used in various ways over the
centuries. Early in this century, there was a good deal of activity
in the use of sample investigations to obtain information concerning
social and economic problems. Bowley (1913) made, perhaps, the initial
serious effort at applying statistical theory to guide the design and
interpretation of results from such sample surveys.

As early as 1924 and before, the major international association
of statisticians, the International Statistical Institute (ISI),
recognized that the time had come to urge wider use of the representative method (the term was used for both random and purposive
selections) in statistics and to provide guidance for those who
would use it. It is interesting to note that while intuition led
to the development of both good and poor survey designs, there was
little directly applicable statistical theory, at that time, that
could be used (1) to evaluate survey designs, (2) to formulate a

methodology that could be applied with confidence if correctly used
and implemented, and (3) to suggest the creation of new survey designs.

In the paper prepared by A. L. Bowley (1926) as part of the 1926
report of the ISI Commission,* there were at least two important
technical points that should be mentioned

1. The insistence that, for a sample to be selected, there
    should be a population defined in the form of a list or
    directory consisting of elements that would be accessible
    if selected, i.e., a frame of numbered or indexed or labeled
    elements.

2. The definition of and theory for stratified proportionate
    random sampling, of elements, not clusters. (In the main
    body of the report, concern was expressed about using random
    rather than purposive selection of clusters, primarily, we
    surmise, because they seemed to be thinking in terms of
    large clusters rather than small clusters.)

Bowley also developed a theory for purposive sampling of clusters.
Although he did not strongly advocate the use of purposive samples,
the ISI report essentially recommended random selection of elements
and very small clusters such as households and purposive selection
of larger clusters.

The ISI report strongly advised that any report of a sample
survey include a detailed statement of the steps taken in selection
of the sample, that the survey be made so that there could be "a
mathematical statement of the precision of the results, that with
these results should be given an indication of the error to which
they are liable," and that the "strict rules" required by the
statistical theory be carried out in practice. In the almost half-
century since this advice was given, it has been followed more and
more in practice.

---

*In May 1924, the International Statistical Institute appointed
a commission "for the purpose of studying the application of the
Representative Method in Statistics." The Commission consisted of
A. L. Bowley, C. Gini, A. Jensen, L. March, V. Stuart, and F. Zizek.
The Commission's report was published in the *Bulletin of the Inter-
national Statistical Institute* 22, liv. 1 (1926), 359-451, followed
by a separately paged memorandum by Bowley: "Measurement of the
Precision Attained in Sampling," 6-62.

By the mid-1930s, the needs of governments and research had led
to much greater recognition that relatively* fast and inexpensive
methods of gathering information on the state of society were essential.
Political polling, under the leadership of Gallup and others, was
beginning to call public attention to sample surveys as a means of
obtaining measures of public opinion. Beginning in about 1933, many
(and some large-scale) sample studies were supported by or done in
the Federal Emergency Relief Administration, and by its successor, the
Work Projects Administration, to measure employment and unemployment,
and other characteristics of people and families, and of communities,
a review of which is given by Stephan (1948). Sample studies still
had relatively little guidance from the methodology that has come to
be the theory and practice of sample surveys.

The Pioneering Contributions of J. Neyman

In the mid-1930s an event occurred that had a special impact in
stimulating the recognition of the role of statistical theory in
sample surveys of finite populations; that event stimulated additional
development of sample-survey theory and methods--this was the paper
by J. Neyman (1934), "On the Two Different Aspects of Representative
Method: The Method of Stratified Sampling and the Method of Purposive
Selection," presented to the Royal Statistical Society. Both in theory
and practice, Neyman's paper served for many as their initial course
and long-time guidance in developing sample-survey methodology.
Although the paper was related to the earlier ISI report, its origin
went back to work done by Neyman in Poland in the early 1930s.

In this paper, Neyman:

1. Emphasized the importance of random rather than purposive
    selection of the defined sampling units and the need for

---

*The fast and inexpensive methods of gathering information that
were needed by governments and research in the mid-1930s could only
be partly achieved. Good surveys are likely to be expensive and often
take time. There is still much to be done.

using samples based on a relatively large number of relatively small sampling units when the units are to be whole clusters of the elements of analysis

2. Defined the problem of optimum allocation of the sample among the strata and solved the problem of choosing the allocation that minimized the variance of the estimator subject to a fixed total size of sample*

3. Extended stratified simple random sampling to the sampling of groups, or, as we would call them today, clusters, including the use of ratio estimators as a necessary consequence of the fact that with cluster samples both numerators and denominators of averages of interest are random variables

4. Introduced the concept of the confidence interval

5. Proposed that where appropriate, the Markoff method of obtaining best linear unbiased estimators be used in sample surveys

6. Stated a statistical model for purposive sampling that today would be said to yield balanced samples; determined conditions that must be satisfied for purposive sampling to have desirable properties; raised the question of the effects on bias, consistency, and variance when these conditions are not satisfied; and gave illustrations of the lack of robustness of purposive sampling

7. Remarked on the approximate normality[†] in large samples of linear estimators and the consequent use of

$$\frac{\theta' - \theta}{s_{\theta'}}$$

---

*It was learned later that the theory for this principle had been published earlier by Tschuprov [*Metron* 2, 646-680 (1923)]. However, the earlier theoretical results had been effectively lost, and had no practical impact on sampling theory or practice until it was independently developed and presented in an appropriate setting in this paper by Neyman.

[†]Although there has been much research on the limiting distribution of estimates based on samples from finite universes since the first paper of Madow (1948), it is perhaps noteworthy that the first paper to give a remainder term for finite sampling comparable to the Berry-Esseen theorem was published by A. Bikeles (1969); also, the first demonstrations of normal limiting distributions when samples are selected with varying probabilities were described in papers by Hájek (1964) and Rosén(1972).

as approximately normally distributed, where θ' is the estimator of the population characteristic θ, and $s_{\theta'}$ is the estimator of the standard deviation of θ'.

The impact of Neyman's 1934 paper on United States sampling practice was not immediate. There was still the need for communication, understanding, acceptance, and the adaptation and extension of the results he had presented. This process was hastened for those in the United States, and especially in the Washington, D.C. area, when W. Edwards Deming arranged for some lectures and consultations by Neyman in April 1937. Deming was head of the statistical program in the U.S. Department of Agriculture Graduate School, and often took the lead in encouraging and promoting the dissemination of statistical innovations. The lectures and consultations resulted in development of the theory of double sampling, as another illustration of the principles of optimum design in sample surveys [see Neyman (1938)].

The contents of Neyman's two papers (1934 and 1938) on sampling were important not only for their contributions to survey design and theory, but also for relating theory to practice. Especially in the latter paper, Neyman stressed that "the advantages of methods are rarely universal...," and the rational decisions on what survey design to use "are possible only if some previous knowledge of the population is available."*

SOME DEVELOPMENTS IN VARIOUS ORGANIZATIONS

The Contributions at Rothamsted

Beginning in the 1920s, and led by R. A. Fisher and his students and successors at the Rothamsted Experimental Station, fundamental ideas and advances that had substantial relevance to sample-survey methodolog

---

*We may remark that gaining such information and deciding what designs to recommend on the basis of that information is one of the more exciting aspects of participation in surveys. Although Bayesian approaches are not discussed in this paper, their relevance to the utilization of this prior knowledge is obvious.

were emerging in connection with statistical inference, experimental designs and their analysis, the analysis of variance, and related topics. However, the recognition of the applicability of those methods to sample surveys, at least beyond a small group of statisticians at Rothamsted and elsewhere, had relatively little impact on operating sample studies until the mid-1930s and later.

In 1935, Yates and Zacopanay published an extensive analysis of sampling and subsampling of agricultural fields. This paper included analyses of labor costs as well as effects on variance and variance components, that pointed up the strong relevance and importance of this work, and that gave results of widespread applicability [see also Yates (1946)]. Yates has been a leader in the development and application of sample-survey methodology over the years, with wide-ranging contributions to many aspects of sample-survey methodology.

DEVELOPMENTS AT IOWA STATE UNIVERSITY STATISTICAL LABORATORY

After some work at Rothamsted, Cochran came to the Statistical Laboratory at Ames, Iowa in 1939. Among other things, he worked in experimental design, and took the lead on the theoretical side on the extensive work in sample-survey methodology that was being encouraged and supported by the U.S. Department of Agriculture (USDA). Arnold King and Raymond Jessen were already providing leadership in studies of sample-survey methodology at the Laboratory, under USDA sponsorship, especially in the definition of sampling units for farm surveys, the efficiency of sampling units of varying size, the use of supplementary information in stratification and estimation, and numerous other aspects of sample-survey design. Jointly with the U.S. Census Bureau, in the early 1940s, they developed the Master Sample to facilitate farm and household sampling in rural areas. Ames has continued to be one of the leading centers of sampling methodology in the United States. Cochran made a number of contributions while there, including some important work jointly with Jessen on the optimum allocation of the sample and estimation procedures for sampling on

each of two occasions [see Jessen (1942)]. Their research was an
initial step in later work (by others) on sampling on successive
occasions, for estimating time series. Among the numerous other
sample-survey contributions by Cochran, we mention especially the
development of regression estimation [see Cochran (1942)] as a means
of making effective use of supplementary information available on a
comparable basis for individual units in the sample, and for all units
in the population. The regression estimator provides a means of
utilizing supplemental information such that, at least in the first-
order approximations, the estimator is unbiased and has variance equal
to or less than that of the best linear unbiased estimator or ratio
estimator--sometimes substantially.

Cochran's productivity in sampling theory and methods continued
after he left Ames for North Carolina State College, then Johns Hopkins,
and later Harvard. He made contributions to the theory of systematic
sampling, extending the original work in this area by the Madows, which
is referred to later.

Ames continued its strong leadership role in sampling by bringing H. O. Hartley to the Laboratory. Hartley made substantial contributions in widely varying aspects of sample-survey methodology--his
work on unbiased ratio estimation, the use of multiple frames in
sample selection, and the unbiased estimation of variances when units
have been selected with varying probabilities [Hartley and Ross (1954);
Hartley and Rao (1962)]. Hartley left Ames to establish and direct
the Institute of Statistics at Texas A & M University.

Many statisticians who began their work at Ames have made
continuing contributions to sample-survey theory and practice, including Earl Houseman, Dan Horvitz, J.N.K. Rao, Wayne Fuller, and others.

Developments in India

P. C. Mahalanobis had an especially interesting role in the historical
and continuing development of sample theory and practice. He created
and directed the work of the Indian Statistical Institute, which, along

with other activities, developed a strong program of sample-surveys. Shortly after the end of World War II, Mahalanobis visited Europe and the United States; it was then that we learned of extensive parallel but independent developments in sample surveys, beginning with surveys of acreage of jute in the 1930s [Mahalanobis (1940, 1944, 1946)].

On one of his early visits to the Census Bureau, Mahalanobis surprised and impressed us with the validity of his statement that the Census Bureau and the Indian Statistical Institute were, perhaps, the two organizations in the world that were most similar in the way they dealt with the total systems-design aspects and operational approaches to sample surveys. The similarity became more apparent as Mahalanobis described the nature of the studies done to improve the efficiency of design through empirical studies, to resolve theoretical problems as needed, but especially to achieve effective operational control in the actual execution of surveys through design, control, and feedback procedures in data collection and data processing.

Among the numerous contributions made by Professor Mahalanobis and his colleagues, one that deserves special mention is the extensive use of interpenetrating samples from initial sample selection through the successive stages of data collection, data processing, and analysis. The purpose was to provide measures of important sources of nonsampling errors, and to facilitate more effective control and interpretation. Special mention should also be made of both theoretical and empirical studies, such as the definition of efficient areal sampling units and their application in agricultural surveys. Later developments include the initiation in India of large-scale sample surveys in an effort to provide information needed for economic and social planning in India, covering wide-ranging subjects, including demographic studies, social and economic characteristics of households and small production units, and related subjects.

Professor Mahalanobis retained his vigor and charm in communicating his ideas, his methods, and his accomplishments until his death in

1971.* His international prestige was great, and he appeared at most
international statistical meetings and at the Statistical Commission
of the United Nations--often with an important paper that was almost
ready, but to be finished after its presentation.

A related development in India took place at the Indian Council
for Agricultural Research (ICAR), under the leadership of P. V. Sukhatme
including, again, numerous theoretical contributions as well as practical work. Among some of the well-known studies at the ICAR is a series
of agricultural crop-cutting surveys designed to estimate agricultural
production in India, with interesting supporting evaluative studies by
Sukhatme (1935, 1946) and others. Similar studies have been made at
the Indian Statistical Institute, in the United Kingdom, the United
States, and elsewhere.

Some controversy developed in India owing to differences in
methods and results, the extent to which some substantial differences
in results were the consequences of border biases, crop-harvesting
losses, the distinction between biological production and harvest for
use, and the effectiveness of control of the sampling and crop-cutting
operations. The state of the art and understanding of issues was substantially advanced as a result of the intensive work done by Sukhatme,
Mahalanobis, and their colleagues.

Later, Sukhatme left India for the Food and Agriculture Organization in Rome, where he took the responsibility for the statistical
programs. With the colleagues he assembled there, notably S. S.
Zarkovich, he has continued to extend methods and theory, focusing
on agricultural food production, consumption and nutritional applications; on the methods appropriate to the needs in developing countries;
and on response or measurement errors in surveys.

---

*The Indian Statistical Institute, under the direction of C. R.
Rao, has continued to be a major source of contributions to statistical theory and practice.

A DEVELOPMENT AT THE WORK PROJECTS ADMINISTRATION

Toward the end of the 1930s and following the experience gained in
measuring unemployment in connection with early work at the Census
Bureau, a pioneering study of special significance was initiated in
the Work Projects Administration (WPA) to provide continuing measurements of unemployment through a repetitive national-population sample
survey. Frankel and Stock (1942) did the innovative work on the design
of a two-stage sample with counties as the first-stage sampling units,
and clusters of households within counties selected in a quasi-random
manner. The sample consisted of 64 counties. They also remarked that
as few as 27 counties could have achieved the same degree of stratification using a Latin square design. The survey design represented
a major methodological advance at that time. However, because
probabilities of selection of households within counties were not
known and could not be reflected in the sample estimates, the design
was revised after the survey was transferred to the Census Bureau,
with consequences that will be discussed subsequently.

DEVELOPMENTS AT THE U. S. CENSUS BUREAU

In the mid-1930s C. L. Dedrick and Stuart Rice, working within the
Census Bureau, and also as members of staff of the Committee on
Government Statistics and Information Services, urged that the Census
Bureau explore and initiate studies concerning the contributions that
sampling methodology might make to the statistical program of the
U.S. Census Bureau. Other leaders in urging and advising on such
developments included Samuel A. Stauffer and Frederick F. Stephan.
In those days, sampling was neither trusted by the public nor by the
members of Congress; the idea was both innovative and perhaps "far
out" that the Census Bureau, which based almost all of its work on
complete coverage or censuses, might make use of sampling methods.
A common view at that time was that others could take such risks with

loose methods, and thereby undermine the basis for confidence in
their data, but not the Census Bureau.

A large-scale sample study was not long in coming as a part of
the so-called 1937 "Census of Unemployment." The United States was
in the grip of the Great Depression, and there was urgent need for
current information on the unemployed. But estimates of the number
of unemployed varied by many millions of persons and the next Decennial
Census would not occur until 1940. Actually, the "Census of Unemployment" was to be a nationwide voluntary registration of the unemployed
and partially unemployed. The voluntary registration was to be
conducted through the post offices. The post offices distributed
registration forms to households, to be completed by partially or
totally unemployed persons and mailed back.

Lack of confidence in the ability to control the accuracy of the
unemployment registration led to the idea of an enumerative check
census, to be taken on a sample basis. The Enumerative Check Census
was a large-scale sample survey, using what have come to be known as
probability sampling methods designed to evaluate the voluntary
unemployment registration and to provide independent estimates of
unemployment and employment status. The Enumerative Check Census
involved an enumeration of a sample of the total population, including
all households in a 2% sample of postal delivery routes. The postal
carriers canvassed their routes. Dedrick had general responsibility,
Stephan was a principal adviser in the design of this first large-scale
national area sample, and Hansen was primarily responsible for the
analysis of the results, including preparation of estimates from the
sample, and measures of their precision [Dedrick and Hansen (1938)].
The national registration and the check survey were done in November
1937, preliminary reports began by January 1938, and the final published
reports were completed in 1938. And although some $5,000,000 had been
allocated, only $2,000,000 were spent.

Along with the estimates of various characteristics, confidence
intervals were computed and published. They were estimated by a
random-group or interpenetrating-subsample procedure. The complete

registration returns were segregated and summarized by individual postal delivery routes. Ratio estimates were introduced utilizing the independent registration information as a means of improving the precision of the estimates. Although the ratio estimates were subject to minor biases, the biases were such that the combined effects of minor biases and variances produced estimates with smaller mean-square errors than would a best linear unbiased estimation procedure.

The Enumerative Check Census achieved the recognition, in the Census Bureau and elsewhere, that large-scale sample surveys could make substantial contributions, and under appropriate design and control, could produce timely information that was more accurate than complete censuses or national registrations.

The next substantial development in the Census Bureau was the introduction of sampling as an important part of the data-collection method in the 1940 Population Census. Extensions in the use of sampling were made in subsequent censuses until the great bulk of the subject-matter coverage in the recent Censuses of Population and Housing was obtained through samples of 5 to 25%. The principal sample-design work in the 1940 Census was by Stephan, Deming, and Hansen, (1940) with strong support and administrative leadership from Philip M. Hauser. Sampling was still a novel method, and not generally accepted as trustworthy. The two major successes supported new significant sampling efforts at the Census Bureau.

A particularly significant step in both the development of sample-survey methodology and in contributions to theory at the Census Bureau was the transfer in 1942 (after the United States had entered World War II, and the WPA was to be abolished) of the Labor Force Survey, mentioned earlier, from the WPA to the Census Bureau. In 1943, the Census Bureau placed the Labor Force Survey on a full probability-sampling basis, and introduced some innovative procedures and theory. These were presented in papers by Hansen and Hurwitz (1942, 1943) that not only gave the formulation of the cluster sampling variance in general terms, including the definition of the cluster-sampling variance in terms of the intraclass correlation coefficient,

but also introduced sampling with probabilities proportionate to
measures of size, with stratified multistage sampling and ratio
estimates. To a large degree, the 1943 paper by Hansen and Hurwitz
provided the theory that is now basic to the sampling methods used
in multistage surveys, and especially interview surveys in which
relatively small samples have been drawn from relatively large
populations.

A significant practical result of placing the Labor Force Survey
(now known as the Current Population Survey, or the CPS) on a
probability-sampling basis was that trends in employment by major
sectors of the economy, especially trends in agricultural and non-
agricultural employment, since the 1940 Census were sharply different
when measured by the revised probability sample as compared with the
earlier design in which probabilities of selection were not known and
not appropriately reflected in the sample estimates. The differences
were substantial and were large enough to affect national manpower
policy in the war effort.

One conclusion we have drawn from this and some related situations
was that, while nonrandom samples, or other methods that depended on
assumptions of the continuity of relationships over time or of models,
could often be introduced into a sample-survey design to reduce
variance, the bias that resulted from such assumptions (at least in
relatively large samples) was likely to be larger than the variance.
This was especially true in periods of instability and upheaval, such
as a war, a depression, or the current energy crisis, which could not
ordinarily be foreseen in time to finance and design a new survey.
Consequently, a national statistical system that provided important
measures had to be designed to withstand such crises; a system that
produced results for which the precision could be evaluated from the
sample itself, and which did not depend, for the results to be valid,
on the continuity of observed past relationships. Lesser standards
could and often should be observed in studies that did not provide the
basis for decisions of great concern to the public interest and welfare.

The Current Population Survey, improved and enlarged over the
years, now serves basic and wide-ranging needs. It has become a model

for survey designs used all over the world that embody the elements
of stratified multistage sampling of clusters, probabilities of
selection proportionate to measures of size, use of auxiliary information in estimators, a high level of quality control and measurement,
and control of nonsampling as well as sampling errors.

Although selection with unequal probabilities used straightforward estimators of aggregates and functions of the aggregates such as
averages or ratios, the corresponding variance estimators as developed
by Hansen and Hurwitz relied upon the assumption of sampling with replacement. In a significant development Horvitz and Thompson (1952)
obtained unbiased variance estimators for selection of two or more
units from a stratum with varying probabilities without replacement.
Unbiased variance estimators when samples are selected with varying
probabilities has become a topic of considerable interest, with contributions by Yates and Grundy (1953), by Rao, Hartley and Cochran (1962),
by Brewer (1963), by Fellegi (1963), by Durbin (1953), and by others.

Another development by the Census Bureau staff that deserves
special mention was the introduction of rotational sampling in sample
surveys for estimation of time series, with supporting theory to guide
sample rotation and estimation [see Bershad et al. (1953)]. It is a
method that requires more work, but that has great potential for
application in sampling over time. Rotation sampling has been
effectively applied in several census surveys, including the CPS, and
in surveys of retail trade and service establishments, in measuring
monthly changes, changes over varying other periods, and monthly,
annual, and other aggregates. This theory is an extension of the
work by Jessen and Cochran, mentioned earlier (1942), for sampling on
two occasions, and the method has great power and utility under certain
circumstances. Patterson's paper (1950) on sampling on successive
occasions, provided important basic and relevant theory. However, his
paper does not adequately deal with some practical problems of an
ongoing survey, in which estimates must be made currently, using
prior available information, but which cannot continue to be revised
in successive periods as additional information becomes available.

As was suggested earlier, effective survey design involves not
only models or theory concerning the sources of errors and costs, but
also requires information on the approximate magnitudes of the parameters in the error and cost models. Hence, empirical information is
needed on the components of sampling variances, biases, and other
sources of error as well as on unit costs, and how they vary for
survey-design alternatives. Appropriate design and analysis of
earlier surveys, supplemented by experimental studies, constitute
the major source of such information.

Over the years, many empirical studies of sampling variances
and biases and of response variances and biases in surveys have been
made, a number of which have measured the effects of interviewers
and other sources of error. In addition to the work by Mahalanobis
with interpenetrating samples, we mention a study by Stock and
Hochstein (1951), an earlier reinterview study by Gladys Palmer
(1943) that measures respondent effects, and a paper by Deming (1944).

A response-error model that has proved particularly helpful was
described in a paper by Hansen et al. (1951) and later elaborated by
Hansen et al. (1961). The model has served as the basis for guiding
improved census and sample-survey design as well as some extensive
experimental studies of response errors, and especially of interviewer
contributions to response errors that were included within the 1950
Census of Population and Housing, and subsequent censuses. The 1961
paper also included a summary and integration of the results of the
1950 Census studies with rather striking implications. The studies
showed that in a massive decennial census operation the contributions
of enumerators to the errors in census results were considerable, and
that a self-enumeration approach, in conjunction with a large-scale
sample for many items in the census, would yield results that were
generally of equivalent or improved accuracy, at lower cost, and on
faster time schedules. The consequence was major redesign of the
1960 and 1970 Censuses, depending far more on sampling, and on self-
enumeration rather than on direct interviewing as the principal means
of data collection in the Censuses. A Canadian study by Fellegi (1964),

confirmed and extended these results. [For relevant illustrative
material see Waksberg and Neter (1964), Hansen and Tepping
(1969), and Hansen and Steinberg (1956)]. Such studies also
demonstrated, again, the fact that well-designed and controlled
continuing sample studies, such as the CPS, had the potential for
yielding more accurate information at the national level than did
complete censuses, although complete coverage or much larger samples
may have been needed to yield small-area data of sufficient precision.

Although much remains to be done, a relatively advanced state
has been achieved in the control of sampling errors through appropriate
design and size of sample, as compared with the remaining problems in
the measurement and control of response or measurement errors.

Census Bureau activities have been extended to sample-surveys in
many subject areas in varying circumstances, and the staff has been
expanded to become perhaps the strongest sample-survey staff that has
been assembled anywhere. Many developments and extensions of theory
and methods have been accomplished through the work of this group, in
solving problems encountered under a wide range of circumstances.
Notable names on the staff, in addition to Hansen and Hurwitz, include
Bershad, Daly, Marks, Nisselson, Pritzker, Ogus, Steinberg, Tepping,
Waksberg, Woodruff, and numerous others. Some of the results are
published in the scientific journals. Many are not, but appear in
the survey methodology and descriptions for the particular surveys
involved.

Mention should be made of an advisory group that worked with the
Census Bureau from 1955 to 1968. This was a group known as the Panel
of Statistical Consultants, under the chairmanship of William G. Cochran,
and included such members as Stephan, Hartley, and Madow during this
period. It included Nathan Keyfitz, formerly at the Dominion Bureau
of Statistics in the early years, and later Ivan Fellegi, from the
same institution (now Statistics Canada). The contributions of this
group through interaction with the Census staff were very substantial.

A special word is needed on the role of William N. Hurwitz, who
died in 1969. Hurwitz's leadership, insistence on high standards,
personal magnetism, and contributions in the definition and solution

of problems cannot be sufficiently emphasized [see Hansen et al. (1969)]. Our personal debt and affection for him are very great, and this paper is dedicated to his memory.

Other Early Contributors

We cannot list all or most of the early contributors, but in this section we wish to mention briefly a few additional contributions and contributors not primarily associated with the institutions we have identified above, but who are among those who made early and substantial contributions. Their contributions are neither more nor less than those of others that have been mentioned, but they do not readily fall into one of the other sections in the approach we have taken.

As indicated earlier, Frederick F. Stephan was one of the earliest pioneers in the United States in the development and application of sample-survey methods. He was an advisor and participant in many early developments; his contributions and impact in many different areas were substantial and continued until his death in 1971.

W. Edwards Deming has been a contributor to sample-survey theory and methodology from the time he participated in the 1940 Census developments. He has been a world-wide consultant and has been a leader in the area of nonsampling errors, in the use of replicated samples, and in the promotion of high standards generally, as well as in other areas.

William G. Madow had been a member of the Census staff in the early 1940s, and as an advisor continued to have a close continuing tie to that staff. We mention two particularly relevant contributions.

Systematic sampling had long been part of the practice of surveys, and its usefulness in providing implicit proportionate stratified random sampling had been noted in the 1926 ISI reports. Empirical investigations had shown it was often more efficient than random or stratified random sampling. In 1944, W. G. Madow and L. H. Madow published the first theoretical investigation of single and multiple random-start systematic sampling and discussed the efficiency of systematic sampling compared to other designs for various types of population. The first use of a superpopulation having concave upward correlogram was in 1946 when Cochran showed that single random-start systematic sampling could be more efficient than stratified sampling with equal-size strata from each of which one element was selected.

With the development of multistage, multiphase, and other complex aspects of survey designs had come the need for techniques of reducing the work of deriving expected values and variances of estimators for such designs. The intuitive use of conditional probability approaches for deriving such survey characteristics had occurred at the Census Bureau by 1944 or 1945, and the formal statement and proof of the conditional variance theorem had occurred at least as early as 1946 in lectures by W. G. Madow.* This has come to be such a powerful and widely used theorem for simplifying otherwise difficult variance developments for complex designs that it deserves special mention.

Tore Dalenius of the University of Stockholm and Brown University has contributed to sample-survey theory and methods, and has served as a consultant on sample surveys in many nations around the world. One of his early contributions was concerned with the principles for the optimum definition of strata boundaries.

Leslie Kish, at the Survey Research Center at the University of Michigan, was one of the pioneers, and has focused on the use of analytical measures from survey results. One of his early contributions jointly with Roe Goodman, that has received a great deal of attention and widespread application, is controlled selection in which dependence between sample selection across strata is introduced in an effort to achieve fuller stratification and thereby reduce sampling errors to a low level [Goodman and Kish (1950)].

---

*Madow, W. G. (1950). *Teoria Dos Levantamentos por Amostragem*, Publicacoes do Centro de Estudos Economicos, Instituto Nacional de Estatistica, Lisbon, Portugal, p. 277 [in Portuguese]. This is an account by V. P. de Magalhães of lectures given in Rio de Janeiro, Brazil, in the winter of 1946-1947. The theorem was also published in Madow (1949).

E. K. Foreman, K. R. W. Brewer, and others in the Australian
Bureau of Census and Statistics have done extensive work in a number
of areas, including especially the use of superpopulation models to
examine the effectiveness of alternative sample selection or estimation
procedures in finite-population sampling.

Nathan Keyfitz was a leader in the early application of statistical methods at the Dominion Bureau of Statistics, Canada and as a consultant throughout the world. One of his contributions that has been
of great practical utility was concerned with the updating of samples
when units have been selected with varying probabilities [Keyfitz
(1951)]. For example, the Census Bureau Current Population Survey
and other surveys ordinarily utilized a fixed set of primary sampling
units (PSUs) for a period of years, and then steps were taken to update the sample when new information became available from the decennial censuses. There was considerable investment in trained manpower
and in other ways in a particular set of sampled PSUs. Keyfitz's substantial contribution was a technique to update the samples so that the
desired new probabilities of selection were appropriately reflected in
the updated sample, but in such a manner as to yield maximum feasible
overlap with the PSUs in the prior sample. This powerful and ingenious
technique has been extended by the U.S. Census Bureau staff, and others
to apply to more complex sample designs and revisions than those
originally covered in the Keyfitz paper, but the basic principles are
the same.

The work at Statistics Canada under Ivan Fellegi deserves special
mention. It has dealt with many types of theoretical and practical
design and sample-execution problems. We mention specifically the
work on nonsampling errors, on variance estimation when samples are
selected with varying probabilities, and on a mathematical theory for
confidentiality of statistical information in censuses and surveys,
as illustrative examples.

SAMPLING FOR EVALUATIVE AND ANALYTICAL STUDIES

In drawing inferences about a cause system, probability sampling of the cause system is not possible, but probability sampling of an existing population that is a consequence of the cause system often is feasible. Such probability sampling may be quite helpful in assuring representations of varying observed conditions, although other factors may also be relevant or operative. Models can be exceedingly helpful in the design and drawing of inferences in evaluative and analytical studies.

In recent years, sample surveys have served increasingly as a method of evaluation of government or other programs or activities. Sometimes the measurement of the outcomes of activities may be fairly direct; for example, to determine whether or not informational activities have led to increased knowledge of a program in a specified population may be measured by a sample survey. In some evaluative studies of important programs, it has been feasible to make random allocations of elements of the sample to treatment and control groups. In others, it has been impractical to assign innovative approaches or some kinds of treatments in a random design. In these studies, a combination of observational sampling with regression, matching, or other models may be particularly useful. Review of this topic is beyond the scope of the present paper, except to remark that research in these areas may be increasingly important in the future.

RECENT ISSUES

We have discussed probability-sampling methods for finite populations, in which probabilities of selection are determined in the sample-selection process and reflected in the estimation process. In either selection or estimation, supplementary information may have a role in increasing the precision of results per cost unit. This approach is

to be distinguished from the use of supplementary information to make
purposive selections, a method that received much attention in earlier
years, and its use in other types of models where the finite-population
estimate is inconsistent or where the bias may tend to zero more slowly
than the variance.

Sometimes the values of variables for the elements of the population are taken to be realizations of random variables, i.e., a superpopulation is assumed to exist, possibly depending on auxiliary variables. Then, sampling designs and estimators may be evaluated on the
assumption that the superpopulation has certain properties; in this
way, much useful information may be obtained, as has been referred
to earlier.

Recently, there has been research in which the properties or
structure of the superpopulation (we will call this a model) are used
to derive sampling designs and estimators that have desirable properties when the model is correct [Royall (1970, 1971, 1973a,b); Madow
(1953)]. Such designs sometimes involve purposive selections of the
units included in the sample, or the use of estimators that do not
reflect the probabilities of selection if probability-selection methods
have been used. Such estimators may have variances that are much
lower than the variances of consistent estimators that utilize the
available information with reasonable effectiveness, but without
assuming the model. However, when the model does not hold, estimators
based on it may be inconsistent or may have biases that tend to zero
more slowly than the variance as the size of sample increases; hence,
with fairly large samples, the bias will dominate the mean-square
error. Consequently, if high precision is needed and the sample is
large enough to yield such precision, the use of such model-based
estimators, or of other than probability samples, may result in substantially larger mean-square errors than with the alternatives.
Studies of robustness based on such models and use of small samples
are subject to the same limitations if the estimated mean-square
error based on the model does not reflect the biases that occur
whenever the models used in the analysis of robustness are not correct.

For these reasons, it may be unwise to generalize from tests of such estimators using small- or moderate-size samples, or even large samples if stable relationships are assumed. It also may be poor policy to depend on the validity of such assumptions when conditions are unstable or the cost of being wrong is substantial. The presence of measurement errors always complicates this situation, but can only add to the problem. On the other hand, there are many circumstances in which stakes are not so high, samples are small, and where substantial economies can be made from assumed models that are likely to hold approximately. Under these circumstances, model-based methods may be quite effective.

Therefore, in making inferences from samples to finite populations, samples can be selected and estimates made by probabilistic methods that only depend on limited assumptions concerning the elements of that population or its distribution. If estimates of high precision and with reliable measures of mean-square error are needed, even under conditions of stress and substantial change in the population or universe for which measurements are made, it may be dangerous to make possibly unrealistic assumptions. In such cases, it seems wiser to depend on large samples and asymptotic and limiting distribution approaches to obtain point and confidence-interval estimators. This is the general philosophy and approach of the finite-population sampling theory, the history of which has been discussed in this paper. Although we emphasize the large-sample randomized approach, we also believe that methods that depend on assumptions concerning distribution or models can be useful in appropriate circumstances, and especially for analytical, research, or small-scale studies.

REFERENCES

Bershad, M., Tepping, B., and Woodruff, R. S. (1953). Regression estimates, double sampling, sampling for time series, systematic and other sampling methods. In *Sample Survey Methods and Theory*, M. Hansen, W. Hurwitz, and W. Madow, Vol. 1, Chapter 11, Section 7, Wiley, New York.

Bikeles, A. (1969). On the limiting distribution of estimates based on sample from finite universes [in Russian]. *Studia Sci. Math. Hung.* *4*, 345-354.

Bowley, A. L. (1913). Working class households in reading. *J. Royal Statist. Soc.* *76*, 672-701.

Bowley, A. L. (1926). Measurement of the precision attained in sampling. *Bull. Int. Statist. Inst.* *22*, Div. 1, 6-62 [a separately paged memorandum].

Cochran, W. G. (1942). Sampling theory when the sampling units are of unequal size. *J. Am. Statist. Assoc.* *37*, 199-212.

Dedrick, C. L., and Hansen, M. H. (1938). *Census of Partial Employment, Unemployment and Occupations: 1937, IV. The Enumerative Check Census.* U.S. Government Printing Office, Washington, D.C.

Deming, W. (1944). On errors in surveys. *Am. Sociological Rev.* *9*, 359-369.

Fellegi, I. (1964). Response variance and its estimation. *J. Am. Statist. Assoc.* *59*, 1016-1041.

Frankel, L. R., and Stock, J. S. (1942). On the sample survey of unemployment. *J. Am. Statist. Assoc.* *37*, 77-80.

Goodman, R., and Kish, L. (1950). Controlled selection--a technique in probability sampling. *J. Am. Statist. Assoc.* *45*, 350-372.

Gurney, M., and Daly, J. (1965). A multivariate approach to estimation in periodic sample surveys. *Proc. Social Statist. Sec.* American Statistical Association, Washington, D.C.

Hájek, J. (1964). Asymptotic theory of rejective sampling with varying populations from a finite population. *Ann. Math. Statist.* *35*, 1491-1523.

Hansen, M. H., Daly, J. F., Bershad, M. A., Tepping, B. J., Waksberg, J., Pritzker, L., Deming, W. E., and Frankel, L. R. (1969). Washington Statistical Society Memorial Meeting for William N. Hurwitz. *J. Am. Statist. Assoc.* *64*, 1121-1153.

Hansen, M. H., and Hurwitz, W. N. (1942). Relative efficiencies of various sampling units in population inquiries. *J. Am. Statist. Assoc.* *37*, 89-94.

Hansen, M. H., and Hurwitz, W. N. (1943). On the theory of sampling from finite populations. *Ann. Math. Statist.* *14*, 332-362.

Hansen, M. H., Hurwitz, W., and Bershad, M. (1961). Measurement errors in censuses and surveys. *Bull. Int. Statist. Inst.* *37*, Part II, 359-374.

Hansen, M., Hurwitz, W., Marks, E., and Mauldin, W. (1951). Errors in surveys. *J. Am. Statist. Assoc.* *49*, 147-190.

Hansen, M. H., Hurwitz, W. N., Nisselson, H., and Steinberg, J. (1955). The redesign of the census Current Population Survey. *J. Am. Statist. Assoc.* *50*, 701-719.

Hansen, M., and Steinberg, J. (1956). Control of errors in surveys. *Biometrics 12*, 462-474.

Hansen, M., and Tepping, B (1969). Progress and problems in survey methods and theory illustrated by the work of the U.S. Bureau of the Census. In *New Developments in Survey Sampling*, N. Johnson and H. J. Smith (Eds.). Wiley, New York.

Hartley, H. O., and Rao, J. N. K. (1962). Sampling with unequal probabilities and without replacement. *Ann. Math. Statist. 33*, 350-374.

Hartley, H. O., and Ross, A. (1954). Unbiased ratio estimates. *Nature 174*, 270-271.

Horvitz, D. G., and Thompson, D. J. (1952). A generalization of sampling without replacement from a finite universe. *J. Am. Statist. Assoc. 47*, 663-685.

Jensen, A. (1926). The representative method in practice. *Bull. Int. Statist. Inst. 22*, Part I, 381-439.

Jessen, R. J. (1942). Statistical investigations of a sample survey for obtaining farm facts. *Iowa Agric. Exp. Statist. Res. Bull. No. 304*, 1-104.

Keyfitz, N. (1951). Sampling with probability proportional to size; adjustment for changes in probabilities. *J. Am. Statist. Assoc. 46*, 105-109.

Madow, W. (1948). On the limiting distribution of estimates based on sample from finite universes. *Ann. Math. Statist. 19*, 535-545.

Madow, W. (1949). On the theory of systematic sampling, II. *Ann. Math. Statist. 20*, 333-354.

Madow, W. G. (1953). On the theory of systematic sampling, III. Comparison of centered and random start systematic sampling. *Ann. Math. Statist. 24*, 101-106.

Mahalanobis, P. C. (1940). A sample survey of the acreage under jute in Bengal. *Sankhyā 4*, 511-530.

Mahalanobis, P. C. (1944). On large-scale surveys. *Phil. Trans. Roy. Soc. London Ser. B, 231*, 329-451.

Mahalanobis, P. C. (1946). Recent experiments in statistical sampling in the Indian Statistical Institute. *J. Roy. Statist. Soc. 109*, 326-378.

Neyman, J. (1934). On the two different aspects of representative method: The method of stratified sampling and the method of purposive selection. *J. Roy. Statist. Soc. 97*, 558-606.

Neyman, J. (1938). Contributions to the theory of sampling human populations. *J. Am. Statist. Assoc. 33*, 101-116.

Palmer, G (1943). Factors in the variability of response in enumerative studies. *J. Am. Statist. Assoc. 38*, 143-152.

Patterson, H. D. (1950). Sampling on successive occasions with partial replacement of units. *J. Roy. Statist. Soc. Ser. B 12*, 241-255.

Rao, J. N. K., Hartley, H. O., and Cochran, W. G. (1962). On a sampling procedure of unequal probability sampling without replacement. *J. Roy. Statist. Soc. Ser. B 24*, 482-491.

Rosén, B. (1972). Asymptotic theory for successive sampling with varying probabilities without replacement, I. *Ann. Math. Statist. 43*, 373-397.

Royall, R. M. (1970). On finite population sampling theory under certain linear regression models. *Biometrika 57*, 377-388.

Royall, R. M. (1971). Linear regression models in finite population sampling theory. *Foundations of Statistical Inference,* V. P. Godambe and D. A. Sprott (Eds.), pp. 259-279. Holt, Rinehart and Winston, New York.

Royall, R. M., and Herson, J. (1973a). Robust estimation in finite populations, I. *J. Am. Statist. Assoc. 68*, 880-889.

Royall, R. M., and Herson, J. (1973b). Robust estimation in finite populations, II: Stratification on a size variable. *J. Am. Statist. Assoc. 68*, 390-393.

Seng, Y. P. (1951). Historical survey of the development of sampling theories and practice. *J. Roy. Statist. Soc. A 114*, 214-231.

Stephan, F. F. (1948). History of the uses of modern sampling procedures. *J. Am. Statist. Assoc. 43*, 12-39. See also, *Rev. Int. Statist. Inst.* III, A, 81-112.

Stephan, F. F., Deming, W. E., and Hansen, M. H. (1940). The sampling procedure of the 1940 Population Census. *J. Am. Statist. Assoc. 35*, 615-630.

Stock, J., and Hochstein, J. (1951). A method of measuring interviewer variability. *Public Opinion Quart. (Summer)*, 322-334.

Sukhatme, P. V. (1935). Contributions to the theory of the representative method. *J. Roy. Statist. Soc., Suppl.*, 253-268.

Sukhatme, P. V. (1946). Bias in the yield of small-size plots in sample surveys for yield. *Nature 157* (3393), 630.

Waksberg, J., and Neter, J. (1964). A study of response errors in expenditures data from household surveys. *J. Am. Statist. Assoc. 59*, 18-55.

Woodruff, R. S. (1963). The use of rotating samples in the Census Bureau's monthly surveys. *J. Am. Statist. Assoc. 58*, 454-467.

Yates, F. (1946). A review of recent statistical developments in sampling and sampling surveys. *J. Roy. Statist. Soc. 109*, 12-45.

Yates, F., and Zacopanay, I. (1935). The estimation of the efficiency of sampling with special reference to sampling for yield in cereal experiments. *J. Agric. Sci. 25*, 543-577.

5.

*A PERSONAL PERSPECTIVE
ON STATISTICAL TECHNIQUES
FOR QUASI-EXPERIMENTS*

NATHAN MANTEL was born February 16, 1919, in New York City. He
received a B.S. degree from the City College of New York in
1939 and an M.A. degree from The American University in 1956.
He is a Fellow of the American Statistical Association, the
Institute of Mathematical Statistics, and of the American
Association for the Advancement of Science. He has been on
the Editorial Board of the *Journal of the National Cancer Institute, Cancer Research,* and of *Biometrics.* From 1939 to 1974
he held various positions with the U.S. government, principally
with the Public Health Service and the National Cancer Institute.
He is a member of the International Statistical Institute. He
is the author of over 200 publications. Presently he is at the
Biostatistics Center, George Washington University, working
under a National Cancer Institute-supported grant, after having
retired from federal employment.

A PERSONAL PERSPECTIVE
ON STATISTICAL TECHNIQUES
FOR QUASI-EXPERIMENTS

Nathan Mantel

Biostatistics Center
George Washington University
Bethesda, Maryland

Observational or quasi-experimental data have been with us for a long time, and for nearly all of human history it has been just such data that mankind has used as the basis of learning. The use of designed experiments as a source of data has only recently come into vogue and has come to be considered ideal, so much so as to give the impression that any other kinds of data are unfit. It is likely, however, that observational data will continue to play an important, possibly domi-nant, role in learning and research--what experiments can be performed and what can be learned from experimental data are subject to limita-tions, perhaps not generally appreciated, and preclude any exclusive role for the experimental approach. In some areas, e.g., astronomy or geology, the observational approach may be the main road to learn-ing. The observational approach is at work wherever we seek to determine if there is some relationship or interrelationship between specified characteristics in some defined population. The population may be the atomic elements or members of some class of molecules, or oils of different kinds or from different sources, with two or more different kinds of measurement taken for each population member. The answers to some problems can only be sought by the observational approach. If, for example, we are concerned with whether or not some disease displays clustering or how it is occurring in the population, we must first get our data by observation. Subsequent analysis is directed to learning what experiments or experimental design nature used in generating the data, the null hypothesis usually relating to some random experiment by nature.

My exposure to problems of observational data analysis has been largely concerned with identifying factors important in disease occurrence, with particular emphasis on retrospective disease studies. Still, there has been considerable tempering of my points of view through handling of data from designed experiments, from having read or reviewed the works of others, from innumerable discussions and consultations, and, in general, I cannot specifically identify those influences today. Any particular identifications I give in this discussion are intended only as personal to me with no necessary implication that they played any key role in statistical methodology. Other practicing statisticians will have been tempered by their own personal experience.

Preparing this perspective causes me some regret. Anecdotal and other material that I have effectively used in lectures and discussions as a seemingly spontaneous display of my wit or wisdom will now be put down in fixed manner. An example is my joking advice to statisticians not to get called in at the start of an investigation. Things are bound to go wrong anyway, and because he is uninvolved, the statistician cannot be faulted. Besides, if he is called in early, the statistician is barred from making the familiar plaint that the investigator should have seen him first. Amusing as this is I have a case in point where in my early years at the National Cancer Institute I helped a leading investigator with the design of a complex experiment. After that I did not hear from him, and about 6 years later I asked him what had happened. All the mice in the experiment had died from toxicity and the investigator had been too embarrassed to seek me out after that. In later years I did come to learn the need for defensively designed experiments with built-in hedges against

wasting the experiment if the investigator's basic information had
not been quite correct. Actually, being called in after the fact of
an experiment has not generally proved an important bar to my finding
some way of being of help to the investigator. In a sense, when this
happens what the investigator takes to be experimental data are, to me,
observational data.

I can recall having a brief discussion at one time with Cornfield
in which I joined with the position that evidence from retrospective
studies of disease should be less persuasive than that from prospective studies (and presumably even less so than evidence from a designed
study). Cornfield would not concede that the retrospective study was
in any way deficient. Today, I would more fully agree with Cornfield,
and if anything, I would be more enthusiastic than he in espousing the
retrospective approach. This enthusiasm I indicated in a presentation
in San Diego in January 1973 on the use of naturalistic data at which
Joe Fleiss and Ardie Lubin were the other statisticians who participated. Preceding medical speakers did give examples that could make one
mistrustful of the data that happened to arise in their studies. To
my mind, this only indicated that one should not be too ready to use
unnaturally occurring data. As an attention-getter, I started my talk
by asserting that, as we had just seen, the retrospective study was
the design of choice in the investigation of etiologic factors in
disease, particularly diseases of low incidence. It has the advantage
of speed and low cost, and if some ambiguity attends the interpretation
of results from a retrospective study, parallel ambiguities would attend those from almost any other study. Then I went on to give some
examples that I will repeat below.

In my publication with Haenszel (Mantel and Haenszel, 1959), my
participation was particularly encouraged by Peter Armitage, and it
most identifies me with the retrospective study problem. Yet, I have
always thought of the statistical methods we give there [and in the
related follow-up paper by myself (Mantel, 1963)] as general methods
for combining information from independent experiments. In a sense,
I have been making no distinction between how I would handle the observational data from a retrospective study and those from a designed

experiment. Recently, I suggested that when a prospective study of
disease has been conducted and has yielded relatively few positive
outcomes, it could be advantageous to convert it in principle to a
retrospective study by the device of taking all the positive respondents, but only a random sample of the negative respondents
(Mantel, 1973).

   As an introduction in this work to the synthetic retrospective
approach, I reviewed three possible approaches to an epidemiologic
study. First, there was the designed experiment approach in which
individuals were assigned at random to treatment groups, with perhaps
an eye to orthogonality, then followed in time; next, in the prospective study approach, the self-selected treatment of individuals was
determined, the individuals then being followed in time; finally, in
the retrospective study, samples of individuals presently with or
presently free of disease were taken and their self-selected treatments
were ascertained. In each case, appropriate analysis could be made,
but there would be varying assumptions required for interpretations
to be valid. I noted at the time that for both the designed experiment approach and for the prospective study approach, the statistical
analysis should take into account both the time and the fact of response, but did not say how to accomplish this. Various techniques
for analyzing time-to-response data are now in the literature, but I
give a rather interesting technique as an extension of my 1959 work
with Haenszel (Mantel, 1963, 1966). (This extension has been cited in
recent work by D. R. Cox, the Petos, and Norman Breslow who recognized certain germinal values in it.) For retrospective studies the
time-to-response problem is resolved, I believe, if age, not too
coarsely subdivided, is taken as a stratifying variable.

   A difficult problem for any of the three approaches is that of
covariates. If for each individual in the study I have measurements
on many factors, I can be concerned with just which ones to consider
as requiring specific attention, which as covariates for which adjustment is to be made, with the added problem of how to make such an
adjustment. If there were many variables, the stratifying approach
of Mantel and Haenszel may be inoperative. If a designed study is

orthogonal on the main study factors, it may become nonorthogonal
when auxiliary measurements are taken. Covariance procedures cannot
be used with reliability if we cannot safely assume linearity or some
other specified regression form. (I have encountered situations in
which reasonably good analyses could have been made if only a few
basic measurements had been taken. But more becomes less when added
measurements make the potential analyses so complicated that the investigator, in desperation and unwilling to be selective about analyses
to be done, makes only simple one-variable-at-a-time analyses.) I
have brought up this issue here particularly to make a point of which
I have only recently become consciously aware, that the simplicity of
analysis of data from designed experiments is largely illusory. We
can see from the foregoing discussion that the covariate problem exists
both for experimental data and for observational data. If an initial
variable in a designed epidemiologic study is a person's blood type,
we might wish to consider that as a study factor in itself and also
as a covariate in evaluating the influence of, say, cigarette smoking.
This kind of thing does not arise in laboratory experiments only because we have failed to make a host of initial observations on each
test animal. Instead of assessing the role or adjusting for each of
these initial conditions, we have chosen to ignore them and to include
the variation attributable to them as part of our error--by so doing,
we make things simple, but in studies involving humans, we are very
conscious of the many ways in which people may differ. (In one clinical trial, the investigator found a significant difference between two
therapies. It was pointed out to him that on one form of therapy
patients were largely female and younger; on the other therapy, they
were largely male and older, and that an analysis adjusted for either
sex or age or both might be reasonable; his reply was that he had
randomized his patients and so saw no need to alter his analysis.
In application, truly random experiments are frequently not conducted--
if an identifiable characteristic attaches to each test subject, the
investigator may well discard a badly lopsided randomization. I can
think of a parallel in the random-pattern laying of floor tiles. The
manufacturer cannot put the tiles into the boxes fully at random else

they might come out in a lopsided arrangement, or even in some sug-
gestively systematic pattern.)

During the preparation of my work on synthetic retrospective
studies, I had consulted with Max Halperin on the theoretical valida-
tion of the principle I proposed there. Halperin came up with some
near-proofs of validity, but need for them disappeared when I suggested
a simple concept. I considered the outcome for an individual to fall
into a four-cell multinomial population defined by whether or not the
individual was in the sample, and whether or not the individual had
the study disease. Our observation is limited to the case in which
the individual is in the sample. The desired adjusted probability for
a sample individual having the disease is given by the initial probability of the individual both having the disease and being in the
sample divided by the sum of the probabilities for the two observable
outcomes, from which the desired result followed. Actually, I had
used this truncation principle in earlier work on incomplete contingency tables (Mantel, 1970a) in which I had taken as the adjusted
probability for a cell in an r × c table when some cell outcomes were
unobservable as the unadjusted probability for the cell divided by
the sum of the probabilities for the observable outcomes. At the time,
I thought of this only as straightforward application of the truncation
concept, but later, in connection with commenting while we were both
briefly in London, to Ted Colton on his Bayes' principle illustration
in the draft of an elementary text he was writing, I came to realize
that the truncation concept was the same as, or rather subsumed,
Bayes' principle.

Certain points discussed in my synthetic retrospective studies
paper had their origins dating back to the early 1950s. Cornfield
had been presented with the problem of estimating the probability of
death of fire victims, using the percentage of skin surface burned as
a basis for prediction. He took the novel approach of proposing that
there were two populations of individuals, each with its own distribution of skin-surface injury, the populations being those that had died
and those that had lived. He then equated the probability of death
to the probability that the individual belonged to the dying population

Cornfield had not proposed this as something intended actually to be realistic, but rather just as an easy way of coming up with answers. In fact, he with Tavia Gordon and Willie Smith elaborated on this at the 1960 I.S.I. meetings in Tokyo, extending it to the multivariate case. Cornfield, in discussions with me, suggested that in the multivariable case there could be a mixture of causative and discriminatory variables, and that it would be just as reasonable to treat them all as discriminatory as to treat them all as causative. In any case, Cornfield's discriminatory approach has been accepted with such enthusiasm as to have become, he says, a Frankenstein monster. My bias for the causative, hence logistic regression, approach is partly premised on the fact that if we are looking for causative factors in disease we should analyze the data as though the study variables were, in fact, causative--if some of the study variables are, in fact, only discriminatory, the logistic regression approach remains valid. Those points are discussed a little further in Mantel and Brown (1973), whereas in Mantel and Brown (1974) we follow through on a suggestion that logistic regression analysis can be validly used as a replacement for univariate or multivariate normal analysis with lessened dependence for validity on the assumption of normality.

On the matter of relating the risk of death to the percentage of skin surface burned, a discussion I had with a representative of an organization concerned with the problem is of interest. We had discussed possible methods for fitting data on death from burns, but then I ascertained that the organization had data on many thousands of cases. The information available was somewhat detailed covering characteristics of the individuals and specific identification of what the burned areas were. In that case I suggested, better estimates of death probabilities could be obtained by categorizing the data than would be yielded by a fitting approach where the fitted model might be wrong and the input data oversimplified. The statistical methods we had been discussing might have been suitable if we had only sparse data. A reasonable graduation model even if it is not quite correct can nevertheless provide useful predictions. Without some graduation model, the data could seem useless and chaotic. [I suggest this idea somewhat in Mantel (1970) where I indicate that probability estimates for the cells of a two-way contingency table when row-column independence is falsely assumed could still be an improvement over use of the observed probabilities.] Even the usual analysis of assumably normal observations classified on two axes corresponds to the use of a simple graduation model, as I have heard Neyman say. The finding of a significant interaction should mean only that a particular simple linear model is false and not necessarily that the data are suggestive of something extraordinary.

Statistics, in general, is characterized by this trickery of making false assumptions yet coming out with good answers, a simple illustration of which I like to make is that relating to setting limits on the population average. In the class of arbitrary distributions no sample, however large, can let us set limits on the population average (but we can set limits on the population median), as there is always the possibility that some small, finite proportion of the population is at arbitrarily extreme values. In setting limits on the mean by normal procedures, what we are really doing is asserting that our population is a well-behaved one, then taking the well-behaved normal distribution as a close guide to what should obtain for the true one. To the extent that the statistical art consists of the use of tricks, it would be important for the statistician not to take those tricks so seriously that he himself is fooled. In a way, the use of such tricks can even become a block to learning. Thus, when the statistician makes the normal assumption, he replaces the infinitely parametered arbitrary distribution about which there may be much to learn by one about which he thinks all he can learn are the mean and variance.

Suppose I review some simple examples I have encountered. The first I owe to John Walsh whom I was visiting at Systems Development Corporation in Santa Monica in 1960. Walsh and his associates had endeavored to use clinical records for getting at prognostic factors in bacterial endocarditis. A curious result of their complex multiple regression analysis was that it demonstrated that prognosis worsened as level of treatment with penicillin or other antibiotics was increased. Their hopes for using clinical data as a substitute for

experimental data were frustrated because treatment was to a large
extent a variable dependent on the patient's condition--patients in
worse shape got more antibiotics--rather than an independent variable
that was a determinant of a patient's outcome. A parallel example
comes from Sid Cutler of the Biometry Branch, National Cancer Institute. Suppose you want to compare the outcome experience at two
medical institutions in the treatment of cancer patients. It may
seem reasonable to compare radiation-treated patients at one institution with similarly treated patients at the other, with a corresponding comparison being made for surgically treated patients. But such
comparison is, in fact, unreasonable because choice of treatment is a
variable dependent on the institution involved. Suppose institution
A selects the 20% most hopeful cases for surgery, leaving the rest
for radiation therapy, whereas institution B is willing to treat as
much as the 50% most hopeful cases by surgery. (In European institutions, I believe, it is the radiologists who get first pick of patients, with surgeons getting the remainder.) A consequence is that
the average surgical patient at A will be better material than that
at B, the same obtaining for average radiotherapy patients. A valid
comparison would have been to ignore treatment and to compare overall
outcomes at the two institutions--the difference in rules for patient
assignment is just one of possibly many ways in which the two institutions differ.

   This confusion of dependent and independent variables seems to
be a rather common error. In the Walsh example, the difficulty is
easily recognizable after the fact, but one can be concerned about
more subtle situations in which it would not readily be recognized
that an incorrect analysis has been made. Nor is the confusion confined to observational data. In a laboratory experiment, it was found
that rats on treatment A had larger tumors than did rats on treatment
B when body weight was taken as a covariate, which suggested that
treatment A involved a tumor growth stimulating substance. An alternative interpretation, suggested by the fact that overall the average
tumor weights were about the same for the two groups, was that treatment had no influence on tumor weight but did cause body weight losses

for the rats, body weight being also a dependent variable. Apparently, treatment A consisted of feeding the rats a diet that included ground-up tumor (which the investigators hoped they were showing to include the growth stimulating substance)--this the rats found sufficiently distasteful as to inhibit their dietary intake so that weight gains were reduced, but there was minimal influence on the tumors because of the tumors being parasitic on the host rat.

Failure to recognize the limitations or even improprieties of a particular regression model is another cause of misleading statistical analysis. A blatant example was one in which the investigators fitted a linear regression of tumor volume against tumor length and tumor diameter. Such a regression was clearly functionally incorrect. Furthermore, the two regressor variables were so highly positively correlated that, not surprisingly, the fitted coefficients came out of opposite sign--in consequence, the fitted volume for a tumor with fixed diameter could vanish or become negative if only the length were increased sufficiently. The responsible investigators were intolerant of any advice on this matter since they had scrupulously followed least squares procedures.

Ed Gehan has provided me with another example in which intercorrelations were involved, but in which the question of the functional form was not. For various drugs, Gehan had data on the optimal dose level in man and in various mammalian species. When Gehan fitted a least-squares regression relating the optimal level in man to that in the other species, owing to the high intercorrelations, one of the fitted coefficients was negative, suggesting that the higher the mouse optimal dose the lower should be the human optimal dose. Rather than defending the least-squares fit as such, in this case I would have preferred to make a prediction for humans separately from each mammalian species, then taking some kind of average of them so that all coefficients would effectively be positive.

Such more obvious misuses or less-than-perfect uses of regression methods in statistical analysis do not seem to me altogether dangerous. The blind use of regression methods without awareness by the investigator of the dependence on implicit assumptions can, however, be frightening. One example I headed off before publication was brought to me by Manny Landau. His multiple regression showed a significant time effect with everything else insignificant. I showed him how if only data for a single year were involved, a seasonal pattern could show up as significant linear time regression but with residual errors so large, since the linear regression was in fact invalid, that the effects of other variables could not be demonstrated. In a published example on which Bernie Pasternack and his associates at the New York University Medical Center sought my comments, the investigator claimed to have demonstrated a role for the acute effects of non-extreme air pollution in causing death in humans when temperature effects were removed, but not vice versa. The report was not entirely specific as to just what had been done, but taking clues from what was given, it was my interpretation that the investigator had found a significant partial regression coefficient for one of the air pollutants considered, but not for temperature. I am sympathetic to uncovering environmental human hazards and so should like to see that such demonstrations are in fact valid. In the present instance, I could recognize a strong dependence of the demonstration on the assumed linear model with the existence of the possibility that a true nonlinear model with only temperature as the causative variable might have been responsible for the findings that were made. This possibility was heightened when I found a casually made point in the report that the temperature used was the absolute deviation from some ideal temperature--with this modification, insofar as temperature was involved, the assumed linear model became an assumed V-shaped model, which if imperfect could have misleading effects. (Nothing came of my report back to the Medical Center on this--they had wanted me to say that the original report was wrong, not simply that it might not be right.)

These last instances hark back to my admonition that the statistician not let himself be deceived by the procedures he uses. In the air pollution study, it would have been feasible to get at the separate influence of any one study variable holding the others constant without the need to postulate a specific multiple regression form. I indicated this possibility in a published letter to *Technometrics*

(Mantel, 1971) in which I responded to a criticism of my earlier paper
on variable selection (Mantel, 1970b). In the letter, I indicated
that the multiple-regression variable-selection problem is an artificial one, because true multiple linear regressions ordinarily do not
occur--fitted partial regression coefficients may be misleading if
multiple regressions are not, in fact, linear. My suggestion was
that if data were abundant, it would be possible to get at the partial
effects of a variable free of an assumed linear model with regard to
the remaining variables by the use of stratification. [In essence,
although without mentioning it specifically, I was recommending the
procedures of Mantel and Haenszel (1959) and Mantel (1963).] In the
summer of 1971 I had occasion to visit with Ed Batschelet in Zurich,
shortly after his return to Switzerland from Catholic University in
Washington, D.C. He was rather excited at his own interpretation that
the Mantel-Haenszel procedure would be effective in such a manner. I
confirmed to him that he was correct, but I cannot say that Haenszel
and I were initially consciously aware of this--maybe we just had
good instincts.

As examples in which a statistical analysis is leading, say rather
misleading, instead of guiding the statistician, I give two instances,
one relating to experimental data, the other to observational data.
In the first case, I was reviewing a paper that involved a bioassay
procedure. The authors were claiming an unusually good index of precision for the assay since in one portion of the response curve the
ratio of the standard deviation to the slope was quite small. I pointed
out that this came about artificially--the data did not necessarily show
such a steep slope, and there was no real basis for postulating that the
slope was so steep. It was only that the regression form fitted called
for a steep inflection point; the high index of precision was a characteristic of the postulated regression form, not of the assay. Perfect
apparent precision would have resulted had the fitted regression form
been one that was vertical at the inflection point. I used my second
example as discussant at a session on variable selection during the
1973 annual meetings of the American Statistical Association in New
York City. Statistical analysis had shown that height was the most

important single determining factor, of about 18 considered, in the
occurrence of coronary incidents. This unusual finding came about
because the data had been fitted, according to a casually stated point,
without an intercept parameter (more correctly, with an intercept
parameter specified to be zero). Such fitting would produce a bias
in favor of finding positive roles for factors with low coefficients
of variation. Human height has a coefficient of variation of only a
few percent and so was destined to appear the most positively influential variable in the analysis made--this would have occurred even if
coronary incidents and height were actually negatively associated.
In an alternate analysis employing an intercept parameter, height
proved to be the least influential of the 18 factors.

Two large-scale observational studies have received particular
attention by statisticians. The American Cancer Society's prospective
study of cigarette smoking has been challenged by Ted Sterling, one
basis of his challenge being that the study group was not representative of the general population. That this should not be a proper basis
for challenge should be self-evident to statisticians, but, in any
case, I have incorporated in (Mantel, 1973) a demonstration that only
an unbiased condition, not a completely random sampling condition,
must be met in order for the results of such a study to be valid. The
retrospective study of the effects of halothane used as an anesthetic
in surgery has been of particular concern to me. I had been called on
by Robert J. Taylor, then with the Food and Drug Administration, to
comment on a published preliminary report on halothane. In my memorandum to Taylor, I brought out that the report quite sensibly emphasized the occurrence of liver necrosis as the endpoint of investigation, as liver necrosis seemed to be a specific toxic effect of anesthetics while death of surgical patients was a rather nonspecific
effect. A puzzling aspect of the analysis that was made was that for
purposes of simplicity the various surgical procedures had been divided
into three groups--high, middle, and low--but on the basis of their
associated death rates, not their necrosis rates. The logic of a
stratification approach is that the various strata should be, if not
homogeneous, at least only mildly or moderately heterogeneous on the

endpoint of study. Such limited heterogeneity on necrosis is, of
course, not achievable if stratification is made on the basis of mortality (heterogeneity within a mortality stratum was rather great).
In fact, no such stratification was needed at all because each surgical procedure could have been kept distinct in getting observed and
expected numbers of cases of necrosis under halothane use. An additional point in my memorandum was that although the analysis showed
reasonable agreement between observed and expected for halothane, this
was partly because the expectations were based on the data for all
anesthetics, including cyclopropane, which was distinctly hazardous--
a direct comparison would have shown halothane more hazardous than
ether. In any case, in my memorandum I suggested that an appropriate
reanalysis of the data could reverse the conclusions reached in the
first analysis.

Halothane appeared again as the subject when I attended the International Biometrics Conference in Sydney in 1967. Mosteller presented
a report on what his associates at Harvard were doing in analyzing the
halothane data. Cornfield as discussant congratulated Mosteller on
the ingenuity and variety of statistical techniques that had been em-
ployed--I raised the question, ineffectually, of why mortality had
been substituted for liver necrosis as the study endpoint and could
this substitution not be largely responsible for the apparently negative findings. The emphasis on ingenuity of statistical techniques
was brought out more forcefully in the final 1969 report (The National
Halothane Study) in which many leading statisticians displayed their
talents, but little was accomplished in the way of useful direct analysis of the data. (In a published review of the 1969 report Peter
Armitage's attempt to be gently critical of it was somewhat vitiated
by a typographical error.) As to the fate of my own initial critical
memorandum which had been urgently requested, Taylor transferred out
of the Food and Drug Administration shortly after receiving it, and
it just got forgotten.

The cigarette smoking-lung cancer issue was the subject of talks
given by R. A. Fisher and Joe Berkson at a program in pharmacology in
Brussels in the summer of 1958 concurrent with the meetings of the

International Statistical Institute (and the World's Fair). Fisher, if I recall correctly, gave a genetic predisposition argument for disputing the association. Berkson's argument consisted of a demonstration that a wide variety of death rates seemed to be increased by cigarette smoking (although generally not proportionally so much as for lung cancer), in consequence of which he could not believe cigarette smoking to be causal of lung cancer. Apparently, he could not accept that one agent would have so wide a variety of effects--this is not so puzzling to me since it could have as many effects as there were target organs. From the floor, I asked if it was not strange that Berkson was using evidence that cigarettes were indeed the coffin nails they were once termed in this way. Berkson then thundered back at me--he didn't care if cigarettes were good for you or bad for you-- he just didn't want them labeled as a cause of lung cancer. It seemed to me that he then turned to Fisher for support, but got none--if they were on the same side of the fence, it was for very different reasons.

(In 1960 at Tokyo, things were amicable between Joe and me. We had a Japanese dinner together with Mrs. Berkson and George Nicholson. There, also, George demonstrated to me his ability to establish rapport with the Japanese bar-girls by teaching them the game of Nim. For myself, I demonstrated, at a Chinese restaurant in Tokyo, to Cornfield my dexterity at eating jellied sea slugs with chop sticks. At the occasion of the 1961 meetings my wife and I found Fisher wandering the streets of Paris one morning--we took him to an American Embassy cafeteria so that he could breakfast on those wonderful cereals that went snap and crackle. There is much to remember from those various meetings, including in the Brussel's year David Cox's frantic activity at running the Royal Statistical Society meeting in St. Andrews, a presentation by Willie Feller at the mathematics congress in Edinburgh, and finally, chasing after the Hotellings from city to city throughout Europe with an over-coat Harold had left behind, only to learn that it was not his.)

The brief review I have given lends emphasis to the important role that the handling of observational data has had in my career. What about experimental data? One thought I have had about my function as a statistician, whether the study is observational or experimental, is that I am a packager. The end product of research is a manuscript, and so I try to visualize what kind of paper the investigator's work can be packaged into. The statistician is likely to find himself in such a role after awhile because he will have had more experience putting things together than will a new, young investigator. The investigator may be hampered from putting together a manuscript because he is conscious of many fine details and because he is aware that his investigation has still left many questions unanswered. I would concentrate on what has been accomplished positively and how effectively to present the data, preferably free of complex manipulations—a good graphic presentation could bring out clearly both the data and any indications therein for statistical significance or insignificance. Fixed rules for statistical analysis of data have to go by the board and statistical practice becomes something of an art. Thus, the use of t tests and the comparison of arithmetic averages may be reasonable in the evaluation of one set of data, but I would not always want to use them. Nor would it be any more reasonable to use nonparametric tests routinely just because they might sometimes resolve otherwise difficult situations. In general, I would prefer to use "cut and try" to "cut and dried" statistical analysis, with perhaps the following justification: the investigator already is convinced that he has found something real and is seeking some persuasive aspect of the data to which to direct the reader. As between the investigator and the statistician, if the investigator is too readily persuaded by his data, the statistician can advise him just how significant or insignificant the results actually are. Incidentally, in classroom lectures I have suggested that the concept of statistical power requires that the statistician take an active part in the preparation of manuscripts, in consequence of which he needs to develop skills in saying things clearly, in simple English and eschewing statistical jargon. Good statistical power implies that our efforts will not be wasted, whether waste occurs because the experiment is too small, or badly designed, or poorly analyzed—waste because the investigation has been written up poorly is just as bad. (What effectiveness I have at clear writing

I believe I owe in large part to Jerry Cornfield as a result of a
brief lesson he once gave me. One day, I was brimming over with ideas
and results I wanted to show him, but this time Jerry would have none
of it. Write it down, he said. No talking it over or anything, just
write it down. So there I found myself having to put everything down
that I wanted to say in some systematic manner, and in a way in which
even Jerry could not misunderstand. And I've been writing ever since.)
A last point here is that I can generally work best with an investigator when we get down to specific data. This is not always so simple.
Instead, the investigator may approach me with an extremely broad
question, which can be thought of as how to deal with the relationship between the generalized variables X and Y. The burden of hedging
and qualifying everything I say is extreme, but is reduced as the problem is stated less broadly or if I am told specifically what the X
and Y variables are--specific X and Y values make life easiest because
then I can see where to go.

Experimental data are not always so perfect that the observational
data statistician has to feel apologetic about the quality of the material he is handling. My direct laboratory experiences are an example
in point. During my varied army service, I was detailed to Ed Knip-
ling's group with the Bureau of Entomology and Plant Quarantine in
Orlando, Florida, doing research on insecticides. (As a statistician,
I continued to carry the occupational designation, clerk nontypist,
and did not succeed in getting it changed to sanitary technician which
my associates carried. As a former New York City Sanitation Department clerk I suppose I qualified for both titles.) My principal
activity was in a team with Jack Williams, a budding pharmacist. Jack
would spray into a testing chamber the insecticide solution he had
prepared, having previously evacuated from the chamber the air that
had been contaminated by the preceding test. He would then swing
back and forth for a specified time period on a long wooden arm two
cages, one with flies, one with mosquitoes, which I had loaded--the
cages had been previously cleaned with solvent. The individual flies
and mosquitoes were all volunteers, not selected at random--they
would have risen toward the light and into the cage I had ready to

receive them. The number of insects entering the cage was variable,
but about 30 was the desired value. After exposure, I would keep the
cages under observation noting, whichever was easier, the number of
insects knocked down at the bottom of the cage or the number still
flying about or on the sides of the cage. My observations were for
something like 5, 15, and 30 minutes after exposure. Some high school
students made later observations for 2 or 3 hours after exposure.
Finally, the next morning, I would make a last observation, then heat-
kill the insects and make a count of the total number of insects in
the cage. What uncertainties attended these observations? Well,
temperature had a very definite influence, and we kept a temperature
record, avoiding testing whenever it was too cold. The high school
students were quite eager to get finished each day and they would rush
through their observations at a breathtaking unreliable speed, putting
down, it seemed to me, any value that took their fancy--no matter,
perhaps, if it was final kill the next day that counted. Overnight,
the cages were kept on an antproofed table, but our antproofing was
not perfect. Somehow the ants would manage to get in and would walk
off with the corpses. Cases of wholesale pillage could easily be
recognized, as could instances where fewer mosquitos were left than
had been seen flying about the day before. Well, I put down the numbers and somewhere someone else analyzed them, hopefully not by some
contingency table method premised on fixed marginals. Something on
this order, maybe not quite so extreme, is true of a good deal of
experimental data, if the statisticians but knew it. (A fringe benefit
of my work at Orlando was that I got some notions on testing agents in
combinations, which I later made use of in my work at the National Cancer Institute. Other chores at Orlando included demonstrating scalp
treatment with DDT for an army training film, but I never saw the
final product.)

(I have found incidentally acquired information useful in my work,
getting much of it from my wife who is the knowledgeable member of the
family. Tom Burch from Arthritis catches me in the parking lot one day
and says he wants to survey diabetes in an American Indian group, how
should he sample. I suggest getting dental records of the population,

forming strata with dental condition one of the stratifying variables, then taking 100% samples of people without teeth, reduced percentages for other strata. Here I am using information I acquired earlier from Howard Schwartz, then at Montefiore Hospital in New York City, while I was on assignment to the New York University Medical Center sharing my time between Bernie Pasternack's and Max Woodbury's departments. Howard had the idea that dentists could be recruited into a campaign for earlier diabetes detection since diabetes patients were typically, or became, edentulous. Edentulous--that's a good word; it signifies my learning to use a few well-chosen terms from the jargon of my scientific colleagues. It can impress fellow statisticians, but care has to be taken to use them correctly lest I appear foolish. A particularly useful scientific term is endpoint, which can variously signify parameter, statistic, observation, result of a calculation, result of a laboratory determination or whatever--laboratory people simply do not have the hassle of statisticians in distinguishing between parameters and statistics.)

An outline of statistical life at the National Institutes of Health (NIH) may be of historical interest, in particular that of the so-titled mathematical statisticians. Mathematical statistics at NIH probably had its beginnings when Harold Dorn brought in for work at the National Cancer Institute five statisticians, over the period 1946-1948--Jack Lieberman in 1946, Cornfield and myself in 1947, Sam Greenhouse and Marvin Schneiderman in 1948. Essentially, none of us had had statistical experience outside the field of economics--what else was available?--and so we were all going to learn this new work on the job. What unusual sense Dorn had for making this selection of people I cannot say, but we all had the characteristic that you could not shake us loose from NIH (save for a brief lapse by Cornfield to take a department chairmanship at Johns Hopkins), and so we all stayed on to retirement, with Schneiderman and Greenhouse still on duty, although already eligible for retirement. (Displaying this same tenacity are two nonmathematical statisticians, also selected by Dorn, Sid Cutler and Bill Haenszel, whom Dorn chose as his replacement at the Cancer Institute when he left for the Heart Institute.) Subsequent

additions to the close-knit group of statisticians were Max Halperin,
recruited by Cornfield and Felix Moore for the Heart Institute and
Seymour Geisser, brought by Greenhouse into Mental Health. Max is now
once again with the Heart (now Heart and Lung) Institute though he had
a longish lapse, and Seymour has defected, but he may yet return.
(Geisser had two peculiarities when he joined us--he wrote on both
sides of the paper and he insisted on giving us a set of statistical
exercises as a test of our competence. Our revenge was in demonstrating the falsity in the proof of Geisser's then pet theorem.)

Anyway, the original five of us were housed together in one small
room, with Dorn next door. And our favorite way of teaching and learning from each other must have been the shout-the-other-guy-down method
if I accept the testimony of the disquieted people elsewhere in the
building. You would score a victory if you could induce a case of
laryngitis in your opponent, for then he would be at your mercy, a
shorn Samson. This wild behavior would continue on into lunch with
our numbers augmented by the later recruits and a few sympathetic
souls. Twenty and even 25 years later, there are cafeteria workers
in Building 1 who can recall our lunch sessions.

One thing that characterized our early work was that essentially
we were not publication conscious. We felt our primary function was
to provide service to the laboratory investigator. He might use our
results with or without acknowledgment in his own publications, and
on various occasions when our work was above and beyond the ordinary
call of duty, might offer one of us a coauthorship. During this
period, if we raised some problem and came up with a solution we would
not recognize it as something extraordinary--we had just learned something that was anyway only a direct consequence of statistical thinking. A particular case is where I showed Cornfield an interesting
idea I had--the analysis of variance could be used to test for parallelism of two regressions against time without the need for specifying
the regression form if one used the animal-time interaction as the
error term. That was interesting, but we took it as just an immediate
consequence of what the analysis of variance was about--the next year
Box published the same idea in his now famous paper on growth and wear

curves in *Biometrics*. (I have since been suggesting to students that they be alert to the possible novelty of seemingly obvious ideas they have. A further piece of advice I give is that they not let slip by the early years they have on a job--this is the time to publish in statistical publications because later on they might just be too swamped with responsibilities.)

At one time, a less bellicose activity we had was poring through the chapters of Kendall's encyclopedic volumes on advanced theory. One of us would have the responsibility each week of previewing a portion of Kendall and explaining it to the others. One discussion I led eventually culminated in the publication by Cornfield and Tukey of an article on the expectations of mean squares in the analysis of variance. Once we had the sense to understand, we could appreciate the preciseness of writing in R. A. Fisher's texts--but then wonder how mysterious it could still be to the novice. (Actually, we found one place where Fisher was clearly wrong on the covariate-adjusted comparison of means. We were gratified to find the matter corrected in a later edition.) Somewhere in the period 1939-1942 I had taken various courses with Abe Girshick at the Department of Agriculture Graduate School (my classmates had included Bill Madow, Morris Hansen, Bill Hurwitz, and Ed Deming; some of my colleagues at the National Cancer Institute probably also studied under Girshick, but I do not know when), but I must say that little of what I studied with Girshick meant much to me at the time. It was only after I began working at the National Cancer Institute that things began falling into place; that I could literally see an F test as relating to direction tangents to points on the uniform-density surface of a generalized sphere. (Earlier at the City College of New York I had taken statistics courses with John Firestone, who must have been the instructor for a very substantial percentage of the Washington area statisticians in the early 1940s, and some with Herbert Arkin. Another teacher I had at Agriculture was Clem Winston. Military service interrupted my part-time graduate studies, and I did not resume them until the 1950s at American University in Washington, D.C., taking my coursework frequently together with Marvin Schneiderman. My instructors there included

John Smith, Bill Connor, Eugene Lukacs, Milton Abramowitz, and Dave
Blackwell. The classroom demonstrations Blackwell put on were so
beautiful that a few of them could carry one through a lifetime. It
was Blackwell who prepared my game theory comprehensive, including as
a question that I solve, in outline, the game of chess—my solution,
I suspect, is not the one he was seeking. My friendly Master's Thesis
Committee at American consisted of John Smith, Jerry Cornfield, and
Max Halperin.)

Statistical activity at NIH has grown enormously since its
earliest beginning which Dorn initiated. A full listing of statisticians whose careers were intimately related to the NIH program, such
as Ed Gehan, Bob Stearman, Don Morrison, Don Richter, Mary Fox, Lila
Randolph, Bob Taylor, and Mort Raff, or who had beginnings at NIH
through some short-time activity, such as Alan Gross, Ted Colton,
Paul Kanofsky, Miles Davis, would be surprisingly long. We have had
Peter Armitage come by for a visit of extended duration, Jerzy Neyman,
George Barnard, H. O. Lancaster, Clive Spicer, and Allan Birnbaum for
visits of only a few weeks or a few months. The careers of our initial statistical group have grown with NIH. Haenszel remains in an
expanded version of the position at the National Cancer Institute to
which he succeeded Dorn, while Schneiderman is technically Haenszel's
superior. Halperin has replaced Cornfield as successor to Dorn in
the top statistical position in the National Heart Institute, Cornfield
now in retirement from NIH, working as Statistics Department Chairman,
George Washington University, Washington, D.C. Lieberman, also in
retirement, remains active as a statistical consultant. The highest
administrative position of any statistician at NIH is that held by
Greenhouse, who is Associate Director, National Institute of Child
Health and Human Development. In my position as replacement for
Cornfield at the National Cancer Institute, I have, in turn, been
replaced successively by Marvin Zelen, George Weiss, and now John
Gart. Somehow I managed to remain in an enviable but untitled position which gave me the greatest freedom to set my own directions.

For the record, and so Cornfield will stop making false attribu-
tions to me, let me state that it was Hy Goldstein, that perennial

winner of slogan contests, who put it so aptly--there is nothing
wrong with NIH that less money could not cure. In those idyllic days
of less money, Dorn would set the pace for us as we pitched horseshoes
during the lunch period. Or we could lie under the apple trees watching the softball game umpired by Bozicevich from Tropical Diseases.
Now I have escaped, by retirement, my awesome burden of irresponsibil-
ity and lack of administrative duties and chores to join Cornfield at
George Washington University. Greenhouse will be along in a few months.
We remain loyal to NIH--that is where our research money comes from.

(I give a final parenthetic addition on how the statistician can
work with the investigator seeking guidance in his work. The trick is
to determine what it is that the investigator really wants to do, and
then, within limits, telling him to do it. Sometimes, however, because of his range of experience a statistician can give helpful
advice on the direction in which a study should continue. It is
always unique whenever I am consulted by a new client, and often these
consultations begin for me with a feeling of dismay. I just don't
know what the client is talking about; the client thinks I know all
about his subject and about what he is doing, and his attempts to fill
me in only seem to make things more confused; what can I possibly do
or say that can make any sense? Somehow or other things come out
right. I pick up a thread, manage to ask a few questions, the answers
to which help things fall into place. I end up with something I can
do for or say to the client; often it is something that is as great a
surprise to me as it is to him.

For greater completeness, let me say how the statisticians at
NIH are indebted to the laboratory investigators there. Before we
came on board, these laboratory men, like Harry Eagle or Ray Bryan,
were already using statistical procedures of which we were in ignorance.
We had to get our learning from them initially, even if we subsequently
refined it. The newer breed of investigators with M.D. degrees that
we have been getting at NIH are a fantastic lot--maybe it's because
they find things easy after the ordeal of their medical training.
Those, for example, who come to the Biometry Branch of the National
Cancer Institute, suddenly become mathematical statisticians, proving

theorems and preparing papers for submission to *Biometrics* or
Biometrika or even the *Annals of Mathematical Statistics*--our early
robin of this M.D.-mathematical statistician species was John Bailar,
who is now M.D.-Ph.D. And they get things done--while you've stepped
out of the room, they've got the problem solved, the solution programmed, and the data analyzed. Their weakness seems to be that they
take the statistical methods seriously and may be unaware of their
pitfalls. In the face of their technical know-how, I assume a visage
of wisdom to conceal my ignorance, bring in a smattering of knowledge
of many little details, and, if needed, effect some corrective changes
in what they are doing.)

ACKNOWLEDGMENT

This work was supported by Public Health Service Research Grant No.
CA-15686 from the National Cancer Institute.

REFERENCES

Box, G. E. P. (1950). Problems in the analysis of growth and wear
    curves. *Biometrics 6,* 362-389.

Breslow, N. (1974). Covariance analysis of censored survival data.
    Biometrics 30, 89-99.

Cornfield, J., Gordon, T., and Smith, W. W. (1961). Quantal response
    curves for experimentally uncontrolled variables. *Bull. Int.
    Statist. Inst. 38,* Part 3, 97-115.

Cornfield, J., and Tukey, J. W. (1956). Average values of mean
    sources in factorials. *Ann. Math. Stat. 27,* 907-949.

Cox, D. R. (1972). Regression models and life tables (with discus-
    sion). *J. Roy. Statist. Soc. Ser. B 34,* 187-220.

Kendall, M. G. (1948). *The Advanced Theory of Statistics* (in 2 vols.),
    Charles Griffin and Company, London.

Mantel, N. (1963). Chi-square tests with one degree of freedom.
    Extensions of the Mantel-Haenszel procedure. *J. Am. Statist.
    Ass. 58,* 690-700.

Mantel, N. (1966). Evaluation of survival data and two new rank order
    statistics arising in its consideration. *Cancer Chemotherapy
    Reports 50* (3), 163-170.

Mantel, N. (1970a). Incomplete contingency tables. *Biometrics 26,*
    291-304.

Mantel, N. (1970b). Why stepdown procedures in variable selection. *Technometrics 12*, 621-625.

Mantel, N. (1971). Letter: More on variable selection and an alternative approach. *Technometrics 13*, 455-457.

Mantel, N. (1973). Synthetic retrospective studies and related topics. *Biometrics 29*, 479-486.

Mantel, N., and Brown, C. (1973). A logistic reanalysis of Ashford and Sowden's data on respiratory symptoms in British coal miners. *Biometrics 29*, 649-665.

Mantel, N., and Brown, C. (1974). Alternative tests for comparing normal distribution parameters based on logistic regression. *Biometrics 30*, 485-497.

Mantel, N., and Haenszel, W. (1959). Statistical aspects of the analysis of data from retrospective studies of disease. *J. Natl. Cancer Inst. 22*, 719-748.

Peto, R., and Peto, J. (1972). Asymptotically efficient rank invariant test procedures (with discussion). *J. Roy. Statist. Soc. Ser. A 135*, 185-206.

Sterling, T. D. (1972). The effect of self-selection factors in the study of smoking and lung cancer. Meeting of the American Statistical Association, Montreal, 1972. (Abstract 2069, *Biometrics 28*, 1183, 1972.)

6.

HISTORY OF THE EARLY
DEVELOPMENTS OF MODERN STATISTICS
IN AMERICA (1920-1944)

BOYD HARSHBARGER was born February 15, 1906, in Weyers Cave,
Virginia. He received an A.B. degree from Bridgewater College
in 1928 and an honorary D.Sc. degree in 1950, also from Bridgewater College. In 1931 he received the M.S. degree from Virginia Polytechnic Institute and in 1935 the M.A. degree from
the University of Illinois. He received the Ph.D. degree from
George Washington University in 1942. He joined the Virginia
Polytechnic Institute faculty in 1931 and has been Professor
of Statistics there since 1941. He organized one of the first
Departments of Statistics in the United States at Virginia
Polytechnic Institute and State University, and was its head
from 1948 through 1972. He was a consultant to the U.S. Army
for 20 years and to various other agencies. He was the winner
of the J. Shelton Horsely Research Award in 1944 for Meritor-
ious Research. He was President of the Biometric Society, ENAR,
in 1957 and of the Virginia Academy of Science in 1949, and
first Chairman of the Committee of Statistics, SREB, in 1956.
He is a Fellow of the American Statistical Association, The
Institute of Mathematical Statistics, and the American Association for the Advancement of Science. He is the founding Editor
of the *Virginia Journal of Science* and has numerous publications
in various journals.

HISTORY OF THE EARLY
DEVELOPMENTS OF MODERN STATISTICS
IN AMERICA (1920-1944)

Boyd Harshbarger

Department of Statistics
Virginia Polytechnic Institute
and State University
Blacksburg, Virginia

The subject of Statistics* as we now think of it is of recent origin in the United States. In his paper, "Our Silver Anniversary," which appeared in *Annals of Mathematical Statistics* (1960), Professor Allen T. Craig stated that "prior to 1920 a scant half-dozen American colleges and universities had, as member of the department, any one who was seriously interested in the newly developing methods of scientific inference called mathematical statistics. ... The august American Mathematical Society took a very dim view of the whole business and looked upon these mavericks (statisticians) with a suspicion of quackery."

Professor H. L. Rietz, at the University of Iowa (1918), instituted courses in Statistics with an option in Mathematical Statistics. In addition, he gave the country a monograph *Mathematical Statistics* (1927) that was the forerunner of modern Statistics. How successful

---

*The word "Statistics" is capitalized throughout to distinguish the broad subject matter area from the technical use of the word in statistical inference.

he was in this area can be appreciated by observing the students who studied under his direction and those who completed their degrees in this field at the University of Iowa. Among some of these protégés were Professor Frank Weida (Ph.D., 1923) of George Washington University, Professor Samuel Wilks (Ph.D., 1931) of Princeton University, Professor Allen Craig (Ph.D., 1931) of the University of Iowa, Professor John H. Curtiss (1930) of Cornell University, and Professor Herbert Meyers of the University of Florida.

Closely following the development of Statistics at the University of Iowa was the program in the Mathematics Department at the University of Michigan. Professor H. C. Carver was given the title of Assistant Professor of Mathematics and Statistics in 1918 after joining the faculty as an instructor in 1915. In 1930 he founded and then published *Annals of Mathematical Statistics* until it was taken over by the Institute of Mathematical Statistics (1938). It was at the University of Michigan in September 1935, that a scholarly group of statisticians met and organized the Institute of Mathematical Statistics with Professor H. L. Rietz of the University of Iowa as its first president. Other faculty in the Department of Statistics at the University of Michigan during this early period were Professor C. C. Craig (Ph.D., 1931) and Professor Paul Dwyer (Ph.D., 1936).

Following this early development of Statistics at the University of Iowa and the University of Michigan, other colleges and universities introduced courses in Statistics into their curricula.

Courses in Statistics or Statistical Methods were offered in the Mathematics Department at Iowa State College as early as 1915. Dr. E. A. Smith, the head of the Mathematics Department, had obtained his doctoral degree in Germany and had some knowledge of probability and mathematical statistics. In 1924, Mr. Henry A. Wallace, assisted by Professor George W. Snedecor and Dr. C. F. Sarle, delivered a series of lectures on multiple regression. The booklet, *Correlation and Machine Calculations*, by H. A. Wallace and G. W. Snedecor, was published in 1925. Snedecor also published the famous text *Statistical Methods Applied to Experiments in Agriculture and Biology* (1937).

Professor Snedecor and Dr. A. E. Brandt in 1927 established the
Statistical Laboratory which, with funding from the U.S. government
continued to grow both in staff size and in the scope of its activities to include research. With appointment in 1939 of Professor
W. G. Cochran and Professor C. P. Winsor, the Statistical Laboratory
acquired both strength and prestige in the theoretical aspects of
Statistics. Courses in Statistics were still given in the Mathematics
Department. The Master Sample Project in the early 1940s, sponsored
by the Bureau of the Census, made important contributions to statistical sampling procedures. The Department of Statistics was organized
separately in 1947.

In the early 1930s William F. Callander, Head of the Crop Reporting Service, U.S. Department of Agriculture (USDA), proposed that
scientific sampling procedures be used for forecasting the production
of various crops. He arranged for the personnel of the Crop Reporting
Service to receive statistical training at Iowa State. Some of the
earlier staff at Iowa State University who took part in developing
the theory and application of sampling techniques were Professor
W. G. Cochran, Mr. Arnold King, and Dr. Ray Jessen.

Early developments in medical Statistics and related fields were
made by Professor Raymond Pearl and Professor Lowell T. Reed at Johns
Hopkins University. The Department of Biostatistics was established
in 1918 and was the first department in the United States with the
word "Statistics" in its title. The influence of these two men was
so strong that Johns Hopkins University is still a leader in medical
statistics.

Professor Frank Weida of George Washington University organized
a Department of Statistics in 1936. He was later joined by Professor
Solomon Kullback (1939). That institution has exerted a continuing
influence for the use of quality statistics in U.S. government agencies.

An important development in the field of Statistics took place
in 1940 at North Carolina State College. Mr. Callander approached
Dr. L. D. Bavor, Director of the Agricultural Experiment Station,
and arranged for North Carolina State to provide training in Statistics

for the staff of the Crop Reporting Service of the USDA. This program
was similar to one then being conducted at Iowa State University.

Dr. Bavor, in looking for a person to initiate his program and
to direct the work at Raleigh, turned to Dr. A. E. Brandt, then at
USDA in Washington. Dr. Brandt suggested Miss Gertrude Cox of the
Statistical Laboratory at Iowa State. Miss Cox's training in psychology, together with her work under Professor Snedecor for the
M.S. degree and her experience in the Statistical Laboratory, made
her well qualified for the program that Dr. Bavor envisaged for North
Carolina State. Things began to happen as soon as Miss Cox arrived
at North Carolina State. She was able to get sizable grants from the
General Education Board for a program in Statistics and immediately
set about to organize the Department of Experimental Statistics with
programs in training as well as consulting. Professor Jack Rigney,
Dr. R. L. Anderson, and Mr. A. L. Finkner were the first faculty to
join her staff.

One of Miss Cox's earliest undertakings was a summer session held
in Raleigh, North Carolina in 1941. The attendance at this session
provided a roster of forthcoming leaders in Statistics who were later
to develop many of the programs throughout the South. Among the
visiting professors for this session were Dr. Harold Hotelling of
Columbia University, Professor C. I. Bliss of Yale University, and
Professor Snedecor of Iowa State University. Some of the students,
including members of other faculties and government, were Walter A.
Hendricks, D. R. Embody, R. J. Monroe, T. A. Bancroft, A. J. King,
Boyd Harshbarger, Margaret Hagood, C. C. Li, J. H. Torrie, C. H.
Hamilton, Paul G. Homeyer, J. G. Osborne, J. F. Kenney, W. J. Youden,
D. B. Delury, C. F. Kossack, F. M. Wadley, and George Tyler. Later
Miss Cox brought Dr. Hotelling to the University of North Carolina
(1946).

In 1945 Miss Cox organized the Institute of Statistics which
combined training in Statistics at the University of North Carolina
and North Carolina State College. Courses in statistical theory were
to be given at the University and the methodology courses at State
College. At the University, the department was called Mathematical

Statistics and at State College it was called Experimental Statistics, but both were part of the Institute of Statistics of which Miss Cox was the director.

Princeton University was a nucleus in Statistics under Professor Samuel Wilks (1936). He was an early president (1940) of the Institute of Mathematical Statistics and also editor of the *Annals of Mathematical Statistics*. One of his early colleagues was Dr. John Tukey.

Columbia University made progress in the area of Statistics in the Department of Economics under the leadership of Professor Harold Hotelling (1931). He was later assisted by Professor Abraham Wald. Professor Helen Walker was active in Statistics (1929) in the Department of Education. Professor Frederick C. Mills of the Business Department and author of a widely used text, *Statistical Methods* (1923), was a leader in economic statistics.

Virginia Polytechnic Institute offered courses in Statistics as early as 1927. In 1934, a course was given in Mathematical Statistics, but it was not until 1942 that a degree option in Statistics and a Statistical Laboratory were initiated, under the direction of Professor Boyd Harshbarger, by joint arrangement with the Mathematics Department and the Agricultural Economics Department. Statistics and the Statistical Laboratory were made a separate department in 1949, with instruction offered in the College of Arts and Sciences and with the Laboratory a part of the Agricultural Experiment Station.

There was activity in Statistics at Stanford University under the direction of Professor Holbrook Working (1925) and Professor Eugene Grant (1938). Both were later associated with the War Production Board. Professor Jerzy Neyman was placed in charge of the Statistical Laboratory, University of California in 1938 and he was to influence the programs of Statistics throughout America. Prior to 1946, he directed two doctoral dissertations with emphasis in Probability and Statistics.

Professor Arthur R. Crathorne and later Professor J. L. Doob were able to continue the development of Statistics at the University of Illinois after Professor Rietz moved to the University of Iowa.

Washington University at St. Louis had a program in Statistics
under the leadership of Professor Paul Rider. It was he who produced
one of the earlier books, *Modern Statistical Methods* (1939), in mathematical Statistics. In addition to his profesorship at Washington
University, Professor Rider made outstanding contributions to
Statistics through his work with the U.S. Air Force, the U.S. Army,
and in private industry. He was one of the earlier presidents of
the Institute of Mathematical Statistics.

Cornell University, besides courses in Statistics in the Math-
ematics Department, had an early development of Statistics in the
Economics Department under Walter Francis Willcox and under H. H.
Love of the Genetics Department.

Harvard University had little activity in Statistics except for
a course in Economics Statistics in the Business School and in the
School of Education where Professor Phillip Rulon and Professor
Truman Kelly made contributions.

Dr. Chester I. Bliss was active in Statistics at the Connecticut
Agricultural Experiment Station and at Yale University.

Carnegie Institute of Technology, under the direction of Professor
Edwin Olds, also deserves recognition for its early program in Statistics. Among some of his early graduates were Professor Frederick
Mosteller, David L. Wallace, and several statisticians in the applied
field. Professor Olds was also secretary of the Institute of Mathematical Statistics (1941-1943), later to become president.

In New England, Statistics made progress at Wesleyan University
in the early 1930s. Professor Burton H. Camp of the Department
of Mathematics published the text, *Mathematical Part of Elementary
Statistics* (1931). He was also the third president of the Institute
of Mathematical Statistics (1938). At nearby Dartmouth College, Dr.
C. H. Forsyth gave a course in mathematical Statistics and in 1924
published a book, *An Introduction to the Mathematical Analysis of
Statistics*.

Other early texts in Statistics were *Mathematics of Statistics*
(1939) by John F. Kenney of Northwestern University, *Elementary*

Mathematical Statistics (1938) by W. D. Baten, Michigan State College, and *Industrial Statistics* (1942) by H. A. Freeman, Massachusetts Institute of Technology.

In Canada during the 1930s there were developments in Statistics by D. B. Delury of the University of Toronto and Dr. E. S. Keeping of the University of Alberta. Dr. C. H. Goulden was an applied statistician at the Rust Research Laboratory of the Dominion Experimental Farm in Manitoba. Dr. George L. Edgett of Queens University influenced a number of students, by his teaching, to choose Statistics as a career. Dr. John W. Hopkins was the principal research officer of the Research Council in Ottawa, Canada and contributed to Statistics.

An impact on Statistics and statistical research was made by several groups at Princeton and Columbia during World War II. The following three groups were described by Professor Fred Mosteller in correspondence with the author.

> In Princeton during the War, in addition to the Statistical Research Group of Princeton University, there was also a branch of research under the direction of Merrill Flood called Fire Control Research. This branch had working for it John W. Tukey, P. S. Dwyer, George W. Brown, Albert Tucker, I. E. Siegal, and sometimes W. J. Dixon, and also Charles Winsor. It was here that Tukey was weaned away from topology into statistics.

> The group at Columbia University was ultimately under the direction of W. Allen Wallis, who had as his executive director the now famous economist, Milton Friedman. Among the stars there were Churchill Eisenhart, M. Girschick, L. J. Savage, A. Bowker, Edward Paulson, J. Wolfowitz, Abraham Wald, Kenneth J. Arnold, Herbert Solomon, Myra Levine and Harriet Levine.

> Princeton had a branch of its Statistical Research Group in New York City under the direction of John D. Williams in the same building with Wallis's group. Olaf Helmer, Cecil Hastings, and I worked with John, and L. J. Savage spent a great part of his time with us. I was exchanged with SRGP Columbia in part payment to work on their *Sampling Inspection* plans for the Navy, which created the book *Sampling Inspection* by Harold Freeman, Milton Friedman, Frederick Mosteller, and W. Allen Wallis. (On the Navy's side, Commander John H. Curtiss and Lieutenant Commander Joseph F. Daley made technical contributions.) Another nice book coming out of the Columbia Statistical Research Group was *Selected Techniques of Statistical Analysis* by Eisenhart,

Hastay, and Wallis, both published by McGraw-Hill. These, of course, are in addition to the *Sequential Analysis* work by A. Wald.

The main group at Princeton, according to Professor Cochran, was under the direction of Professor S. S. Wilks. The research staff included, among others, Professor W. G. Cochran, Dr. R. L. Anderson, Dr. Phil McCarthy, and Professor Alex Mood.

A great impetus during World War II was given to Statistics and quality control by the program initiated by Dr. W. Edwards Deming, Professor S. S. Wilks, Dr. Walter Shewhart, and Professor Eugene Grant. They presented plans for a series of 10-day courses to be given throughout the United States so that leaders in industry could learn of Statistics and quality control and how it could be used by them for increasing production and quality in manufacturing. Dr. Deming took part in 23 such courses. Among the additional faculty for such courses were Dr. Holbrook Working, Professor Edwin Olds, Col. Leslie Simon, Professor Martin Brumbaugh, and the author. In addition to this program, the War Department sponsored courses in acceptance sampling under the direction of Mr. George Edwards and Mr. Harold Dodge. These courses marked a serious beginning of quality control charts, acceptance sampling, and the applications of Statistics to engineering problems.

One of the important developments in early Statistics was not in the colleges and universities but in the Bureau of the Census of the United States. Sampling and sampling theory commenced with the big census of unemployment in 1936. Dr. Calvert L. Dedrick and Mr. Morris Hansen were able to persuade the authorities to take a sample consisting of one postal route in 50 and to make it complete with inspections. As it turned out, the only part of the census of employment that was worth tabulating was the 2% sample. This great success certainly helped to put sampling on the map. The sampling activity of the census was further accelerated by the addition of Dr. Phil Hauser in the program. Dr. W. Edwards Deming was another one of the earlier contributors to Statistics in the Bureau of the Census.

The National Bureau of Standards was among the early government
leaders in Statistics. Major contributors here were Drs. W. J. Youden
and Churchill Eisenhart. From that early date to the present the
Bureau of Standards has continued to have a prominent statistical staff.

The National Institutes of Health at an early date sponsored
statistical research contracts and grants, and, at a later date,
fellowships for statistical training. Some of the early statisticians
with the Institutes were Dr. Harold F. Dorn, Professor Jerome Cornfield, and Dr. E. Cuyler Hammond.

In a paper entitled "Statistics in Army Research Development
and Testing," Dr. Clifford Maloney of the Division of Biological
Standards names several important contributions to Statistics during
the period 1920-1944. He notes the contributions of Dr. L. S.
Dederick, the author of "Relation Between the Probable Error of a
Single Shot and the Bracket of a Salvo or Series," and also contributions of Messrs. R. H. Kent and Joseph R. Lane.

During World War II, General (then Lieutenant) Leslie Simons,
along with Mr. Harold Dodge, introduced modern sampling plans in U.S.
Army testing. At this same time Dr. (then Captain) Frank Grubbs
began his outstanding contributions to Statistics and its applications
to Army research and testing. Also during this period punch cards
and electronic computers came into use in Army research.

Mrs. Charles E. Mauss (then Bessie Day) of the U.S. Forest
Service, and later with the U.S. Navy, introduced Statistics (1943)
into the Engineering Experiment Station of the U.S. Navy, Annapolis.
The entire Navy was influenced by her contributions to their programs
whenever Statistical methods were involved.

The organized use of experimental statistics in the USDA during
the period 1920-1945 was limited to the work of Dr. A. E. Brandt.
The research organization of the Forest Service also established a
small statistical consulting staff during this period at each of its
Regional Research Experiment Stations.

Prior to World War II and during that period, the U.S. Weather
Bureau had on its research staff a number of prominent statisticians.

The Bureau fostered Statistics both by research and by support in the training of statisticians.

There were other groups of statisticians located in industry and government. Dr. T. C. Fry, a member of the technical staff of Bell Laboratories, published a book, *Probability and Its Engineering Uses* (1928). Dr. Walter Shewhart of the Western Electric Company and Bell Laboratories was a leader in developing Statistics for industry and in promoting its applications in the engineering field. He was the author of a statistical text, *Economic Control of Manufactured Products* (1931), that was to influence the engineering curriculum throughout America. Dr. Robert W. Burgess of Western Electric Company was a leader in Statistics among industrialists as well as the author of a widely used Statistical text, *Introduction to The Mathematics of Statistics* (1927). Dr. W. E. Deming developed a program in Statistics for the evening college that was located in the South Building of Agriculture in Washington, D. C., and was also the author of a text, *Statistical Adjustment of Data* (1938). Dr. Bradford Smith and Dr. Mordecai Ezekiel made important contributions to the use of punch card equipment in regression analysis. Dr. Frank Wilcoxon developed methods for testing fungicides and insecticides with new statistical procedures. He initially worked with the Boyce Thompson Institute and later with American Cyanamid Company.

Dr. F. R. Immer, the Associate Director of the Agricultural Experiment Station, University of Minnesota, and Professor Sewell Wright of the University of Chicago were contributors to Statistics and genetics.

A major and successful unifying force for Statistics, and an organization that changed with the times, was the American Statistical Association. In the 1920s and 1930s its objectives were more the collecting and combination of data, with little or no probability theory or statistical inference. With the passing of time and an awareness by its membership of statistical inference, the society modified its objectives. Instead of carrying only descriptive articles, the *Journal* began accepting more analytical papers.

Another important contribution to Statistics by the American Statistical Association was the encouragement of two new journals, *Biometrics* by the Biometric Society and *Technometrics* by the American Statistical Association and the American Society for Quality Control. The working arrangements with the Institute of Mathematical Statistics in meetings and in publications were commendable. More recently, the American Statistical Association in its publication of the *American Statistician* has been a major news medium for all statistical members. There is a large overlapping of membership of all other statistical societies with the American Statistical Association. Many of the presidents, editors, and executive officers of the Association have been members of the Institute of Mathematical Statistics, the Biometric Society, and the American Society for Quality Control.

There was a great expansion in the use of Statistics by the government and industry following World War II. A number of new Departments of Statistics were soon organized, and increases occurred in the number of courses in Statistical Methods offered by other departments. A few Mathematics Departments in universities and colleges not having separate Departments of Statistics continued to be the major departments offering courses and awarding degrees in Statistics.

ACKNOWLEDGMENTS

I am indebted to the following individuals for making suggestions during the preparation of this paper. I assume responsibility for omissions and other shortcomings of the manuscript. T. A. Bancroft, Iowa State University; A. E. Brandt, University of Florida; W. G. Cochran, Harvard University; Gertrude Cox, Institute of Statistics, North Carolina State College (retired); Allyn Kimball, Johns Hopkins University; Solomon Kullback, George Washington University; Clifford Maloney, National Institutes of Health, Division of Biological Studies; W. Edwards Deming, Washington, D.C. (formerly with the Bureau of Census, Budget Bureau); Frank Grubbs, Dept. of the Army, USA Ballistic Research Lab., Aberdeen Proving Grounds; Robert V. Hogg, Iowa

University; Frederick Mosteller, Harvard University; Paul R. Rider,
Laguna Hills, Calif. (formerly Washington University); J. A. Rigney,
North Carolina State University.

REFERENCES

Baten, W. D. (1938). *Elementary Mathematical Statistics*. Wiley,
    New York.

Burgess, R. W. (1927). *Introduction to the Mathematics of Statistics*.
    Houghton Mifflin, Boston.

Camp, B. H. (1931). *The Mathematical Part of Elementary Statistics*.
    D. C. Heath, New York.

Craig, A. T. (1960). Our silver anniversary. *Ann. Math. Statist.*
    31, 835-837.

Dedrick, L. S. (1926). Relations Between the Probable Error of a
    Single Shot and Bracket of a Salvo or Series. Aberdeen Proving
    Ground Report.

Deming, W. E. (1938). *Statistical Adjustment of Data*. Wiley, New
    York.

Eisenhart, C., Hastay, M. W. and Wallis, W. A. (1947). *Selected
    Techniques of Statistical Analysis*. McGraw-Hill, New York.

Forsyth, C. H. (1924). *An Introduction to the Mathematical Analysis
    of Statistics*. Wiley, New York.

Freeman, H. A. (1942). *Industrial Statistics*. Wiley, New York.

Freeman, H. A., Friedman, M., Mosteller, F., and Wallis, W. A. (1948).
    Sampling Inspection. McGraw-Hill, New York.

Fry, T. C. (1928). *Probability and Its Engineering Uses*. Van Nostrand,
    New York.

Kenney, J. F. (1939). *Mathematics of Statistics*. Van Nostrand, New
    York.

Maloney, C. J. (1926). Statistics in Army Research Development and
    Testing. U.S. Army CMLC Biological Laboratories Report.

Mills, F. C. (1924). *Statistical Methods*. Holt, New York.

Rider, P. R. (1939). *Modern Statistical Methods*. Wiley, New York.

Rietz, H. L. (1927). *Mathematical Statistics*. Open House Publish-
    ing Co., La Salle.

Shewhart, W. (1939). *Statistical Method from the Viewpoint of Quality
    Control*. U.S. Department of Agriculture, Washington D.C.

Snedecor, G. W. (1937). *Statistical Methods Applied to Experiments
    in Agriculture and Biology*. Collegiate Press, Ames, Iowa.

Wald, A. (1947). *Sequential Analysis*. Wiley, New York.

Wallace, H. A., and Snedecor, G. W. (1925). *Correlation and Machine Calculation*. Iowa State College Press, Vol. 23(35), Ames, Iowa.

THE EMERGENCE OF
MATHEMATICAL STATISTICS

JERZY NEYMAN, of Polish ancestry, was born April 16, 1894, in Bendery, Russia. In 1916 he earned a degree of Candidate in Mathematics at the University of Kharkov and from 1917 to 1921 was Lecturer at the Institute of Technology, Kharkov, Russia. From 1921 to 1923 he was a statistician at the Agricultural Research Institute, Bydgoszcz, Poland. After receiving a Ph.D. degree in Mathematics from the University of Warsaw, he was Lecturer at that University from 1923 to 1934. Also he was head of the Biometric Laboratory of the Nencki Institute. In 1926-1927 he was a Rockefeller Fellow in London and Paris. From 1934 to 1938 he served at the University College, London, first as Lecturer and then as Reader in Statistics. From 1938 to present he is Professor and Director of the Statistical Laboratory, University of California, Berkeley. Neyman's scholarly publications are in the following domains: set theory, probability, statistics, astronomy, biology, weather modification, and philosophy of science. He has received five honorary doctorates, from the University of Chicago; University of California, Berkeley; University of Stockholm, Sweden; University of Warsaw, Poland; and the Indian Statistical Institute. He is Honorary Fellow (and also medalist) of the Royal Statistical Society (London) and of the London Mathematical Society. He is a member of the International Astronomical Union and Honorary President of the International Statistical Institute. He is a member of the U.S. National Academy of Sciences and Foreign Member of the Swedish and Polish Academies of Sciences. He has been Editor and Coeditor of the Proceedings of Berkeley Symposia on Mathematical Statistics and Probability held in 1945, 1946, 1950, 1955, 1960, 1965, and 1970-1971. He edited the Copernican Volume of the U.S. National Academy of Sciences, 1974. In 1968 he received the U.S. National Medal of Science.

7.

THE EMERGENCE OF
MATHEMATICAL STATISTICS

*A Historical Sketch
with Particular Reference
to the United States*

Jerzy Neyman

Statistical Laborabory
University of California
Berkeley, California

GENERAL BACKGROUND

Problems of Science as a Breeding Ground of Novel Mathematical Disciplines

As a general rule, novel mathematical disciplines have their roots in the problems of science, that is, in man's efforts to understand the mechanics of the universe. Arithmetic has its ultimate roots in the awareness of our distant ancestors of the distinction between "few" and "many." Geometry originated from notions of "near" and "far" and, at a more advanced stage, from the desire to measure areas of land. Differential calculus stems from the empirical notions of velocity and acceleration. These examples involve a direct connection between mathematical disciplines and empirical studies. The connection with science of some other mathematical disciplines, such as set theory, also exists but is indirect. Their origins are in the snags in logic discovered in the earlier disciplines, which

Jerzy Neyman

were more closely connected with the empirical world. After a start
from the treatment of a class of natural phenomena, a mathematical
discipline usually diverges from its "mother domain" of science
and begins its own life as an abstract theory, with only occasional
occurrences of "feedback." Therefore, modern geometry has little to
do with measuring the areas of fields, which is a separate domain of
surveying. The feedback of modern geometry to science can be exemplified by non-Euclidean geometries, first considered as fancy logical
structures, but now assumed by many astronomers to represent real
characterizations of the space we live in.

The origins of probability theory and of mathematical statistics
are also empirical. Now, both have reached the stage of maturity
and live their own lives. However, because of certain circumstances
recounted below, we witness a seemingly endless sequence of feedbacks
to empirical sciences and, simultaneously, a comparable sequence of
stimuli from empirical sciences to mathematical theories of both
probability and statistics. Also, other mathematical disciplines
supply probability and statistics with novel analytical tools.

As just described, the whole process would appear to be very
harmonious. However, the impression of harmony disappears just as
soon as instead of the various disciplines we turn our attention to
practitioners. The stream of novel problems, of novel ideas, and
novel techniques overwhelms the older people who would like to rest
on their laurels. But the energetic and mathematically better
equipped "youngsters" would not leave them in peace. Unavoidably,
the conflict of "fathers and sons" continues, frequently causing
some bitterness. This is illustrated by the following quotation
from William Feller (1), the recently deceased most brilliant scholar:

> It has been claimed that the modern theory of probability is too
> abstract and too general to be useful...the argument could be
> countered by pointing to the unexpected new applications opened
> by the abstract theory of stochastic processes, or to the new
> insights offered by the modern fluctuation theory which once
> more belies intuition.... However, the discussion is useless, it
> is too easy to condemn. Only yesterday the practical things of
> today were decried as impractical, and the theories which will
> be practical tomorrow will always be branded as valueless games
> by the practical men of today.

Empirical Backgrounds of Probability Theory

There are two classes of phenomena calling for the development of
mathematical theories that might be called theories of probability.
One class is the class of apparently stable relative frequencies
which generated the so-called "frequentist" theory. The other class
is the class of psychological phenomena connected with the feelings
of confidence and diffidence. The related theory of probability is
called "subjective." This paper is concerned only with the frequentist theory.

The basic notion of the frequentist theory of probability is
that of chance mechanism. We call a mechanism a chance mechanism
if (1) its operation can result in any one of several possible outcomes, say A, B, ..., C, and it appears impractical to predict which
will appear in a given instance, and (2) when the frequencies of
these outcomes in many repeated operations of the mechanism seem
predictable. Coin tossing, dice throwing, and extracting balls from

a bag are familiar examples of primitive chance mechanisms. The
origin of the notion of chance mechanisms is difficult to trace.
Quite possibly, the honor of authorship belongs to the first crook
who loaded his dice. This must have happened in distant antiquity.
I am told that in the burial chambers of certain Egyptian kings,
several sets of dice are customarily found, some untampered and
others obviously loaded. Prior to loading his dice, the crook must
have become aware of the important phenomenon that the long-run frequency of throwing an "ace" with one die D_1 is not necessarily equal
to that corresponding to another die D_2. Here, then, the frequencies
of a die falling in the six possible ways appeared as this die's
measurable properties, comparable to its volume and weight. Having
noticed this, the crook must have thought of the possibility of
manufacturing dice with more-or-less preassigned frequencies of falling this way or that way.

All of the above refers to the awareness of an empirical phenomenon of chance mechanisms. The origin of the frequentist theory of
probability goes back to the question of whether one can compute the
long-range frequency of some event E from known frequencies of some
related events A, B, ..., C. With an unavoidable degree of oversimplification, one might say that the theory of probability started
in 1713, with the publication of the book, *Ars Conjectandi*, by Jacob
Bernoulli (2).

With this empirical background, the frequentist theory of probability could be most appropriately called the calculus of frequencies,
or some such. The reference to "probability" is likely to have arisen
from the intuitive association between "frequent" and "probable."

There were many attempts made to formulate the foundations of an
abstract, purely mathematical theory of probability as a conceptual
model of the empirical realm of relative frequencies. The most
successful work in this direction is due to A. N. Kolmogorov (3).
The title of this work is *Grundbegriffe der Wahrscheinlichkeitsrechnung*
published in 1933 by Julius Springer, in Berlin. This work is purely
mathematical. However, the author points out that its relationship

with the empirical world is frequentist as contemporarily advocated
by Richard von Mises (4). The fundamental question is "how frequently?"

"Individualistic" and "Pluralistic" Subjects of Research in Science

When Newton, and later, Laplace worked on celestial mechanics, their
studies were "individualistic." The typical problem was to calculate
the orbit of a given planet, say, Mars, under the Newtonian attrac-
tion by the sun. In several studies of this kind, Laplace found the
orbital planes of all the planets studied to be rather close together.
The reader will notice that the last statement is qualitatively different from the various conclusions regarding any particular planet,
say, Mars. The phrase *orbital planes of all planets are close together* applies not to any individual planet, but to the whole category
of celestial bodies labeled "planets." Here, then, we encounter a
novel subject of study: a category of objects all satisfying a certain
definition but varying in their individual characteristics. Such a
subject of study is called "pluralistic."

When, at the turn of the eighteenth century, Laplace (5) asked
whether *comets,* like *planets,* are members of the solar system, this
question was also *pluralistic.* At that time, observations of several
comets were available sufficient for calculation of their orbits and
Laplace noticed that the planes of these orbits were far from being
as close together as those of the planets. It is this circumstance
that suggested to Laplace the further pluralistic question of whether
the *category* of objects called *comets* is different from another *category* called *planets.*

Subjects of study in modern scientific research are mainly
pluralistic. For example, in astronomy there is the question of
whether elliptical galaxies (a category of galaxies) are more massive
than the spirals (another category of galaxies). In medicine, there
are important questions about the internal disorders of patients exhibiting a specified set of observable symptoms (that is, of a
specified *category* of patients). In transportation engineering, one
of the important questions is whether the automobile traffic on a

two-lane highway (a *category* of traffic) involves more accidents
than that on a three-lane road, etc.

In modern terminology, a category of objects, the subject of a
pluralistic study, is called a "population." Hence, we customarily
speak of "populations" of people satisfying certain definitions, of
populations of elliptical or spiral galaxies, of populations of molecules of gases in a container, etc. Here, it is important to note
a connection with probability theory. This connection is through the
omnipresent question, "how frequently?" that occurs in all pluralistic
studies. Laplace's suspicion that comets are not regular members of
the solar system was based on the frequency distribution of the angles
of their orbital planes with the ecliptic. The question as to the
masses of elliptical and of spiral galaxies is really one of the frequency distribution of mass in the two populations. The medical question mentioned above is a question about the degree of association
between external symptoms and the physical disorders of patients; for
example, how frequently do the patients who have these particular
symptoms suffer from cancer?*

Mathematical Statistics as the Mathematical Discipline Developed for Pluralistic Studies of Nature

Whereas single pluralistic studies kept recurring from Laplacian
times and on, it took several decades for the research workers to
become aware that they represent a new category and that this new
category of problems of science splits itself into several subcate-
gories, each requiring a novel mathematical discipline. The totality
of mathematical disciplines developed to satisfy these needs is what
we call mathematical statistics (or occasionally "statistical
mathematics").

---

*The relevance of this question should not be confused with that
of the following: How frequently are the victims of cancer found to
have been smokers?

Descriptive Statistics

Logically, the first subdiscipline of mathematical statistics (but not historically the first) is what we now call "descriptive statistics." The original definitive term to describe it was coined in the nineteenth century by G. T. Fechner (6). The term is *Kollektivmasslehre*. It is just a little long, but very descriptive. If the subject of our study is a population of individuals each characterized by some trait X varying from one individual to another, then this study requires a mathematical discipline especially adjusted to describe what we now call the distribution of X in the population. As is well known, the realization of the need for means to describe empirical distributions generated several systems of theoretical frequency curves and frequency surfaces, those attributed to Bruns, to Gram, to Charlier, and to the most successful of all, Karl Pearson (see ref. 7). Subsequent generalizations are due to Romanovsky and many other authors.

Problem of "Goodness-of-Fit"

One of the subdisciplines of mathematical statistics that follows immediately is that of goodness-of-fit. It is connected with the fact that despite the principal interest of a study being centered on a population of some objects, say, elliptical galaxies, this population is not open to exhaustive study. Instead of the entire population of interest, we are forced to deal with a sample drawn from it. Given a function $f(x)$, supposed to represent the population distribution of a variable X, the question arises as to whether the deviations from f found in the sample could be ascribed to sampling variation. The first familiar method of treating this problem, that of the χ^2 test, is attributable to Karl Pearson. It was published in 1900 (8). A few decades later many alternative methods were proposed of which the best known are those of Harald Cramér in Sweden

(9), of Richard von Mises in Germany (10), and of A. N. Kolmogorov
(11) and N. V. Smirnov in Russia (12): quite an international effort!
The problem of goodness-of-fit ties in with further statistical
problems, those of estimation and of testing statistical hypotheses,
to be discussed presently (13).

Model Building: Problem of Chance Mechanism
Capable of Generating a Given Distribution

Depending on the predominant interest of the investigator, the
studies listed below belong to two distinct categories: (a) substantive and (b) theoretical.

1. The studies of the substantive category deal with the ques-
    tions of the following type: What is likely to be happening
    within this specified population (say, of galaxies, or of
    cells forming a human body, or of bus drivers in the streets
    of London) which could result in this particular distribution
    of the observable X?

    Not infrequently, this scientific question is combined with
    a utilitarian one: Can anything be done to modify the pro-
    cesses in the population so as to remove a specified unsatis-
    factory feature of the distribution of X?

    Studies of this substantive kind can be exemplified by the
    work of G. Udny Yule conducted in the 1920s jointly with M.
    Greenwood and Miss E. M. Newbold, concerned with accidents
    incurred by bus drivers in London (14). The basic concept
    of the model is that the number of accidents per unit time
    incurred by a particular driver is a Poisson variable with
    expectation λ, termed this driver's "accident proneness."
    The value of λ varies from one driver to the next following
    a gamma distribution. Thus, according to this model, the
    observed distribution of the number of accidents per driver
    per year is a gamma mixture of Poisson distributions which
    happens to be a negative binomial. The "utilitarian" question
    was whether the population of drivers could be modified (by
    appropriate selection) to diminish the number of accidents.
    The underlying scientific question was whether the hypothet-
    ical λ of each given driver does or does not change markedly
    in time.

    In more modern times, a tremendous number of similar model-
    building problems arose and have been treated. Here are a
    few examples, intentionally selected to be heterogeneous:

    a. Is the mechanism of cancer development a one-stage or
        a multistage mechanism (15)?

b. Does the mechanism behind the observed close associa-
    tions of galaxies consist in the gravitational "capture"
    of "vagabond" galaxies (Lemaître) or in cataclysmic
    explosions within nuclei of very big galaxies (e.g.,
    Ambartsumian) (16)?

c. Does the spectacular variation in the amount of rain-
    fall in a given locality (remember the biblical "seven
    fat years followed by seven lean years!") depend on the
    variability in the amount of ice nucleation particles
    in the clouds? [If so, could one diminish the severity
    of droughts by cloud seeding? (17).]

2. The alternative approach to model building is characteristic
of the scientists with predominantly mathematical interests
and abilities who, however, are capable of becoming interested
in the substantive questions of one kind or another. The line
of thinking is, roughly, as follows. Here is a phenomenon
that worries certain "substantive people," say, the phenomenon of contagion. Also, here is a familiar chance mechanism
M, which offers certain possiblities of modification. The
question is whether or not some modification of M can produce
a distribution of the observable X comparable to that charac-
terizing the real phenomenon, say of accidents among bus
drivers, or cases of influenza, etc.? More generally, can
the modifications of M produce distributions characterized
by all of the Pearson curves, which appear to fit an astonishing variety of empirical distributions? This was precisely
the subject of a brilliant study (1930), essentially probabilistic, of George Pólya (18), then in Zürich, now at
Stanford.

One of the interesting results of Pólya was that one of his
"contagion" models, presuming that accidents incurred in the past
increase probabilities of more accidents in the future, produced the
negative binomial distribution of the number of accidents per driver
per unit time, identical with that of the Yule-Greenwood-Newbold
model. As described above, this latter model may be termed a
"mixture-no-contagion mechanism." This particular finding by Pólya
demonstrated a phenomenon which was unanticipated--two radically
different stochastic mechanisms can produce identical distributions
of the same variable X! Thus, the study of this distribution cannot
answer the question which of the two mechanisms is actually operating.
Now this phenomenon is labeled phenomenon of nonidentifiability. In
a number of cases, notably in econometrics and in the psychological
and other factor analyses, the phenomenon of nonidentifiability proved

most embarrassing and stimulated many mathematical studies. Luckily,
it was found that a family of chance mechanisms that are nonidenti-
fiable by the distribution of one variable X can become identifiable
by the joint distribution of the same X in company with some other
variables Y, ..., Z. There are some very interesting results on
identifiability in econometrics, which are due to O. Reiersøl of Oslo
(19).

After World War II, the study of stochastic models of natural
phenomena advanced substantially both (1) from the substantive point
of view and (2) from the mathematical point of view. This is illustrated by terms current in mathematical literature which are obviously
borrowed either from biology or physics, such as "birth-and-death
processes," "branching processes," "renewal processes," etc. An
excellent source of information on such subjects is the book by
William Feller (1). Another excellent book, by Theodore E. Harris
(20), provides a wealth of information both on substantive sources
of a variety of mathematical concepts and on mathematical results
accumulated through the efforts of a very international galaxy of
outstanding authors.

Theories of Estimation and of Testing Statistical Hypotheses

When a stochastic model of a natural phenomenon is constructed,
there immediately arise two distinct problems both connected with
the question of whether the model is realistic or not.

One problem is how best to fit the model to the available observations. Ordinarily, a model yields only a family of distributions
that the observable X must follow, but not the values of the parameters
("nuisance parameters") that identify the members of this family. For
example, although both the Yule et al. and also the Pólya models of
accident proneness lead to the conclusion that the number X of accidents must follow a negative binomial distribution, the two parameters
of that distribution, say α and β, are not specified by either model.
Thus there is the problem of estimating these parameters from the
available data: given the observations X_1, X_2, ..., X_n, what functions

say $\hat{\alpha} = \hat{\alpha}(X_1, X_2, \ldots, X_n)$ and $\hat{\beta} = \hat{\beta}(X_1, X_2, \ldots, X_n)$ should one use
to obtain estimates of α and β that are in some sense "best"?

The other problem, that of testing a statistical hypothesis, follows immediately. Even if the hypothetical model is identical to the
one that operates behind the phenomenon being studied, and even if the
estimates of all the "nuisance parameters" are free of error (which
is too much to hope for), chance variation involved in collecting the
data will create discrepancies between the empirical and the theoretical distributions. In contrast, the construction of the model may
well be erroneous and it is essential to discover this circumstance.

Here, then, we have two problems of mathematical theory of
statistics: the problem of "point estimation" and the problem of
testing a "statistical" hypothesis (that is, of a hypothesis regarding the distribution of an observable X).

The first proper formulation of the problem of point estimation
is that of Laplace. Later, a somewhat more satisfactory formulation
was given by Gauss (21). Both cases involved the contemplation of a
fixed unknown value, say θ, of a parameter and a certain number n of
its measurements X_i, all subject to a random error. Also, both cases
involved the formulation of what we now call a "loss function" $L(\hat{\theta}, \theta)$,
which is supposed to represent the penalty to be paid by the statistician if he adopts $\hat{\theta}$ as the estimate of the unknown parameter when
its value is θ. Briefly and roughly, the problem was to determine
$\hat{\theta}$ as a function of the observable Xs, which minimizes the expectation
of $L(\hat{\theta}, \theta)$. Laplace used the absolute value of the difference $L_L(\hat{\theta}, \theta) = |\hat{\theta} - \theta|$, whereas Gauss preferred its square $L_G(\hat{\theta}, \theta) = (\hat{\theta} - \theta)^2$. There
resulted the familiar theory of least squares. Because of its simp-
licity and of intuitive appeal, the least-squares method was rapidly
adopted by the "consumers" of the theory, and continues to be widely
used. However, the theoretical background involving the loss function
L has been forgotten for about a century to be revived in the early
twentieth century by A. A. Markov (22) in Russia and, less distinctly,
by Edgeworth in England. Now, some textbooks on statistics include
a theorem, the Gauss-Markov theorem on least squares.

The development of the theory of testing statistical hypotheses was slower in coming. To my knowledge, the first attempt to test a statistical hypothesis is attributed to Laplace (5). As mentioned above, Laplace contemplated the possibility that comets are not regular members of the solar system, but "intruders" from outer space. He reasoned that, if this be the case, then the angles between the orbital planes of the comets and the ecliptic would be uniformly distributed between zero and $\pi/2$. This was the hypothesis that Laplace wanted to test. To do so, he *adopted* the arithmetic mean, say $\bar{\phi}$, of the observed angles as his criterion. (The word "adopted" is italicized in order to emphasize the fact that it was not *deduced* to possess some property of optimality as was the case with the problem of point estimation.) Next, Laplace *deduced* the distribution of the mean $\bar{\phi}$ as implied by the hypothesis tested, and referred to it the actual value computed for the several then investigated comets. As a result of this comparison, Laplace made up his mind to act on the assumption that comets are not regular members of the solar system like planets.

The reader will notice that in constructing a test of his hypothesis, Laplace had to solve a special problem, the distribution problem of his criterion $\bar{\phi}$. The distribution problems are invariably connected with both the problem of estimation and with the problem of testing hypotheses. The Pearson χ^2 criterion for goodness-of-fit also necessitated the solution of a distribution problem. This was solved by Karl Pearson for the case in which the distribution supposed to fit the data is completely specified, involving no adjustable parameters. Originally, Pearson himself and his many followers thought that the presence of adjustable parameters to be estimated did not affect the distribution of χ^2. Using simulation sampling, G. Udny Yule became convinced to the contrary (23). Subsequently, in a series of investigations, beginning with one by R. A. Fisher (24), it was established that the suspicions of Yule are justified--the distribution of χ^2 was found to depend on the so-called number of degrees of freedom and on the method used to estimate the adjustable parameters.

The method of estimation introduced by Karl Pearson, and systematically developed by him particularly with regard to his frequency curves, was the method of moments (7). An alternative method, the method of maximum likelihood (25), was advocated by R. A. Fisher over which there resulted a spectacular dispute. Both methods were advanced on intuitive grounds, somewhat dogmatically. Fisher's idea was based on what he called a new measure of "confidence or diffidence," namely, the likelihood function. However, the arguments themselves were not dogmatic, but were based on the precision of estimation.

The problem of estimation was studied extensively by Fisher who introduced a number of fruitful concepts such as "consistency" of an estimator and its "efficiency." Other important concepts introduced by Fisher are those of "sufficiency" and of the "amount of information." These concepts attracted the attention of many investigators. Early American contributors were Harold Hotelling (26) and J. L. Doob (27). These same concepts are still being investigated from many points of view in somewhat modified form.

The theory of testing statistical hypotheses was initiated in the early 1930s by Egon S. Pearson and the present writer (28). In modern terminology, the problem studied is a two-decision problem—given a hypothesis H concerned with the distribution of an observable X, and given an alternative hypothesis $\bar{H}$, we contemplated two possible *decisions* regarding H: to act on the assumption that it is false (and, therefore, that $\bar{H}$ is true), or to abstain from such action. Each of these decisions may well be false and the consequences of the two errors may be of different importance. The error that a practicing statistician would consider the more important to avoid (which is a subjective judgment) is called the error of the first kind. The first demand of the mathematical theory is to deduce such test criteria as would ensure that the probability of committing an error of the first kind would equal (or approximately equal, or not exceed) a preassigned number α, such as $\alpha = 0.05$ or 0.01, etc. This number is called the level of significance.

When a class, say, $K(\alpha)$, of test criteria is determined, all
ensuring the same low frequency α of errors of the first kind, it is
necessary to consider the possibility of committing an error of the
second kind. Briefly and roughly, the mathematical problem consists
of determining within the class $K(\alpha)$ the particular criterion that
minimizes the probability of the second kind error. The concepts
involved in the theory include "power function" of a test, "similar
regions," "most powerful tests," "uniformly most powerful tests," etc.

For a number of cases, now mostly considered to be "bookish,"
and not likely to be encountered in "live" statistical problems
encountered in modern studies in science, the early theory established
the desired uniformly most powerful tests. In other cases, such
tests were found not to exist, which opened the door to a variety of
"compromise" definitions of optimality, starting with "unbiased" tests.
The general problem is still very much "on the books."

A novel form of the problem of estimation was noticed by the
present author, that of estimation "by interval" or, more generally,
"by a set." Suppose that the distribution of an observable X (usually
a vector) depends on the value of a parameter θ which is unknown,
except that it must lie within a specified interval, perhaps from
zero to unity, etc. The problem is to associate with each possible
value x of X an interval $S(x|\gamma)$ of possible values of θ satisfying
the condition that the probability of $S(X|\gamma)$ covering the true value
of θ is "large," being equal (or approximately equal, or at least
equal) to a preassigned number γ, just as close to unity as desired,
say $\gamma = 0.95$ or $\gamma = 0.99$ or the like. The number γ so chosen is
called the confidence coefficient.

When a class of such intervals ("confidence intervals") is
determined, it is also necessary to determine the one that satisfies
some intelligible condition of optimality. Depending on the circumstances of the problem, the "optimal" confidence interval may be the
"shortest" (in a certain defined sense). However, this is not a
general rule. For example, in a number of problems of science and
technology it is important to be "confident" that the value of the

unknown θ does not exceed a number $\bar{\theta}(x)$ calculated from the observations on X. One particular case of this kind is the case in which θ designates the average number of noxious bacteria per unit volume of drinking water. Here, then, the optimal confidence interval for θ extends from zero to $\bar{\theta}(x|\gamma)$. The routine use of $\bar{\theta}(x|\gamma = 0.99)$ ensures that the assertion that the bacterial density does not exceed $\bar{\theta}(x|\gamma = 0.99)$ will be true in about 99% of the cases.

Whereas the first publication regarding the confidence intervals goes back to 1930 (29), the definitive theory appeared in two papers in 1937 and in 1938 (30,31). More recently, both the theory of testing hypotheses and of estimation by set became particular cases of the general theory of statistical decision functions built by Abraham Wald (32).

Problem of Experimental Design

Perhaps the greatest achievement of R. A. Fisher was his initiation and development of the theory of experimentation with variable material. Although a full day is devoted to this domain at this symposium, this achievement of Fisher is so important in constituting a section of mathematical statistics as to require at least a brief mention in the present paper.

As I see it, the most basic of Fisher's numerous ideas is that, to be reliable, an experiment with variable material must be "randomized." The meaning of this term is as follows. Suppose an experiment is designed to test the effectiveness of several treatments T_1, T_2, ..., T_s. The treatments are to be compared on some "experimental units." In an agricultural trial, these might be plots of an experimental field. In medicine, the experimental units might be similarly diagnosed patients. In weather modification, the experimental units might be days with some specified weather conditions, and so on.

Fisher's principle of randomization requires that the studied treatments be assigned to experimental units not by the choice of the experimenter, but through the use of a well-designed chance

R. A. Fisher as he was in the late 1920s.

mechanism. This is because, without randomization, the apparent
differences among the treatments exhibited by the experiment may, in
fact, be due not to intrinsic properties of the treatments but to
some extraneous causes. In retrospect, one of the frequent causes
of bias of this kind is that the experimenters have emotional attach-
ments to particular treatments studied and, perhaps subconsciously,
tend to assign the preferred treatments to individual experimental
units which appear, in some sense, to be "better." The frequent
results of such procedures is first a degree of self-deception and
later the deception of others.

The randomization of an experiment may be "unrestricted" or
subject to a variety of restrictions. Jointly with Frank Yates,
Fisher produced a number of tools to achieve an effective randomization. Here, three books deserve mention: *Statistical Methods for
Research Workers* (33) and *The Design of Experiments* (34) (many
editions) both by Fisher himself; the third book, by Fisher and Yates,
Statistical Tables for Biological, Agricultural and Medical Research
(35), was published by Hafner, again in many editions.

All three books exercised an immense influence on experimentation in all domains. However, most usually, their effect was not immediate. Quite frequently the initial reactions of the experimenters are characterized by pronouncements such as the following: "Oh, get that Fisher out of my hair--I know about experimentation and about my material, all that I need to know!" Particularly, such protests are voiced against randomization. In due course, attitudes change.

Proliferation of Pluralistic Studies in Science and Technology as the Source of Inspiration for the Development of Mathematical Theories of Probability and Statistics

At the outset of this paper it was mentioned that, even though by now the theories of probability and statistics reached the stage of maturity as mathematical disciplines and began to live "their own lives," they continue to grow and to diversify in many novel directions. The reason for this is the ever-increasing number of pluralistic problems within an astonishing variety of substantive domains which almost invariably involve novel mathematical problems. In the preceding pages, the following 10 examples were briefly mentioned:

Accident proneness of bus drivers

Astronomy: a variety of problems (36)

Attempts to stimulate rain

Bacterial contamination of drinking water

Econometrics

Experimentation with variable material in all domains

Factor analysis in psychology

Mechanism of carcinogenesis

Medical diagnosis

Traffic engineering

To this list it is appropriate to add another list of novel fields of theoretical and substantive activities. The following few must suffice as illustrations.

1. *Population genetics*, from G. H. Hardy (a "purist mathematician" from Cambridge who in 1908 published an eye-opening letter to *Science*), to R. A. Fisher, to Sewall Wright, to modern "neo-Darwinian" studies of evolution (37).

2. *Technology*, from Walter A. Shewhart and "quality control," to acceptance sampling (38), to operations research, to reliability theory.

3. *Enviromental pollution and public health studies* (39).

4. *Demography*, from the "substantive" work of Alfred Lotka in the United States (40) and the highly "mathematical" work of Vito Volterra in Italy (41), both in the 1920s and 1930s, to modern international concern with overpopulation.

These, then, are the reasons for the current widespread research and instruction efforts in the two interrelated domains of probability and statistics.

EARLY DAYS OF MATHEMATICAL STATISTICS IN THE UNITED STATES, ROUGHLY THROUGH WORLD WAR II

Before embarking on an account of the early history of mathematical statistics in the United States, I must stress that I am not a historian and that, therefore, my account must be fragmentary, based primarily on personal observations. Furthermore, I must confess that my information on the subject goes back to the late 1920s which, compared to the material in the first section of this paper is not very early. My excuse for writing with this severe limitation is that the information I do have indicates the absence in the United States of impressive developments before the 1920s. As far as I can gather, in the 1920s and 1930s there were in the United States three leaders endowed with broad vision and energy who generated, each in his own way, the splendid developments we now observe. The statisticians in question are Henry Lewis Rietz, Harry C. Carver, and Harold Hotelling. Although their efforts were not easy and the progress was slow at first, a phenomenon occurred, roughly after World War II, indicating that the period before the war was an early stage of a supercritical branching process--a slow start followed by an explosion. This applies to the number of individuals concerned with mathematical statistics, to the number of fruitful ideas born and developed, and to the number of periodicals, symposium proceedings and to the amount of book literature. Something similar happened in other countries, with a change in emphasis on the directions of study.

In the following, I shall be concerned primarily with Rietz,
Carver, and Hotelling, with only cursory glances at the contemporary
"statistical environment." Quite possibly, in identifying, more or
less, the early history of mathematical statistics in the United
States with the history of activities of the three individuals mentioned, I may be committing oversights and injustices. If so, I
hope to be corrected.

Henry Lewis Rietz

Rietz was a man of a generation preceding that of Carver and Hotelling.
He received his Ph.D. at Cornell in 1902, for a time was teaching in
the University of Illinois at Urbana, and then, in 1918, settled at
the University of Iowa, at Iowa City. In 1937, I visited him for a

H. L. Rietz, first president of the Institute
of Mathematical Statistics, 1935-1936.

few days and had some conversations. Rietz told me about his concern
with the status of mathematical statistics in the United States and,
particularly, with the present lack of opportunities for studies in
actuarial sciences. The young people interested in this field of
activity had to go abroad to study, at least to Canada! I am not
sure that the present situation is much different.

Subsequently, I had the opportunity to see some publications of
Henry Rietz. By and large, they were published in the *Annals of
Mathematics*, the first of which I saw mentioned having appeared in
1912. I am particularly fond of a little monograph *Mathematical
Statistics* (42) published by the Mathematical Association of America.
The first printing appeared in 1927, the fifth in 1947. I like this
monograph because it emphasizes the differences among several domains
of statistical activities: the manipulative work, frequently conducted
without clear understanding of what it is all about, the domain of
descriptive statistics, and both the substantive and the mathematical
domains of model building (see first section of the present paper).
Rietz tried to communicate the underlying ideas to his readers. The
following quotations are illustrative:

> ...a large number of textbooks on statistical methods...(have
> been published) during the past two or three years...although
> the recent books on statistical method will serve useful purposes
> in the teaching and standardization of statistical practice, they
> have not, in general, gone far toward explaining the nature of
> the underlying theory, and some of them may even give misleading
> impressions as to the place and importance of probability theory
> in statistical analysis....

> The exposition of mathematical statistics here given will be
> limited to certain methods and theories which, in their inception,
> center around the names of Bernoulli, De Moivre, Laplace, Lexis,
> Tchebycheff, Gram, Pearson, Edgeworth and Charlier, and which
> have been much developed by other contributors...Borel has ex-
> pressed this somewhat restricted point of view that the general
> problem of mathematical statistics is to determine a system of
> drawings carried out with urns of fixed composition, in such a
> way that the results of a series of drawings lead, with a very
> high degree of probability, to a table of values identical with
> the table of observed values.

In line with the above "somewhat restricted" view of Emile Borel, Rietz published a paper (43) under the title "Urn Schemata as a Basis for the Development of Correlation Theory."

Here, then, we have a reference to the mathematical aspect of model building. To the substantive aspect a special chapter of the monograph is given describing "The Lexis theory" of "normal," "supernormal," and "subnormal" variation. Other publications of Rietz are mostly concerned with descriptive statistics.

The general perspective that one gains from the perusal of all this literature is as follows: (1) Rietz appears as a scholar excellently informed of the early and the contemporary conceptual developments in mathematical statistics, and (2) that the novel developments occurred predominantly in Europe. Of the 68 references in Rietz's monograph, only 10 of which are references to American authors, the remaining 58 being European! The period during which Rietz was active appears to have been the period of transplantation into the United States of the ideas conceived in Europe.

I wish I could establish just how many intellectual descendants Rietz did produce. There must have been quite a few. When a monograph appears in five editions, this means that at least the first four must have been sold out. I visualize that many of those who bought the book, later came to Iowa City to study under Rietz. Of those who became scholars in their own right, I know three—Allen T. Craig (later Rietz's successor at the University of Iowa), Carl F. Fischer (subsequently Professor at Ann Arbor, Michigan) and Samuel S. Wilks (the initiator and developer of courses in mathematical statistics at Princeton). Each of these three men produced his own brood, including some of the most outstanding scholars of today.

Harry C. Carver

Now 83 and retired in Santa Barbara, California, Carver developed as a mathematical statistician possibly under some slight and indirect influence of Rietz. In addition to a number of research papers,

Harry C. Carver as he was in 1930
when he founded *The Annals of
Mathematical Statistics.*

Carver's outstanding contribution to the development of mathematical
statistics in the United States was the founding and development of
our special journal, *The Annals of Mathematical Statistics,* which
grew tremendously and is now replaced by two journals, *The Annals of
Statistics* and *The Annals of Probability.* Still "in addition," Carver'
activities in connection with the *Annals* appear to qualify him as the
father of our professional organization, The Institute of Mathematical
Statistics, which now has an impressively broad international membershi

In order to appreciate the achievements of Carver properly, it
is necessary to take at least a cursory glance at the situation of
mathematical statistics in the United States as it was in the late
1920s and early 1930s. At that time, "statistical life" in the
United States was concentrated in the American Statistical Association
(ASA), founded in 1839, which had published a journal, first called

Quarterly Publication of the ASA and later *Journal of the ASA,* with
its 25th volume appearing in 1930.

The rather crowded annual meetings of the ASA were held in
various cities, as is true today. There were additional regional
meetings organized by the seven chapters of the Association. The
matters discussed at these meetings varied tremendously but were
concerned, predominatly, with contemporary social and economic problems in the United States. The following titles of the six papers
published in a single September 1931 issue of *J. Amer. Statist.
Assoc.* (Vol. 2b) are illustrative (243-318):

1. *An attempt to measure public opinion about repealing the
   eighteenth amendment* (Prohibition), by Walter F. Willcox.

2. *Statistical method from an engineering viewpoint,* by
   W. A. Shewhart.

3. *Enumeration and sampling in the field of the census,* by
   R. H. Coats.

4. *Can we find out how the American income is spent?,* by
   Louis Bader.

5. *The business cycle and accidents to railroad employees in
   the United States,* by C. D. Campbell.

6. *Statistical measures of social aspects of unemployment,* by
   Meredith B. Givens.

Contributions to statistical theory? No. Source of interesting
theoretical problems? Occasionally, yes. This situation was a matter
of concern of Carver.

Carver's own contact with theoretical statistics goes back to
1916 when, as a young instructor in mathematics at the University of
Michigan, he was asked to teach a course billed *Mathematical Statistics.* Earlier, this course was based on the book of W. P. Elderton,
Frequency Curves and Correlation (44). According to a recent letter
of Carver, he adopted a different text, the book of G. Udny Yule, on
theory of statistics (45) (much more conceptual in character than was
the book of Elderton). The material in this book was supplemented
by that in the writings of Karl Pearson, "Student" (Gosset), R. A.
Fisher, and others. The following quotation from Carver's letter
describes some of the subsequent developments:

As interest in mathematical statistics grew in this country, it
became more and more difficult to find editors who would accept
contributions to mathematical statistics. So I brought this
matter up before the Directors of the American Statistical Association at their 1928 Annual Meeting. These gentlemen were very
understanding, but pointed out:

(a) Most of their membership were economists, bankers and census
people whose knowledge of mathematics was very limited, and

(b) Their Association's membership was beginning to fall off as
though a recession were approaching, and our formulae would
require expensive monotype composition, rather than the cheaper
linotype used then by the Association.

I agreed with the difficulties enumerated by the Directors; they
agreed that a mathematical statistical journal should be established when funds were available; I said I was going to continue
working on this project, and they wished me luck. I promised a
progress report for their 1929 Annual Meeting; so we closed our
first meeting on the friendliest of terms.

My first step was to notify others interested in mathematical
statistics that I expected to edit a new Journal called the
Annals of Mathematical Statistics to which I hoped they would
contribute from time to time. Within a few months, I had sufficient material for the first year's issues of four quarterly
numbers....When I showed the Directors (of the ASA) the off-set
version of Vol. I, No. 1, they were favorably impressed and
suggested that we work out a sponsorship that would benefit all
of us. We then worked out an agreement under which

(a) The editorship and financial responsibilities of the *Annals*
would rest completely on my shoulders.

(b) Every member of the Association had his choice of Journals.
The *Annals* would receive $4.00 of the $6.00 annual dues for
every choice of the *Annals*.

(c) Either the Association or the *Annals* could terminate this
sponsorship at any time in the future, with a reasonable notice
to the other party.

The first issue of *The Annals of Mathematical Statistics* appeared
in February 1930. The top of its cover carried an imprint "The
American Statistical Association." The Editorial Committee was described as consisting of H. C. Carver (Editor), B. L. Shook (Assistant

Editor), J. Shohat* (Foreign Editor) and J. W. Edwards (Business

Manager). The first item published in the first issue of the *Annals*

was a two-page statement of Professor Wilford I. King, then Secretary

of the ASA, describing the ASA sponsorship of the *Annals*. The fol-

lowing passages are illustrative:

> For some time past...it has been evident that the membership of
> our organization is tending to become divided into two groups--
> those familiar with advanced mathematics, and those who have not
> devoted themselves to this field. The mathematicians are, of
> course, interested in articles of a type which are not intelli-
> gible to the non-mathematical readers of our Journal. The Editor
> of our Journal has, then, found it a puzzling problem to satisfy
> both classes of readers....This Journal (*The Annals*) will deal
> not only with the mathematical technique of statistics, but also
> with the applications of such technique to the fields of astron-
> omy, physics, psychology, biology, medicine, education, business,
> and economics. At present, mathematical articles along these
> lines are scattered through a great variety of publications. It
> is hoped that in the future they will be gathered together in
> the *Annals*.

Further remarks of Professor King, an economist, indicate the

awareness of the need in the United States for expository literature

on the theory of statistics to stimulate fresh research. I am some-

what regretful that, as witnessed by the above quotation, the thinking

of Professor King involved the idea of theory of statistics as being

identified with "mathematical techniques."

Judging from the contents of the first several volumes of the

Annals, the editorial policy of Carver was two-pronged. One direction

of emphasis was to transplant from Europe to the United States the

conceptual (as contrasted with manipulative) developments in mathe-

matical statistics.

---

*Subsequently, Shohat, a noted mathematician, then living in
France, moved to the United States. I remember his being a coauthor
of the monograph on the problem of moments from which I learned much
(46).

The "transplantation policy" of Carver manifested itself in a
number of articles by European scholars published or republished in
the early issues of the *Annals*. This includes contributions of the
following authors:

S. D. Wicksell, an astronomer-statistician from Lund, Sweden,
who wrote on regression and correlation (47).

J. A. Shohat, a noted mathematician then in France, who wrote
on Stieltjes integrals (48).

L. v. Bortkiewicz of Berlin, Germany, on his work on Lexis ideas
regarding stability and heterogeneity (49). (See section
on H. L. Rietz.)

Charles Jordan of Budapest, Hungary, on approximations by orthogonal polynomials (50).

Some of these contributions were written in foreign languages
and were translated into English. In addition, some 150 pages of
Vol. 2 of the *Annals* (51) were given to the republication of the famous
monograph of T. N. Thiele, first published in 1889 in Danish and then
republished in English in 1903. Here Carver was motivated by
the importance of semi-invariants, introduced by Thiele. Although I
agree with Carver on this particular point, the reading of Thiele's
philosophical discussion of causality, etc., created in me a degree
of amusement ("...the law of causality cannot be proved, but must be
believed; in the same way as we now believe the fundamental assumptions of religion, with which it is clearly and intimately connected.")

Apparently, the University of Michigan went along with Carver's
ideas on transplantation. In particular, Wicksell served there as a
Visiting Professor and managed to generate one of his intellectual
descendents. The person in question is Cecil C. Craig, subsequently
Professor at Ann Arbor and a frequent contributor to the *Annals*. Paul
Dwyer, another professor at Ann Arbor, appears as a descendant of both
Carver and C. C. Craig.

As a mark of Carver's policy of encouraging home efforts at
conceptual statistical studies, the early volumes of the *Annals*
contain contributions of a great number of young people. The three
descendants of Rietz, namely Allen T. Craig, Carl F. Fischer, and

S. S. Wilks, are among them. Then there was a galaxy of Carver's
own students, some of whom served as his Assistant or Associate
Editors, B. L. Shook, R. S. Sekhon, and A. M. O'Toole, to name a few.

The above-mentioned young people were budding mathematical
statisticians. In addition, Carver took care of representatives of
substantive domains. Because of the mathematical background of
engineers, the situation with technology was relatively easy (Walter
Shewhart, Edward Molina). Also, some of the biologists studying
evolution, such as Sewall Wright, may have helped the *Annals* more
than the *Annals* helped them. Because of the notorious distaste for
mathematics among the M.D.s, this is not the case with medicine. At
this point, the paper by Joseph Berkson, an M.D., published in the
first issue of the *Annals*, must be mentioned (52). Its title, "Bayes
Theorem," is most unexpected from a representative of the medical
profession!

In subsequent years, Berkson justified the unusual forecast
suggested by his (I think) first contribution to a conceptual statistical journal. As a leader in statistical work at the Mayo Clinic,
Berkson participated in countless substantive statistical studies in
that institution and I personally learned from him many important
details of this domain. In addition, however, Berkson contributed
considerably to the solution of theoretical statistical problems
through his studies of bioassay, of properties of "logits," and of a
variety of χ^2 problems.

Vols. 2 and 6 of the *Annals* each have a special distinction, due
to papers by Harold Hotelling and J. L. Doob, undisputed American
leaders in their respective domains.

The contents of the paper by Hotelling (53) "The Generalization
of Student's Ratio" correspond exactly to its title. The main result
is the distribution of the so-called Hotelling's T statistic. However,
there is also a discussion that indicates that the ellipsoids provided
by the T statistic have the property of confidence regions for the
population means of several possibly correlated normal variables with
unknown variances and covariances. I believe that this is the first

English-language discussion of the new concepts which are still "on the books."* In particular, there is a distinct ancestor-descendant relationship between the results of Hotelling in 1931 and the modern ideas of Henry Scheffé (55) on the "multiple comparison" problem. In my opinion, the paper (53) in Vol. 2 of the *Annals* is the most important of Hotelling's contributions to mathematical statistics.

As I see it, Joseph L. Doob is what the biologists call a "mutant." A red-blooded mathematician and a Ph.D. from Harvard, Doob suddenly "mutated" into a red-blooded mathematical statistician, with a special double mission: to teach mathematicians theory of statistics (27) and to teach statisticians some other mathematical disciplines which at first sight may seem unrelated (56). At the University of Illinois at Urbana, for several decades now, Doob persists in this pattern of activities with outstanding success, having produced not only excellent scholarly results (57), but also a large succession of intellectual descendents beginning (I think) with my own friend and colleague, David Blackwell.

Regretfully, the originally idyllic relationships between the *Annals of Mathematical Statistics* and the ASA did not last long. As is not infrequent in human relations, the cause of the break was financial. It occurred when, as a delayed aftereffect of the 1929 stock market crash, the membership of the ASA began to decline rapidly, with a consequent decrease in the revenue of the Association. There was an unpleasant meeting of Carver with the ASA Directors at the conclusion of which Carver demanded the dissolution of the arrangements of 1929.

Apparently due to Carver's proverbial "infinite resource and sagacity," the *Annals* were not hurt badly. In fact, the *Annals* may have gained from the divorce. First, the subscription price went

---

*I am grateful to Henry Scheffé for calling my attention to an earlier relevant paper. This is the paper by H. Hotelling and H. Working, Applications of the theory of error to the interpretation of trends (54).

down, from $6 for Vols. 1 through 4, to $4 beginning with Vol. 5 and on. Next, perhaps unexpectedly, the quality of printing went up, from photo-offset with hand-written formulas, to ordinary mono-type settings. This occurred with Vol. 6, the last pages of which carry two somewhat triumphal pieces of information.

The first item, which is very brief, records a meeting of several interested persons arranged at Ann Arbor, Michigan, on September 12, 1935. At this meeting, it was decided to form an organization to be known as the Institute of Mathematical Statistics (IMS). A constitution and bylaws were adopted (totaling not quite six printed lines!); the following officers were elected to serve until December 31, 1936: President, H. L. Rietz, Vice President, W. A. Shewhart; Secratary-Treasurer, A. T. Craig. "A resolution, instructing the officers to investigate the feasibility of the affiliation of the Institute with the American Mathematical Society or the American Statistical Association, was adopted." The annual dues for members were set at $5 which included the subscription to the *Annals of Mathematical Statistics,* the official journal of the Institute.

The other piece of information at the end of Vol. 6 is the list of subscribers to the *Annals.* There were 118 individual subscribers in 1935, including 14 from abroad. Furthermore, a total of 98 librar-ies subscribed to the *Annals* with the surprising number of 33 from abroad, including a few from far-away Soviet Union and China. In each case, there must have been several individuals who wanted the *Annals!* This was the evidence that Carver's journal fulfilled a widespread need, and that the *Annals* had a future.

While in 1935 the *Annals* became the official journal of the IMS, it continued to be Harry Carver's journal. However, the obvious vitality of the periodical was bound to attract the attention of people with broad vision. One of these was Luther Eisenhart, a spe-ialist in differential geometry and a Dean at Princeton. He made an offer to Carver to take over the *Annals*, which would then become a journal of Princeton. Carver declined, but he suggested that Princeton

might find financial support for the IMS, in which case Carver was
prepared to yield his journal to the new professional organization.
After some deliberation, Dean Eisenhart found a foundation willing
to provide $25,000 as a five-year support of the *Annals* to be the
responsibility of the Institute of Mathematical Statistics. The
arrangements took some time to complete, and it was only with Vol. 9
that the *Annals* became the responsibility of the IMS with S. S. Wilks
of Princeton as its first elected Editor. The relevant sentence in
Carver's letter reads:

> The Annals then got a wonderful new Editor, and I then stepped
> down and since then have watched the growth of the Institute
> and the *Annals*, as a father does his children.

We, the mathematical statisticians of all lands, owe a lot to
Carver! How can we repay our debt?

Harold Hotelling

Hotelling had many fruitful ideas which he published in a number of
periodicals. Also, Hotelling produced a very substantial generation
of his intellectual followers and friends, who have now become leaders
in statistical research and instruction. In 1960, a group of them,
headed by Ingram Olkin, S. G. Ghurye, Wassily Hoeffding, William G.
Madow, and Henry B. Mann published a volume under the title *Contributions to Probability and Statistics: Essays in Honor of Harold
Hotelling* (58). In addition to his scholarly achievements, Hotelling
must be credited with an effective effort to regularize the status of
mathematical statistics within the American system of higher education,
that is, within American colleges and universities.

The reader will remember that while writing his monograph (42),
Henry Rietz considered the question of its usefulness in the presence
of many books on statistical methods published in the United States
in the early twentieth century. He found that, although useful in
describing some manipulative techniques, these books were conceptually
below a desirable standard; therefore, he had his monograph completed
and published. Thirteen years later, at the annual IMS meeting at

Harold Hotelling, as he was when he
visualized his "Young Jones."

Hanover, New Hampshire, Hotelling (59) examined more closely a set
of related but more concrete questions. One of these questions is:
What is the category of individuals who currently teach statistics
in American colleges and universities, and what should this category
be? (Hotelling's answer: "The task of leading the blind must not be
turned over to the blind.") The other question examined by Hotelling
refers to the university departments that currently offer courses in
statistics, and what is the appropriate situation? Hotelling noted
the astonishing proliferation of statistical courses in a variety of
academic departments on American campuses, in the departments of
agriculture, anthropology, astronomy, biology, business, economics,
education, engineering, medicine, physics, political science, psychology, sociology, etc., and ridiculed this situation:

> A synoptic picture...would perhaps be something like this. A
> university Department of X, where X stands for economics, psy-
> chology, or any one of numerous other fields, begins to note
> toward the end of the pre-statistical era that some of the out-

standing work in its field involves statistics. The quantity
and importance of such work are observed to increase, while at
the same time its intelligibility seems to diminish. Evidently
students turned out with degrees in the field of X who do not
know something about statistics are going to be handicapped, and
are not likely to reflect credit on Alma Mater. The department
therefore resolves that its students must acquire at least an
elementary knowledge of the fundamentals of statistics. To
implement this principle, it perhaps inserts some acquaintance
with statistics among the requirements for a degree. This situation naturally calls for the introduction of a course in statistics. Accordingly the head of the Department of X, in preparing the next Announcement of Courses, writes:

X 82. Elements of Statistics. An elementary but thorough course
designed to acquaint students of X with the fundamental concepts
of statistics and their applications in the field of X. The
viewpoint will be practical throughout. Second semester, MWF
at 10. Instructor to be announced.

The problem now arises of finding someone to teach the new course.
The few well-known statisticians in the country have positions
elsewhere from which it would be impossible to dislodge them
with the bait to be offered; for though the department wishes
to have statistics taught as an auxiliary to the study of X, it
feels that there must be no question of the tail wagging the
dog, and that economy is appropriate in this connection. The
members of the department of professorial rank do not respond
favorably to the suggestion that they should themselves undertake
to teach the new and unfamiliar course. But every university
department has a bright graduate student whose placement is an
immediate problem. Young Jones has already demonstrated a quantitative turn of mind in the course on Money and Banking, or in
the Ph.D. thesis on which he has already made substantial progress, dealing with The Proportion of Public School Yard Areas
Surfaced with Gravel. He may even recall having had a high-
school course in trigonometry. His personality is all that
might be desired. He is a white, Protestant, native-born American. And so the "Instructor to be announced" materializes as
Jones.

This earnest young scholar now finds that, in addition to complet-
ing his thesis, he must look up the literature of statistics and
prepare a course in the subject. His attention is directed by
older members of the department to some of the research papers
in the field of X involving statistics. He pursues "statistics"
through the library card catalog and in the encyclopedias. He
reads about census and vital statistics, price statistics, statistical mechanics. Perhaps he encounters probable errors.
Eventually he learns that Karl Pearson is the great man of statistics, and that *Biometrika* is the central source of information. Unfortunately most of the papers in *Biometrika* and of
Pearson's writings, while not lacking in vigor, trail off into

mathematical discourse of a kind with which young Jones feels
ill at ease. What he wants is a textbook, couched in simple
language and omitting all mathematics, to make the subject clear
to a beginner. Perhaps he finds the impressive books of Yule
and Bowley, but decides that they are too abstruse. Elderton's
"Frequence Curves and Correlation" is far too mathematical. Jones
decides that a simple book on statistics must be written, and
that he will do it if he can ever succeed in mastering the subject. In the meantime, he contents himself perforce with the
less mathematical writings of Karl Pearson, with applied examples
in the field of X, and with such nonmathematical textbooks as
may have been written by other young men who have earlier trod
the same path as that on which Jones is now beginning. Somehow
or other he gets the class through the course. After doing this
two or three times, Jones is an experienced teacher of statistics
and his services are much in demand. His course expands, takes
on a settled form, and after a while crystallizes into a textbook.
At the same time he may be getting out some research, consisting
of studies in the field of X in which statistical methods play
a part. His promotion is rapid. He becomes a Professor of
Statistics, and perhaps an officer in a national association.
His textbook has a large sale, and is used as a source by other
young men writing textbooks on statistics.

The textbooks written in this way form an interesting literary
cycle. Measures of 'central tendency' and of dispersion are
introduced, and the use of one against another of these measures
is debated on every ground except the criterion that modern
research has shown to be the important one, the sampling stability.
Sampling considerations, indeed, get little attention. The urge
to simplify by leaving out the more difficult parts of the subject, and especially the mathematical parts, is accompanied by
pride in the great number of examples drawn from real life, that
is, actual data that have been collected.

But the most fascinating features of this literary cycle is the
opportunity it offers for research by the standard methods of
literary investigation, tracing the influence of one author upon
another through parallelism of passages, and so forth. This
study is facilitated by the accumulation of errors with repeated
copying.

What Hotelling recommended as a result of the above kind of

analysis was the establishment of properly staffed departments of

statistics. Hotelling's paper was enthusiastically received by the

audience. As a result, at its business meeting of September 11, 1940,

the IMS adopted the following three resolutions:

    1. If the teaching of statistical theory and methods is to be
       satisfactory, it should be in the hands of persons who have

made comprehensive studies of the mathematical theory of
statistics, and who have been in active contact with applications in one or more fields.

2. The judgment of the adequacy of a teacher's knowledge of
    statistical theory must rest initially on his published
    contributions to statistical theory, in contrast with mere
    applications, in a manner analogous to that long accepted
    in other university subjects.

3. These ideas are expressed in detail in the paper *The teach-
    ing of statistics,* by Professor Harold Hotelling, and the
    Institute decides to give both the resolution and the paper
    as wide a circulation as possible.

More than three decades have elapsed since the presentation of
Hotelling's paper and the subsequent passage of the three resolutions
just quoted. As we all know, the subsequent developments went along
with the ideas of Hotelling. Currently, we have separate departments
of statistics staffed by Ph.D.s in our discipline. Research, both
in theory of probability and statistics and in their applications in
many domains, is flourishing in a number of centers. Could all this
have happened without the activities of Hotelling? Probably yes, but
then someone else would have had to come out with an authoritative
and vigorous call for a broad reform. Harold Hotelling did it for us.

Concluding Remarks

In the preceding pages, I tried to sketch the developments related
to mathematical statistics in the United States either caused or
influenced by the activities of three individuals--Henry Rietz, Harry
Carver, and Harold Hotelling. In the process, it was necessary for
me to take account of some chronology and some geography. In particular, I tried to construct a schematic map of the United States with
marked areas symbolizing the several centers of intellectual influence,
and with arrows indicating the wanderings of individuals who appear
to me to be the first generation that followed the three original
"founders." Unfortunately, the multiplicity of novel ideas and of
movements of persons involved in developing them proved so complicated
that the project of the map had to be abandoned. Instead, Fig. 1,

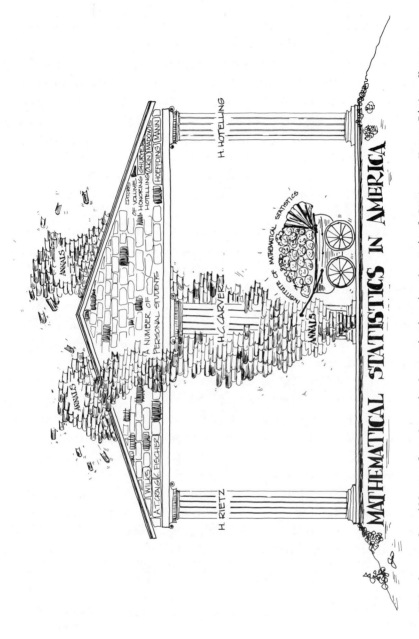

FIG. 1. Early pillars of mathematical statistics in the United States: Henry Rietz, Harry Carver, and Harold Hotelling.

conceived in a cheerful spirit, was drawn to summarize my views on
the activities of Rietz, Carver, and Hotelling.

However, the effort expended in preparing the map was not entirely
wasted. In particular, it demonstrated that, in addition to activities
of some individuals, whether the three "founders" or their progeny,
the early developments of mathematical statistics in the United States
were influenced, very favorably, by a particular locality, without
my being able to point out any particular person involved. The
"locality" in question is the combination of mathematics departments
in the leading universities in the northeast--Cornell, Harvard, Brown,
and probably others.

As mentioned in the account given above, Rietz came from Cornell.
J. L. Doob came from Harvard as a mathematician and "mutated" into a
statistician. Then I noticed that Professor Arthur H. Copeland, a
statistician at Ann Arbor, also came from Harvard. Of the several
modern outstanding leaders in mathematical statistics, Herbert E.
Robbins is from Harvard and Herman Chernoff is from Brown, to name
only two. There must have been quite a strong and broad-minded
mathematical atmosphere at these universities.

These thoughts led me to a succession of others. One is related
to Hotelling's recommendation that, in order to facilitate the development of mathematical statistics, the American universities organize
separate departments of statistics. There are good reasons for this.
However, it seems to me that there are also dangers. The danger lies
in the separateness of statistics from mathematics which, depending
on the attitudes of individuals concerned, may be excessive. All
mathematical disciplines are interconnected to some degree. Mathematical statistics is connected with probability theory so strongly
that one might say that the connection is "indivisible." Next come
measure theory, topology,* functional analysis, and algebra. I am

---

*Note that another of the present-day leaders in mathematical
statistics is a "mutant" from topology. He is John W. Tukey.

convinced that, in our epoch, a young scholar without familiarity
with these domains would be strongly handicapped in his work in mathematical statistics.

It is my considered opinion that those departments of statistics
that are emphatically separated from mathematics can produce "practitioners" of statistics. These practitioners would be capable of
applying such theory as they learned and could do useful substantive
work, but I would be most pleasantly surprised to see one contributing
to theory without first filling in the gaps in his mathematical education. Yes, such fillings of gaps are possible. In fact, I have
seen two such instances in my own department at Berkeley.

Incidentally, along with Harvard mathematics, another phenomenon
must be mentioned which, one might say, marked a new epoch in the
development of mathematical statistics. This is the appearance, in
1946, of the book by Harald Cramér, *Mathematical Methods of Statistics*
(60). For a considerable time at least, this book served many people
interested in statistics who felt a need to fill their "mathematical
gaps." It is still very useful as a textbook.

The conclusion is: Yes, let's have separate departments of statistics, but let them be, so to speak, next door to departments of
mathematics, and let's have some reasonably close intellectual contacts.

Here is another thought suggested by the examination of developments in the 1920s and 1930s. An effort should be made to popularize
mathematical statistics and probability among the general educated
public in the United States. The reader will remember that in wel-
coming the appearance of the *Annals* in 1930, Wilford I. King spoke
and wrote about "mathematical techniques of statistics" as the presumed
material to be published in the new journal. This is the more thought-
provoking, because King was Professor of Economics and at the same
time served as secretary of the ASA. It is not the "techniques" that
are important to the general educated public, but a few basic concepts.

One might think that over the 40-odd years that elapsed, the
situation must have changed. It did, of course, but not sufficiently.

In 1971, we had a special conference on pollution and health as a
part of the Sixth Berkeley Symposium on Mathematical Statistics and
Probability. Because of the national importance of the problem, an
effort was made to assemble the representatives of the various disciplines concerned, such as biology, radiation technology, etc.; statistical practitioners cooperating with them, representatives of the
relevant governmental agencies and their critics, and all had complete
freedom of expression. The contents of the conference *Proceedings*
(39) are most embarrassing.

The problem of "young Joneses" of Hotelling is still with us,
except that it changed (or expanded?) its locale. The *Proceedings*
of the pollution-health conference contain descriptions of many most
interesting laboratory studies on health effects of various pollutants,
including radioactivity, and also sounds of alarm voiced by M.D.s.
Also, there are descriptions of numerous statistical studies.* These
are disconcerting. They indicate that in modern times the "young
Joneses" of Hotelling, or their descendants, are in responsible
positions in a number of research institutions, councils etc., directing extensive studies on problems of great national importance, for
which they are not competent. This circumstance caused me to close
the *Proceedings* with an article of my own (61) pointing out a number
of deficiencies. Because of these deficiencies, there are currently
no reliable answers to many pressing questions relating to the present
efforts to meet the energy crisis. One illustrative example must
suffice: Is the proposed huge soft-coal-burning electric power plant
less dangerous or more dangerous to public health than a nuclear
power facility?

To produce answers to this and to a number of other similar
questions, a large and complicated "multipollutant" and "multilocality"
study is needed. However, to have such a study organized and to

---

*Unfortunately, the authors of these studies did not take part
in the conference.

obtain its adequate financing, we need another Harry C. Carver as he was in 1928. Eventually, he is bound to appear, but when?

How can we speed up the appearance of a Carver, not only a competent and productive statistician, but also a man of action on a broad scale? How can we make his life easier when he appears? It seems to be that here all the existing departments of statistics have a special mission: to raise the level of comprehension of mathematical statistics among the so-called general educated public. To achieve this goal, special statistical courses are needed for university students at large, including such diverse fields as history and business administration, classics and political science, musicology and medicine. Let's teach these students not "techniques," but general concepts of statistical ("pluralistic") subjects of study and a few others. One of them ought to be the phenomena behind the old saying about "lie, damn lie and statistics." One of these phenomena ought to be the phenomenon of spurious correlation.

Discovered by Karl Pearson in the 1890s, this phenomenon ruined a great number of important statistical studies ranging from anatomy of shrimp, to farm economics, to railroad finances, to race relationships, to alcoholic beverages and crime, to air pollution with copper dust and breast cancer in women. Using small-scale Monte Carlo simulation, the subject is not hard to teach and the students enjoy the occasion for having a good laugh. If we offer nontechnical courses including this and some similar subjects, within a relatively short time the underlying ideas would reach the decision-making authorities and our hoped-for Harry Carvers would have easier times in organizing and conducting large-scale studies on important national problems.

These are the current problems, but what about the achievements? These are immense. The epoch of transplantation of mathematical statistics from Europe to the United States is over and, if anything the pendulum appears to swing the other way. Here the priority belongs to J. L. Doob. Although the fundamental ideas of his papers written in the 1930s, exemplified by refs. 27 and 56, are those of

R. A. Fisher, born in England, with mathematical rigor added, Doob soon developed conceptual initiative of his own. In a series of papers published in many countries, culminating with his famous book (57), he developed entirely new fields of study symbolized by terms "martingales" and "semi-martingales." Early in 1975, his international influence earned him the well-deserved very rare honor of being elected foreign member of the French Academy of Sciences.

On a different level, the books of the big John Wiley & Sons statistical series can now be seen on many shelves all over Europe. There are also many translations. For example, I happen to be informed that a book by Lehmann (62) appeared in translations in Japanese, Polish, and Russian. Also, the more recent book by Hodges and Lehmann (63) has been translated into Danish, Hebrew, and Italian. Presumably, there are other occurrences of the same kind.

Naturally, this is not to say that American mathematical statistics dominates the world. We do have many most important foreign books on our own shelves. Also, the names of many probabilists-statisticians from Asia, Australia, and Europe are well familiar to all of us. A realistic appraisal seems to be that starting from the pre-World War II status of a recipient of ideas born in Europe, the United States turned into a co-equal contributor to the current splendid development of our discipline, on a worldwide scale.

A POSTSCRIPT

I am indebted to Professor Herman Chernoff for important information relating to the role of a particular school in bringing up generations of scholars who now take an active part in the development of mathematical statistics in the United States. After examining the above pages of the present essay emphasizing the role of Cornell, Harvard, and Brown Universities, Professor Chernoff sent me a group photograph of 20 young men, some looking like teenagers, who, during the dark years of the Great Depression, were members of the Mathematics Club in the City College of New York (CCNY), an undergraduate school of much less glamour than the three universities I mentioned. Ten of these youngsters are now professors in either pure mathematics or in one of the subdisciplines of mathematical statistics, serving in leading universities all over the country. The statistical subgroup is composed of the following: (1) Kenneth Arrow, now member of the National Academy of Sciences and Nobel Prize winner in econometrics, (2) Herman Chernoff, (3) Milton Sobel, and (4) Oscar Wesler. After reading Professor Chernoff's letter, I found from other sources that the following leading mathematical statisticians, not in the group photograph, also received their basic training at CCNY: Lester Dubins, Ingram Olkin (now Chairman of Statistics at Stanford University), Herbert Solomon, and Jacob Wolfowitz, another member of the National Academy of Sciences.

How did all this happen? There must have been some sources of inspiration at CCNY which attracted talented young men and subsequently fed them a variety of novel ideas. According to Professor Chernoff, two particular individuals played an outstanding role, one of whom is John M. Firestone, still Professor of Economics at CCNY and a Fellow of the American Statistical Association. The other individual is Dr. Selby Robinson currently retired in Laguna Beach, California. Both deserve an appropriate mark of recognition.

ACKNOWLEDGMENTS

This paper was written with the partial support of Public Health Service Grant No. GM-10525-11, National Institutes of Health, Public Health Service, and U.S. Army Research Office-Durham, Grant No. DA-ARO-D-31-124-73-G31.

I would like to thank Terry Jue for the drawing of Fig. 1; Marilyn Hill for the drawings of Fisher, Hotelling, Carver, and Rietz; and Jeanne Lovasich for the drawing of my own likeness.

REFERENCES

1. W. Feller, *An Introduction to Probability Theory and Its Applications,* Wiley, New York, (1st ed., 1950; 3rd ed., 1967.)

2. J. Bernoulli, *Ars conjectandi* (1713: reprinted in *Wahrscheinlichkeitsrechnung*), Engelmann, Leipzig, 1899.

3. A. N. Kolmogorov, *Grundbegriffe der Wahrscheinlichkeitsrechnung*, Julius Springer, Berlin, 1933.

4. R. von Mises, *Probability, Statistics and Truth* [2nd rev. English ed., H. Geiringer (Transl. Ed.)], Allen and Unwin, London, 1957.

5. P. S. Laplace, *Théorie analytique des probabilitiés*, Académie Française, Paris, 1812 (English version: Dover Publications, New York, 1951).

6. G. T. Fechner, *Kollektivmasslehre*, G. R. Lipps (Ed.), Engelmann, Leipzig, 1897.

7. M. G. Kendall, *The Advanced Theory of Statistics* (3rd ed.), Vols. 1 and 2, Griffin, London, 1947.

8. K. Pearson, On the criterion that a given system of deviations from the probable in the case of a correlated system of variables is such that it can be reasonably supposed to have arisen from random sampling. *Phil. Mag. Ser. V 50*, 157-175 (1900).

9. H. Cramér, On the composition of elementary errors. First paper: Mathematical deductions. *Skand. Aktuarietidskr. 11*, 13-74 (1928).

10. R. von Mises, *Wahrscheinlichkeitsrechnung und ihre Anwendung in der Statistik und theoretischen Physik*, Deuticke, Leipzig, 1931.

11. A. N. Kolmogorov, Sulla determinazione empirica de una leggi di distribuzione. *Giorn. Ist. Ital. Attuari 4*, 83-91 (1933).

12. N. V. Smirnov, On deviations of the empirical distribution functions, (Russian). *Mathematicheskii Sbornik 6*, 3-26 (1939).

13. J. Neyman, "Smooth" test for goodness of fit. *Skandinavisk Akuarietidskr. 20*, 149-199 (1937). *See also: A Selection of Early Statistical Papers of J. Neyman*, University of California Press, Berkeley, 291-319, 1967.

14. G. U. Yule and M. Greenwood, An inquiry into the nature of frequency distributions representative of multiple happenings with particular reference to the occurrence of multiple attack of disease or of repeated accidents. *J. Roy. Statist. Soc. 83*, 255-279 (1920).

15. J. Neyman and E. L. Scott, Statistical aspects of the problem of carcinogenesis. *Proc. 5th Berkeley Symp. Math. Statist. Prob.* 4, 745-776 (1967).

16. W. Zonn, Explosive events in the universe. In *The Copernican Heritage: Theories "Pleasing to the Mind,"* J. Neyman (Ed.), MIT Press, Cambridge, Mass., 1974.

17. J. Neyman, Experimentation with weather control. *J. Roy. Statist. Soc. A 130*, 285-326 (1967).

18. G. Pólya, Sur quelques points de la théorie des probabilités. *Ann. Inst. Henri Poincaré 1*, 117-161 (1930).

19. O. Reiersøl, Identifiability of a linear relation between variables which are subject to error. *Econometrica 18*, 375-389 (1950).

20. T. E. Harris, *The Theory of Branching Processes*, Springer-Verlag Berlin, 1963.

21. C. F. Gauss, *Abhandlungen zur Methode der kleinsten Quadrate*, Stankiewicz, Berlin, 1887.

22. A. A. Markov, *Wahrscheinlichkeitsrechnung* (German transl.), Teubner, Leipzig, 1912.

23. G. U. Yule, An application of the χ^2 method to association and contingency tables, with experimental illustrations. *J. Roy. Statist. Soc. 85*, 95-104 (1922).

24. R. A. Fisher, The conditions under which χ^2 measures the discrepancy between observation and hypothesis. *J. Roy. Statist. Soc. 87*, 442-450 (1924).

25. R. A. Fisher, On mathematical foundations of theoretical statistics. *Phil. Trans. Roy. Soc. (London) Ser. A 222*, 309-368 (1921).

26. H. Hotelling, The consistency and ultimate distribution of optimum statistics. *Trans. Amer. Math. Soc. 32*, 847-859 (1930).

27. J. L. Doob, Probability and Statistics. *Trans. Amer. Math. Soc. 36*, 759-775 (1934).

28. J. Neyman and E. S. Pearson, On the problem of the most efficient tests of statistical hypotheses. *Phil. Trans. Roy. Soc. (London) Ser. A 231*, 289-337 (1933).

29. W. Pytkowski, The dependence of the income in small farms upon their area, the outlay and the capital invested in cows, (Polish, English summaries), Monograph no. 31 of series *Biblioteka Pulawska*, publ. Agri. Res. Inst. Pulawy, Poland, 1932.

30. J. Neyman, Outline of a theory of statistical estimation based on the classical theory of probability. *Phil. Trans. Roy. Soc. (London) Ser. A 236*, 333-380 (1937).

31. J. Neyman, L'estimation statistique traitée comme un problème classique de probabilité. *Actual. Scient. Indust. 739*, 25-57 (1939). [Russian transl.: *Usp. Matemat. Nauk 10*, 207-229 (1949)].

32. A. Wald, *Statistical Decision Functions*, Wiley, New York, 1950.

33. R. A. Fisher, *Statistical Methods for Research Workers*, Oliver & Boyd, London. (1st ed., 1925; 12th ed., 1954.)

34. R. A. Fisher, *The Design of Experiments*, Hafner, New York. (1st ed., 1935; 8th ed., 1966.)

35. R. A. Fisher and F. Yates, *Statistical Tables for Biological, Agricultural and Medical Research*, Hafner, New York. (1st ed., 1838; 6th ed. 1963).

36. J. Neyman and E. L. Scott, Field galaxies and cluster galaxies: abundances of morphological types and corresponding luminosity functions. In *Confrontation of Cosmological Theories with Observation*, M. S. Longait (Ed.), D. Reidel Publishing Co., Dordrecht, 119-130 (1974).

37. *Proc. Sixth Berkeley Symp. Math. Statist. Prob., 5: Darwinian, Neo-Darwinian and Non-Darwinian Evolution.* University of California Press, Berkeley, 1972.

38. A. Bowker, *Sampling Inspection by Variables*, McGraw-Hill, New York, 1952.

39. *Proc. Sixth Berkeley Symp. Math. Statist. Prob., 6: Effects of Pollution on Health.* University of California Press, Berkeley, 1972.

40. A. J. Lotká, *Elements of Physical Biology*, Williams & Wilkins, Baltimore, 1925. (Republished, Dover, New York, 1956.)

41. V. Volterra, *Lecons sur la théorie mathématique de la Lutte pour la vie.* Gauthier-Villars, Paris, 1931.

42. H. L. Rietz, *Mathematical Statistics*, Carus Mathematical Monographs, Mathematical Association of America, New York. (1st ed., 1927; 5th ed., 1947.)

43. H. L. Rietz, Urn schemata as a basis for the development of correlation theory. *Ann. Math. 21*, 306-322 (1920).

44. W. P. Elderton, *Frequency Curves and Correlation* (2nd ed.) Layton, London, 1927.

45. G. U. Yule, *An Introduction to the Theory of Statistics*, Griffin, London. (1st ed., 1910; 11th ed., coauthored by M. G. Kendall, 1937.)

46. J. A. Shohat and J. D. Tamarkin, *The Problem of Moments*, Carus Mathematical Monographs, American Mathematical Society, New York, 1945.

47. S. D. Wicksell, Remarks on regression. *Ann. Math. Statist. 1*, 3-13 (1930).

48. J. A. Shohat, Stieltjes integrals in mathematical statistics. *Ann. Math. Statist. 1*, 73-94 (1930).

49. L. v. Bortkiewicz, The relation between stability and homogeneity. *Ann. Math. Statist. 2*, 1-22 (1931).

50. C. Jordan, Approximation and graduation according to the principle of least squares by orthogonal polynomials. *Ann. Math. Statist. 3*, 257-357 (1932).

51. T. N. Thiele, The theory of observations. *Ann. Math. Statist. 2*, 165-307 (1931).

52. J. Berkson, Bayes' theorem. *Ann. Math. Statist. 1*, 42-56 (1930).

53. H. Hotelling, The generalization of Student's ratio. *Ann. Math. Statist.* 2, 360-378 (1931).

54. H. Hotelling and H. Working, Applications of the theory of error(s) to the interpretation of trends. *J. Amer. Statist. Assoc.* [March Suppl.] 24, 73-85 (1931).

55. H. Scheffé, *The Analysis of Variance,* Wiley, New York, 1959.

56. J. L. Doob, The limiting distribution of certain statistics. *Ann. Math. Statist.* 6, 160-169 (1935).

57. J. L. Doob, *Stochastic Processes,* Wiley, New York, 1953.

58. *Contributions to Probability and Statistics: Essays in Honor of Harold Hotelling,* I. Olkin, S. G. Ghurye, W. Hoeffding, W. G. Madow, and H. B. Mann (Eds.), Stanford University Press, Stanford, Calif. 1960.

59. H. Hotelling, The teaching of statistics. *Ann. Math. Statist.* 11, 457-472 (1940).

60. H. Cramér, *Mathematical Methods of Statistics,* Princeton University Press, Princeton, N.J., 1946.

61. J. Neyman, Epilogue of the health-pollution conference. *Proc. 6th Berkeley Symp. Math. Statist. Prob.* 6, 575-587 (1972).

62. E. L. Lehmann, *Testing Statistical Hypotheses,* xii, 369, Wiley, New York, 1959.

63. E. L. Lehmann and J. L. Hodges, Jr., *Basic Concepts in Probability and Statistics* (2nd ed.) 441, Holden-Day, San Francisco, 1970.

8.

*FOUNDATIONS OF PROBABILITY
THEORY AND ITS INFLUENCE ON
THE THEORY OF STATISTICS*

J. L. DOOB was born February 27, 1910, in Cincinnati, Ohio. He received three degrees from Harvard University: a B.A. in 1930, an M.A. in 1931, and a Ph.D. in 1932. He received a National Research Council Fellowship and a Carnegie Corporation grant from 1932 to 1935. In 1935 he joined the faculty of The University of Illinois at Urbana-Champaign, where he rose to Professor in 1945. He is a member of the National Academy of Sciences and past President of the Institute of Mathematical Statistics and of the American Mathematical Society. He is a member of the International Statistical Institute, American Academy of Arts and Sciences, and is Foreign Associate of the French Academy of Sciences. He is a Fellow of the Institute of Mathematical Statistics and the author of numerous papers. He is an internationally recognized authority on probability theory.

8.

FOUNDATIONS OF PROBABILITY THEORY AND ITS INFLUENCE ON THE THEORY OF STATISTICS

J. L. Doob

Department of Mathematics
University of Illinois
Urbana, Illinois

MATHEMATICAL PROBABILITY

Probability as a purely mathematical subject is not controversial. It is a part of measure theory with specialized terminology and emphasis harking back to its provenance. In fact, probabilists studied measure and integration theory and developed their own terminology before these topics were formulated and made precise by Lebesgue and by later mathematicians who did not suspect that they were laying the rigorous foundations of mathematical probability as finally made explicit in 1933 by Kolmogorov. Only culturally retarded writers now contrast "probabilistic proofs" with "analytical proofs," ignoring the fact that mathematical probability is just as much a part of analysis as, say, differential calculus, and that a measurable function is no less sweet if called a random variable.

In summary, mathematical probability concerns itself with a measurable space, that is a space Ω with a distinguished class (σ algebra) F of its subsets. In applications, these sets are interpreted as events. Each set A in F is assigned a measure $P(A)$, "the probability of A."

Here the set function P is countably additive and $P(\Omega) = 1$. (More
generally, it is sometimes appropriate to allow $P(\Omega)$ to be other than 1,
sometimes even to be ∞; in some contexts, probabilists treat situations in which F is an algebra and P is only finitely additive.) If
x is a measurable function on Ω, x is called a "random variable" and
its integral $\int_\Omega x \, dP$, if this integral exists, is called the "expectation of x" and is denoted by $E(x)$. Probabilists have been reluctant
to give up this special terminology which recalls the nonmathematical
origin of their subject. This origin and corresponding terminology
still suggest directions for research and thereby maintain the special
place of probability in measure theory.

Finally, it should be noted that mathematical probability has
both taken and given. It uses the methods and results of other fields
of mathematics and in turn has contributed to these fields. For example, certain parts of probability theory are indistinguishable from
ergodic theory; others have revolutionized potential theory.

NONMATHEMATICAL SIGNIFICANCE OF MATHEMATICAL PROBABILITY

The intrinsic interest of mathematical probability and the power of
its applications to other parts of mathematics are accepted. There
is controversy outside the domain of pure mathematics, where the relation of mathematical probability to the real world is discussed. It
is obvious that certain mathematical theorems have real-life analogs,
but the significance and proper formulation of the relationship between these theorems and their analogs have induced debates as endless
as the infinite sequence of coin tosses so beloved by the debaters.
The nonmathematical significance of mathematical probability will be
illustrated in the contexts of coin tossing and electrical engineering.

We first consider coin tossing. Mathematical probability theory
includes a theorem of pure mathematics, the strong law of large numbers
which goes back to Borel (1909), according to which (making the usual
translation into the language of coin tossing) if s_n is the number of
heads obtained in n tosses of a symmetric coin the event $\lim_{n\to\infty} s_n/n = 1/2$ has probability 1. On the empirical side, everyone who has ever

tossed coins has observed (I) a clustering of s_n/n around 1/2 for moderately large n, say n at least 20. Moreover, (II) a somewhat more sophisticated observation leads to the remark that if not all tosses are accepted, but if the criterion of whether to accept the jth toss only depends on the past before that toss (say, if the jth toss accepted is the first toss preceded by j heads), then the tosses accepted have the clustering described in (I).

Thus there is a mathematical theorem and there are empirical observations, and one wishes to relate the two, in some rational and useful way. Of course, one can say that the strong law of large numbers corresponds to (I) and it is easy to prove a mathematical theorem corresponding in the same way to (II), but the problem is to put this obvious correspondence into operational use. A great deal of cloudy philosophy has poured its rain upon the scientific community in connection with coin tossing and this correspondence. The following points may offer some shelter.

1. The strong law of large numbers is a mathematical theorem. It can be proved only by mathematics not by tossing coins, and it alone implies nothing whatever about actual coins, their tossing, or the fingers that toss them.

2. Actual coin tossing has nothing to do with the limit of s_n/n if for no other reason than because a coin can be tossed only finitely many times. It is, therefore, as useless to debate whether or not the strong law of large numbers holds for actual coin tossing as to debate the table manners of children with six arms. (But, remarkably enough, the statement has appeared in print that it is all right to assume that $s_n/n \to 1/2$ in actual coin tossing because since this limit statement cannot be verified or disproved the assumption is safe!)

3. General principle: Any analysis of what actually happens when a coin is tossed can be ignored unless Newton's laws of motion of a body falling under the influence of gravity are involved explicitly. (This principle is a valuable time saver.) In fact, these are the laws that govern the tosses of genuine coins. If mathematical coins are to be discussed they should be treated like any other mathematical concepts; that is they should be defined and their properties stated in mathematical form. Otherwise they are counterfeit.

4. The observed stability of s_n/n for moderate values of n is impossible to state in precise mathematical form, even though this stability lies at the base of the profits of insurance

companies and justifies the confidence of moralists that the
distribution of male and female births will be such that neither polyandry nor polygamy will be necessary to keep the human race content.

At the opposite extreme from coin tossing, consider the phenomenon of spontaneous "noise" in electric circuits (Johnson effect), which
sets an upper limit to the accuracy of electronic measuring apparatus.
Even if there are no power sources in an electric circuit, there is a
nontrivial flow of current generated by the heat motion of charged
particles in the conductors. The usual mathematical model for a circuit containing resistors, condensers, and inductances, with the usual
interpretation, makes the spontaneous current in a wire at a given time
t a random variable x(t). The family of these random variables as t
varies is Gaussian, and asymptotically stationary, and the covariance
function is easily calculated in terms of the circuit constants. The
theoretical harmonic analysis of sample functions of the x(t) process
can be compared with the observed results, and by adjusting the circuit
constants the Johnson effect frequencies can be tuned to where their
interference is least important, and the temperature can be reduced to
decrease the Johnson effect simultaneously at all frequencies. The
success of the theoretical analysis, that is, of the mathematical model, can be measured by the success in decreasing the Johnson effect
interference, which may simply mean decreasing a background hiss in an
audible signal. In this case one need not sample the signal a large
number of times. The circuit is doing the averaging of the many elementary jolts making up the Johnson effect.

THE ROLE OF MATHEMATICAL PROBABILITY

Mathematical probability can be either at the beginning or at the end
of a discussion of probability in the real world. That is (beginning
of the discussion), one can accept the mathematical discipline as a
model for certain phenomena and use a translation principle to go from
one to the other, or (end of the discussion) one can make an analysis
of certain phenomena leading to a formal calculus which turns out to
be mathematical probability. Whatever the procedure, one cannot hope

that all the subtleties of the mathematical discipline have operational
images in the real world, and even when the operational image is precise, extreme cases should not be taken seriously. For example, consider the equipartition principle of physics as applied to a pendulum
free to oscillate in a fixed plane. Let $x(t)$ be the angle the pendulum makes with the vertical at time t. Even if $x(0) = 0$ and no initial
push is given, $x(t)$ does not remain at 0 but, in fact, because of the
molecular nature of matter (Johnson effect in a mechanical system) $x(t)$
is a random variable, the $x(t)$ stochastic process is Gaussian, and the
covariance function of the process is easily calculated. Because the
normal distribution is unbounded, an observer who has complete faith
in the mathematical model might expect that if he watches long enough
$x(t)$ will reach, say, $\pi/4$, or if he waits still longer $x(t)$ will reach
$\pi/2$ so that the pendulum will swing all the way up and around. But
only nonscientists should have this much faith in science, and even
one with this abiding faith should be wary. A mathematician once remarked that some day, according to statistical physics, a pot of water
on a fire instead of boiling, will freeze, but that anyone told of such
an occurrence by a cook can safely disbelieve the story since the probability of the event is much smaller than the probability that the cook
is a liar. The Bayesian statisticians (see below) will, of course,
note that the first of these two probabilities can perhaps be calculated by an astute physicist willing to make simplifying assumptions, but
that the very existence of the second probability is controversial to
non-Bayesians except, perhaps, to those expert on the truthfulness
of cooks.

This discussion brings us to the problem, which we take up next,
of what principle should be used to translate mathematical probability
theorems into real life.

FROM MATHEMATICS TO THE REAL WORLD

If one starts with mathematical probability theory the obvious general
operational translation principle is that one should ignore real events
that have small probabilities. How small is "small" depends on the

context, for example, the demands of a client on a statistician. Somewhat more precisely, one first makes a judgment on the possibility of
the application of probability in a given context; if so, one then sets
up a model and comes to operational decisions based on the principle
that hypotheses must be reexamined if they ascribe small probability
to a key event that actually happens. (This is, of course, a great
oversimplification.) For example, some years ago, when small sample
theory was first coming into use, a statistician for Bell Laboratories
applied the technique to the testing of telephone poles. Either he
had not chosen his model correctly or he was ignoring events with non-
negligible probability in this context. At any rate, Bell Laboratories
decided it had lost money using his advice. Presumably Bell also had
a mathematical model (perhaps not explicitly constructed and probably
cruder) which, when applied to a large sample, indicated a financial
loss.

SUBJECTIVE PROBABILITY

A popular approach to statistics, inspiring bitter quarrels, is the
Bayesian approach, which makes probability "subjective." This means
that an individual assigns probabilities to events in a consistent
way (for example, as determined by the odds he would be willing to
accept on their occurrence) and there is a mechanism in the theory
which tends to put different individuals into substantial agreement
with each other, and, presumably, with the real world, so that the
subjective approach becomes objective. The final mathematical discipline is, of course, the usual mathematical probability. A Bayesian
assigns probabilities to events the non-Bayesian would not include in
his probability model and the statistical tests of the two may differ.
The non-Bayesians assert that the Bayesians' very principles are con-
tradictory. The Bayesians assert that their opponents are inefficient
statisticians.

The principal character of a Lessing play, when asked which of
the Christian, Jewish, and Muslim religions was the true one, suggested
waiting a few millenia and then identifying the true religion from its

beneficent effects. The decision would be difficult, but less so than
the problem of recognizing the better of the Bayesian or non-Bayesian
statistical advice after millenia of experience. In fact, although
the three religions share enough that they might agree on a few criteria of truth, each statistical camp would probably insist that only
its own methodology can determine its own merits. Statistics has the
advantage that it is both advocate and judge of its worth.

VON MISES COLLECTIVES

For historical reasons, no discussion of the meaning of probability is
complete without some remarks on von Mises' approach. This approach
straddled mathematics and the real world, but its inventor was finally
persuaded to modify it into pure mathematics. In the context of the
coin tossing discussion, a von Mises "collective" was a sequence of
"observations" satisfying (I') $s_n/n \rightarrow 1/2$ together with a second condition (II') corresponding to (II). Since an "observation" is not a
mathematical concept, and since an infinite sequence of coin tosses is
not physically meaningful, collectives should have been suspect from
the start. When von Mises was persuaded to make his theory mathematical, a collective became a mathematical sequence of ones (heads) and
zeros (tails) satisfying (I') and a weakened version of (II'). This
took the intuitive content out of the approach and left only an awkward technique to get by detours to the usual mathematical probability discipline.

PROBABILITY IN THE UNITED STATES SINCE 1930

In 1930, probability was in a primitive state in the United States.
There was no text available that was much more than a collection of
combinatorial problems and a discussion of easy cases of the law of
large numbers and the central limit theorem. There was almost invariably a confusing identification of mathematical probability with probability in real life. In those days, mathematicians used probabilistic
ideas to suggest mathematical problems, and "mathematical probability"
was simply the class of all these problems. This is not to imply that

no deep research was being pursued. Norbert Wiener had put Brownian motion into rigorous mathematics in 1923. In the Soviet Union mathematicians of the caliber of Serge Bernstein were continuing their country's tradition in probability. The connection between partial differential equations and random walks was known to many. But mathematical probability as such was still not quite in the mainstream of mathematics.

There was a radical change with the publication of Kolmogorov's 1933 monograph on the basic concepts of probability. Although probabilistic concepts had been identified with measure theoretic ones by many authors before 1933, Kolmogorov's formalization was complete and definitive and was soon almost universally accepted. However, there was a lag in the United States and it was years before U. S. mathematicians accepted probability as a legitimate mathematical subject and mathematical statistics is still not fully accepted. Perhaps in self-defense mathematical statistics in the United States became purer and purer, adopting the protective coloration of mathematics. A considerable part of *Annals of Mathematical Statistics* has been mathematics that has had only a tenuous connection with applied statistics and only a small part of this research has brought about changes in statistical procedures.

However this may be, mathematical probability and mathematical statistics are now flourishing in the United States. We have had and continue to have our share of world leaders in these fields. And perhaps the state of war between the champions of the different points of view in statistics is a good sign. It indicates that the issues are important enough to fight about.

REFERENCES

Borel, E. (1909). *Rend. Circ. Mat. Palermo 27*, 247-271.

Kolmogorov, A. N. (1933). Grundbegriffe der Wahrscheinlichkeitsrechnun*Ergeb. Mat. 2*, 62 pp.

Wiener, N. (1923). *J. Math. Phys. 2*, 131-174.

9.

THE INTERACTION BETWEEN
LARGE SAMPLE THEORY AND OPTIMAL
DESIGN OF EXPERIMENTS

HERMAN CHERNOFF was born July 1, 1923, in New York City. He received
his B.S. degree from the City College of New York in 1943, his
M.Sc. degree from Brown University in 1945, and Ph.D. degree also from Brown in 1948. He was with the Cowles Commission at the
University of Chicago from 1947 to 1949 and with the University
of Illinois from 1949 to 1952. He was affiliated with Stanford
University from 1952 to 1974 and since then with the Massachusetts
Institute of Technology. He is a Fellow of the American Statistical Association and the Institute of Mathematical Statistics. He
is a member of the International Statistical Institute, The American Academy of Arts and Sciences, and several other organizations
He was President of the Institute of Mathematical Statistics from
1967 to 1968. He is internationally known for the clarity of his
presentations and is the author of over 75 research papers.

THE INTERACTION BETWEEN
LARGE SAMPLE THEORY AND OPTIMAL
DESIGN OF EXPERIMENTS

Herman Chernoff

Department of Mathematics
Massachusetts Institute of Technology
Cambridge, Massachusetts

It is my intention to present a picture of the power of the application of large sample theory with particular emphasis on its applicability to the optimal design of experiments and with references to
some of the important contributions in this development. The perspective is highly personal and undoubtedly ignores important contributors
and contributions which proper historical research would clarify.

I shall discuss how large sample considerations simplify analysis
and results and how they provide clarity and meaningful concepts of
optimality, relative efficiency, and information. These concepts of
relative efficiency and information, in turn, provide the basis for
the definition and derivation of optimal designs of experiments,
which are applicable when there is a choice of alternative experimental setups.

This sort of treatment always seems to require an explanation of
the potential usefulness of a theory that presents optimal procedures
and measures of efficiency with respect to such procedures. In the
applied world, it is not always convenient or practical, or even desirable, to use a procedure that is derived to be optimal. However,

such a procedure and its properties provide a benchmark against which
other more convenient procedures may be compared. If 60% efficiency
means that one can do equally well with 60% of the sample size or with
60% of the total cost by using an efficient procedure, then it pays to
examine alternatives carefully. Conversely, if a simple practical procedure can be shown to have 98% efficiency, and the "optimal" method
is either impractical or sensitive to simplifying assumptions, or both,
then one would naturally be inclined to be satisfied with the simple
procedure.

It should also be pointed out that finding optimal procedures requires more than the simple optimization of some function. Typically,
the major hurdle is to formulate the problem in such a way that there
is a meaningful function to be optimized. In small sample theory, such
formulations seem difficult or impossible to make. In contrast, large
sample theory often provides useful formulations.

The literature of optimal design of experiments does not abound
in an overwhelming number of important practical applications. Nevertheless, the insights to be gained are important and useful, the implications on the philosophy of experimental science are substantial,
and there are many prospects for future applications in a world that
depends increasingly on the use of computer-controlled experimentation
and Monte Carlo simulations.

ASYMPTOTIC THEOREMS PROVIDE SIMPLICITY

The object of a statistical experiment is to obtain data which, when
properly summarized and reduced, provide information about the underlying laws, which we call the *state of nature,* typically described by
a *parameter* θ. The word *"statistical"* is used in contrast to *"deterministic"* to describe the fact that the relationship between θ and the
data is obscured in part by noise or by random fluctuations. The word
statistic applies to a number that summarizes aspects of the data and
is typically assumed to behave randomly according to some law of probability or *distribution,* depending on θ. The relationship between θ
and this probability distribution determines how the statistic may be
used to make inferences with regard to θ.

Thus, the statistician must cope with random variation, and the mathematical statistician is concerned with the probability distribution of possibly complicated functions of the data. These are called *sampling distributions*. Typically, the complexity of the mathematical derivations increases rapidly with the amount of data or with the size of the sample. It is frequently seen that after sample size n = 1 or 2, analytic expressions in closed form are difficult to attain and, if attained, are difficult to interpret. However, two fundamental theorems of probability theory make it possible to obtain and understand the results for the case n = ∞, or, more precisely, the case in which n → ∞. These theorems are the Law of Large Numbers and the Central Limit Theorem, which state that a sample average approaches the population mean and that its probability distribution is approximated, for large n, by an appropriate Gaussian (normal) bell-shaped distribution characterized completely by only two parameters, its mean and variance.

Large sample theory exploits these remarkably neat and simple limiting or *asymptotic* properties. It is often easy for statisticians to use these properties in order to derive effective and easily understood approximations and to compare these neat limiting forms with rather precise distributions derived numerically for specified finite values of n. To repeat, the limiting forms are typically relatively easy for mathematical analysis and for comprehension.

An outstanding application was the derivation of the χ^2 distribution as the limiting form for Karl Pearson's χ^2 statistic for testing goodness-of-fit, i.e., for testing how well the data fit a model of random variation (27).

ESTIMATION

With the advent of decision theory, it has become clear that in principle the task of finding a useful general criterion of optimality in the estimation problem is difficult in the case of small samples. Here the unknown parameter θ is numerical valued; the statistician seeks some statistic T, depending on the data, in order to estimate θ. He desires a procedure or statistic T that is very likely to be close to θ. The difficulty arises in that one can improve performance for some values of θ at the cost of making it worse for others. An extreme case consists of guessing without paying any attention to the data.

Suppose, for example, that θ is the unknown probability of an event that can be sensibly estimated by T, the proportion of times the event occurs in a sample of n trials. It may then be shown that the standard deviation (root-mean-squared error) of T about θ is no greater than $1/2\sqrt{n}$ and is quite small when n is large, irrespective of the unknown value of θ. In contrast, the apparently ridiculous procedure of guessing that θ = 1/3, i.e., using T* = 1/3 and ignoring the data, is better than using T if, indeed, θ is equal to or close to 1/3.

It would seem that this example makes the concept of an optimal estimate unattainable and yet, the estimate T appears to be a desirable one. R. A. Fisher (14) resolved the dilemma by applying large sample theory and by restricting the class of estimates under consideration to those which are *consistent*, i.e., those that have the property that as n → ∞, the estimate is very likely to be close to θ irrespective of the value of θ. Subject to both this restriction and an additional one, Fisher attempted to prove that the method of maximum-likelihood provides asymptotically optimal estimates. To accomplish this, it is necessary to show how well the maximum-likelihood estimate performs and that it is impossible for one of the allowable competitors to do better. The surprising fact is that a slightly strengthened version of the consistency requirement is adequate in the large sample context. A general informal rule in estimating and testing hypotheses is that there are standards of performance that may be attained for large samples, and that one cannot find a procedure to do substantially better for some range of θ without doing worse by an *order of magnitude* elsewhere.

In the estimation context, Fisher associated a measure of information I(θ) for a parameter θ with a statistic or its probability distribution. The information is additive in the sense that the information for two independent random statistics is the sum of the individual pieces of information. Thus, a sample of n independent observations, each with information I(θ), has combined information

$nI(\theta)$ and Fisher demonstrated that the maximum-likelihood estimator behaves asymptotically as a normally distributed variable with
mean θ and variance $1/nI(\theta)$. Thus, if an alternative procedure leads
to an approximate normal distribution with mean θ and variance $1/nI^*(\theta)$,
it would make sense to regard the alternative as having the relative
efficiency $I^*(\theta)/I(\theta)$, for this is the ratio of sample sizes required
to achieve comparable results with the two procedures.

The present understanding of the mathematics underlying the asymptotic results took some time to develop, for the situation was more del-
icate than was originally appreciated. Normality was not crucial, but
the consistency condition required a somewhat unanticipated strength-
ening. First, Cramér (12) and Rao (29) essentially established the
bound $1/nI(\theta)$ on performance of an arbitrary regular estimator requiring conditions of regularity not only on the probability distributions
but also on the estimators. Subsequently, Hodges produced examples of
superefficiency. The simplest example, a slightly modified form of
the $T^* = 1/3$ estimate, consists of simply guessing when T is sufficiently close to $1/3$, sufficiently close depending on n, and using T otherwise. Then this hybrid estimate does as well as T for all other values
of θ and better than T for $\theta = 1/3$. This phenomenon can be extended,
and partially reflects how asymptotic results may be moderately misleading. In his dissertation, Le Cam (21) demonstrated that the superefficiency property is limited to θ sets of measure zero. Meanwhile,
C. Stein exploited the methods of Chapman and Robbins (5), who derived
Cramér-Rao bounds without conditions of regularity on the estimator,
to show that basically these bounds cannot be improved upon everywhere
in any interval. Stein's unpublished result and an independently derived variant by H. Rubin are referred to in ref. 8.

In summary, the maximum-likelihood estimator achieves the asymptotic bound on the variance given by the information $I(\theta)$, and no
estimator can do better in any interval. Thus, maximum-likelihood
estimates are optimal, and information provides a useful measure of
what is attainable. Incidentally, the maximum-likelihood estimates
are not unique in achieving the Cramér-Rao bounds asymptotically.
Neyman (22) introduced the term BAN (best asymptotically normal) to

describe other classes of asymptotically normal estimators that achieve the Cramér-Rao bounds asymptotically.

OPTIMAL DESIGN IN ESTIMATION--ELFVING'S THEOREM

Optimal design in estimation developed along two parallel tracks. Much interest has been centered on the extension of Elfving's fundamental result in regression (13).

Suppose an observation y of a dependent variable depends on a pre-selected value of x according to the linear relation $y = \alpha + \beta x$, except for a random unobserved additive error term u with mean 0 and constant variance. Suppose that x may be selected in an interval, say, $-1 \leq x \leq 1$, and that an efficient estimate of the slope β is desired on the basis of n observations. The experimenter may select n values $x_1, x_2, \ldots, x_n$. Here each x_i, combined with the observation of the corresponding y_i, constitutes an *elementary experiment*, and the collection $(x_1, \ldots, x_n)$ describes an *experimental design*. A little introspection makes it clear that one should divide the available x_i into two groups--half at $x_i = +1$ and half at $x_i = -1$.

This problem is the most transparent of a class that was described and solved elegantly by Elfving. We present the two-dimensional version as follows. Let

$$y_i = \beta_1 x_{1i} + \beta_2 x_{2i} + u_i \qquad\qquad i = 1, 2, \ldots, n$$

where the u_i are unobserved errors or disturbances with mean 0 and constant variance. The experimenter does not know the parameters β_1 and β_2 and wishes to estimate a linear function of these

$$\theta = a_1 \beta_1 + a_2 \beta_2$$

where a_1 and a_2 are specified. He is permitted to select n points (x_{1i}, x_{2i}) from some specified set S and to observe the corresponding values y_i. How might this be achieved?

In the straight-line example mentioned above, $x_{1i} = 1$, $x_{2i} = x$, S is the line segment from $(1,-1)$ to $(1,1)$ and $(a_1, a_2) = (0,1)$. Each point (x_{1i}, x_{2i}) determines an elementary experiment or an experimental *level,* and the collection of n such points or levels is an experimental design.

The elegant Elfving solution consists of constructing the smallest convex set S* containing S and its reflection about the origin -S. A ray is then drawn from the origin through the point (a_1, a_2). The point z where that ray penetrates S* represents the solution in the following sense. If z is a point of S or -S, it corresponds to a single (x_{1i}, x_{2i}); repeating that elementary experiment n times yields an optimal design. If z is not a point of S or -S, it is on a line segment that connects two such points. An optimal design may be constructed by repeating the experimental levels corresponding to those two points in inverse proportion to their distances to z. One consequence of this result is that there is an optimal design requiring, at most, two of the available elementary experiments to be repeated in suitable proportions.

Elfving also demonstrated that the standard deviation of the *least-squares* estimate of θ is inversely proportional to the distance from z to the origin. This result is relevant to the evaluation of suboptimal designs.

The Elfving result was generalized by Kiefer and Wolfowitz (19). Suppose that $y_i = \beta_1 x_{1i} + \cdots + \beta_k x_{ki} + u_i$ and we wish to estimate r linear functions of these k unknowns $\beta_1, \ldots, \beta_k$, $1 \leq r \leq k$. When r = 1, the Elfving result applies. If r > 1, there is a question of what constitutes a measure of the goodness of the design. Kiefer and Wolfowitz study designs that minimize the generalized variance. This is the determinant of the covariance matrix of the r linear functions estimated. The minimization of the generalized variance strikes me as an arbitrary criterion with little to recommend it in general. However, when r = k, it is equivalent to a minimax criterion that is quite meaningful. In that case, the criterion selects the design that minimizes the largest variance of the estimate of $\beta_1 x_1 + \cdots + \beta_k x_k$ over all choices of $(x_1, x_2, \ldots, x_k)$ available to the experimenter. Thus, when r = k, the criterion is relevant for problems involving interpolation.

OPTIMAL DESIGN IN ESTIMATION--LARGE SAMPLE THEORY

An alternative approach, which overlaps Elfving's, applies Fisher's
information. It is not limited to regression experiments as was
Elfving's, but requires large sample theory to the extent that the
relevance of Fisher's information requires it.

As an example, consider the dose-response model where the probability of nonresponse, $1 - p_d$, is exponential in the dose d, i.e.,
$p_d = 1 - e^{-\theta d}$. The information for estimating θ is maximized when
the dose is set such that the probability of response is 0.8, that is,
$d = 1.6/\theta$. It follows that an asymptotically optimal design for selecting dose levels for n subjects consists of applying the above dose
level to each subject. Unfortunately, this dose depends on θ, which
is unknown. Nevertheless, if it is approximately known, the optimal
(although impractical) design may be approximated by selecting all of
the dose levels at or near an estimate of $1.6/\theta$. Hence, the optimal
design is called *locally optimal*.

The use of information makes it clear that as long as the elementary experiments yield data, the probability distributions of which
depend on only one unknown parameter, information considerations will
lead to optimal designs that require, at most, one elementary experiment to be repeated n times.

Suppose now that the outcomes of the available experimental levels
depend on two or more parameters. For example, in the probit model,
the probability of response to a dose level d follows the cumulative
normal distribution with unknown mean μ amd variance σ^2. Suppose we
want to estimate $\mu - 1.96\sigma$, a quantity below which the probability
of response is only 0.05 (a "safe" dose level). Here Fisher's information must be replaced by an information matrix, and the efficiency
of a design corresponds to a function of the inverse of the information matrix.

Chernoff (7) demonstrated that if there are k unknown parameters,
the locally optimal design for estimating a function of these parameters involves at most k of the elementary experiments. This result
extends to the case in which one is interested in several of the k

parameters. Moreover, if each elementary experiment has an outcome the distribution of which depends on only one parameter, the problem of selecting a locally optimal design may be shown to be equivalent to a related regression design problem. The probit model falls in this category.

The mathematics of selecting an optimal design and the interpretation of this and competing designs is ordinarily facilitated largely by the knowledge that one may restrict oneself to k elementary experiments if k is relatively small.

TESTING HYPOTHESES

Fundamental difficulties in understanding hypothesis testing were finally resolved by Neyman and Pearson (24,25) when they introduced the notion of alternative hypotheses and two types of error. The Neyman-Pearson theory depends upon a compromise between the probability α of rejecting a true hypothesis and β, that of accepting a false hypothesis.

The original development did not seem to require much help from large sample theory. The importance of the likelihood-ratio statistic was clear and informative. Some asymptotic methods were used in deriving "type A" tests.

We describe a test of type A. Suppose the hypothesis H_0 specifies $\theta = \theta_0$. The *power function* $\alpha(\theta)$ of a test is the probability that the test leads to the rejection of H_0 as a function of θ. It is desirable to have $\alpha(\theta)$ large for $\theta \neq \theta_0$. A test is categorized as type A if it maximizes $\alpha''(\theta_0)$ subject to $\alpha(\theta_0) = \alpha$ for some specified value of α, and $\alpha'(\theta_0) = 0$. This definition defines a procedure with a power function that rises most steeply in the neighborhood of $\theta = \theta_0$ subject to the restriction that it attain the specified local minimum value of α at $\theta = \theta_0$. This definition, which applies to a local property, permits the use of limiting (asymptotic) theory but does not require large samples. Therefore, at first glance, large sample theory does not seem to offer much for hypothesis testing.

However, deeper analyses raised by substantial theoretical and applied problems ultimately indicated a powerful range of applications

of large sample theory. Wilks (32) derived an asymptotic property of
likelihood-ratio tests, which showed that for the test that the hypothesis that θ lies on a hyperplane, $-2 \log \lambda$ has asymptotically a χ^2
distribution as the sample size approaches ∞, where λ is the likelihood-ratio statistic. Wald (31) presented a highly detailed and elaborate description of the large sample optimality properties of likelihood-ratio statistics. Neyman (23) introduced an alternative family
of tests, C_α tests, designed to be asymptotically optimal and easily
derived.

Fundamental for design considerations was the contribution of
Pitman (26,28) who was interested in comparing the relative efficiency
of alternative nonparametric tests. These tests are frequently very
simple to apply and have the added advantage that they are insensitive
to underlying assumptions.

Consider a test of $H_0: \theta = \theta_0$, which consists of rejecting H_0 if
T is sufficiently large. We are concerned with two error probabilities. The probability of rejecting H_0 when it is true is $\alpha(\theta_0)$ and
can be set equal to a specified level α by setting the threshold of
rejection (on T) appropriately. For $\theta \neq \theta_0$, the error probability is
$1 - \alpha(\theta)$, the probability of accepting the hypothesis when it is false.
For reasonable tests, if $\theta \neq \theta_0$, $\alpha(\theta) \to 1$ as the sample size approaches
infinity. If we seek that value θ_n near θ_0 for which the error probability assumes a specified value β, then $\theta_n - \theta_0$ will ordinarily approach zero at a rate determined by T, α, and β.

Ordinarily, $n(\theta_n - \theta_0)^2$ converges to a limit a. Suppose the same
procedure applied to an alternative statistic T* gives $n^*(\theta_n^* - \theta_0)^2 \to$
a*. It follows that these two test procedures will give equivalent
error probabilities α and β to the same alternatives near θ_0 if the
ratio of the sample sizes is given by

$$\frac{n}{n^*} = \frac{a}{a^*}$$

and hence a/a* is called the asymptotic relative efficiency of the
test based on T* compared to that based on T.

This definition would lose considerable impact were the relative efficiency to depend on α and β. Ordinarily, T and T* are asymptotically normally distributed; it is easy to show, given the asymptotic distribution, that the relative efficiency depends in a simple way on the means and variances of the asymptotic distributions and *not* on the particular choice of α and β.

This measure of asymptotic efficiency, known as Pitman efficiency, can be used to show that the simple Mann-Whitney or Wilcoxon test used to compare the equality of two populations has relative efficiency of $3/\pi = 0.96$ compared to the optimal t test when the data are normally distributed. Chernoff and Savage (10) established the remarkable conjecture of Hodges and Lehmann (16) that another test, the normal-scores test for equality of distributions, has asymptotic relative efficiency of 1 compared to the t test if the distribution of the data is normal and efficiency greater than 1 otherwise.

We return now to the simplest case of hypothesis testing. The best test of a simple hypothesis $H_0: \theta = \theta_0$ as compared to a simple alternative $H_1: \theta = \theta_1$ is the likelihood-ratio test. There seems to be little scope for large sample theory here because, as $n \to \infty$, we can make both error probabilities α and β approach zero. However, the rate at which these error probabilities approach zero requires a theorem of Cramér (11), which extends the central limit theorem. Chernoff (6) demonstrated that the error probabilities approach zero at an exponential rate in n and that this rate could be used to measure certain essential relative efficiencies and information numbers. These numbers can be applied to compare various test procedures based on sums of random variables, likelihood-ratio tests, and different experimental designs. This work gave rise to a rich theory known as *large deviation* theory, as the threshold for accepting or rejecting a hypothesis corresponds to a large number of standard deviations of the statistic from its mean.

We shall discuss this theory and a few of its consequences with regard to the comparison of experiments in testing hypotheses in the

following section. However, we shall digress briefly to mention an
approach of Hoeffding (17), which involves comparison of the likelihood-ratio and χ^2 test statistics for testing $H_0: p = p_0$, where $p =$
$(p_1, p_2, \ldots, p_r)$ and p_j is the probability that the outcome of a
trial falls in the jth of r cells, and p_0 is a specified value of p.

Both of these values are asymptotically optimal in the Pitman
sense. Nevertheless, if we study values of p whose distance from p_0
exceeds $1/\sqrt{n}$ by an order of magnitude, the error probabilities for the
χ^2 test may exceed those of the likelihood-ratio test substantially,
but cannot be substantially less. The proof of this result is related
to the analysis of large deviation theory for multinomial distributions.

COMPARISON OF EXPERIMENTS IN TESTING HYPOTHESES

The decision theory formulation of statistics gave rise to an interesting theory of comparison of experiments. Two experiments are comparable and one is more informative than a second, if for every possible loss function, any risk function attainable with the second
experiment is also attainable with the first. This basis of comparison was suggested by Halmos and Savage (15) in a paper on sufficiency,
which was later developed and exploited by Blackwell (4). Unfortunately, the conditions under which one experiment is comparable with, i.e.,
more or less informative than, another experiment are so stringent that
one rarely finds situations in which two different experimental setups
are comparable by these standards. Rather, Blackwell used this theory
to study a few important special cases in addition to the statistical
concept of sufficient statistics.

In contrast, the measures of information and relative efficiency
for testing that we have discussed provide a means of comparing any
two experiments in a meaningful way when sample sizes are large.

This may be illustrated with an example. In a quality-control
application, the observations X_1, X_2, $\ldots$ are normally distributed
with mean θ_0 and variance 1. If the process goes out of control in
the usual fashion, the data are normally distributed with mean $\theta_1 \neq$
θ_0 and variance 1. The process of measuring the individual X_i is

prohibitively expensive. It is preferred to use a mechanical device (sieve) that efficiently determines if $X_i < a$, $a \leq X_i \leq b$, or if $X_i > b$. One may select a and b at one's convenience. This choice of a and b is at the disposal of the experimental designer. How should they be chosen?

If one were concerned about values of θ_1 close to θ_0 so that error probabilities were bound to be substantial, even with a large sample size, the Pitman efficiency considerations would be relevant. I shall not elaborate on this case.

Suppose now that loss of control implies that θ shifts from θ_0 to some specified θ_1 such that both error probabilities can be made small with the available sample size. Suppose we want to minimize $\alpha + k\beta$ for some constant k. Large deviation theory tells us that both α and β can be made to approach zero at roughly the exponential rate $\exp(-nI)$, where I is the Chernoff Information (6)

$$I = -\log \left[\inf_{0 \leq t \leq 1} \int f_0^t(x) \; f_1^{1-t}(x) \; dx \right]$$

and f_0 and f_1 are the distribution functions under the two alternative hypotheses. Hence, an optimal design consists of selecting a and b to maximize 1. Note that I does not depend on the particular value of k.

A variation of this problem occurs when the incorrect acceptance of H_0 leads to great danger, and β is to be made as small as possible, keeping α fixed at about 0.05. Then the β for the appropriate likelihood-ratio test approaches zero at roughly the exponential rate $\exp(-nI_{01})$, where I_{01}, the Kullback-Leibler Information (20) number for discriminating between H_0 and H_1 when H_0 is true, is given by

$$I_{01} = \int \log \; [f_0(x)/f_1(x)] \; f_0(x) \; dx$$

In this application, we select a and b to maximize I_{01}.

The concept of Bahadur Efficiency (2) is used to measure the rate at which the observed significance level of the test approaches zero when an alternative θ_1 is true, far from θ_0, even though the sample size is very large.

The remarkable thing about all of these measures is the extent
to which the measure that large sample theory makes relevant seems to
be independent of special cost factors of the problem. They depend
mainly on the relevant distributions. This fact seems less remarkable
in the light of Bayesian Theory, which was unpopular during the time
that many of these results were being derived.

SEQUENTIAL DESIGN OF EXPERIMENTS

The concept of sequential analysis is fundamentally related to that
of design. In sequential analysis, one decides whether or not to continue taking observations after each observation. But this is essentially a choice between the trivial noninformative and costless experiment that does not involve observation and the costly experiment that
involves another observation. It is natural to ask how to proceed when
a set of elementary experiments is available at each stage of sampling.
This is the problem of the sequential design of experiments.

Here again large sample theory plays an important role in providing meaningful answers. Large sample theory corresponds to the case
in which the cost per observation becomes small, which encourages observation of a large sample. Chernoff demonstrated the role of a
natural extension of the Kullback-Leibler Information number (9).

Suppose we wish to test H_0 that θ belongs to some set ω_0 as opposed to the alternative H_1 that θ is in a set ω_1. The following
describes, in part, an asymptotically optimal procedure. After the
nth observation, one decides whether or not to continue sampling on
the basis of the likelihood ratio. If one decides to continue and the
maximum likelihood estimate $\hat{\theta}$ of θ is in ω_0, say, then measure the
Kullback-Leibler Information number for $I(\hat{\theta},\phi,e)$ for each ϕ in ω_1 and
each randomized mixture e of the available experiments. Select e
to maximize

$$I(\hat{\theta}) = \inf_{\phi \varepsilon \omega_1} I(\hat{\theta},\phi,e)$$

This is the next experiment.

Although the problem of sequential design of experiments was posed in a Bayesian, decision-theoretic framework, the large sample results show that asymptotically, the optimal procedure and optimal risks are insensitive to the prior probabilities and to the costs of making the wrong decision. Note that in sequential analysis where the sample size is random, large sample theory corresponds to the case in which the cost of sampling approaches zero.

Chernoff's results were generalized by Albert (1), Bessler (3), Schwarz (30), and Kiefer and Sacks (18).

SUMMARY

We have outlined a few of the major results in large sample theory, indicating how large sample methods contributed simplified insights and allowed the possibility of defining optimal procedures and measures of efficiency and information. These measures, in turn, made it possible to develop methods of optimal design of experiments.

ACKNOWLEDGMENT

This work was supported in part by the Office of Naval Research under contract N00014-67-0112-0078 (NR-042993).

REFERENCES

1. A. E. Albert, The sequential design of experiments for infinitely many states of nature. *Ann. Math. Statist.* *32*, 774-799 (1961).

2. R. R. Bahadur, Stochastic comparison of tests. *Ann. Math. Statist.* *31*, 276-295 (1960).

3. S. Bessler, Theory and application of the sequential design of experiments, k-actions and infinitely many experiments: Part I-theory. Part II-applications. Stanford University Technical Reports 55 and 56 (1960).

4. D. Blackwell, Comparison of experiments. *Proc. 2nd Berkeley Symp. Math. Statist. Prob.*, 93-102 (1951).

5. D. G. Chapman and H. Robbins, Minimum variance estimation without regularity assumptions. *Ann. Math. Statist.* *22*, 581-586 (1951).

6. H. Chernoff, A measure of asymptotic efficiency for tests of a hypothesis based on the sum of observations. *Ann. Math. Statist.* *23*, 493-507 (1952).

7. H. Chernoff, Locally optimal designs for estimating parameters. *Ann. Math. Statist. 24*, 586-602 (1953).

8. H. Chernoff, Large sample theory: parametric case. *Ann. Math. Statist. 27*, 1-22 (1956).

9. H. Chernoff, Sequential design of experiments. *Ann. Math. Statist. 30*, 755-770 (1959).

10. H. Chernoff and I. R. Savage, Asymptotic normality and efficiency of certain non-parametric test statistics. *Ann. Math. Statist. 29*, 972-994 (1958).

11. H. Cramér, Sur un nouveau théorème-limite de la théorie des probabilités. *Actualités Sci. Ind.*, 736 (1938).

12. H. Cramér, A contribution to the theory of statistical estimation. *Skand. Aktuar. 29*, 85-94 (1946).

13. G. Elfving, Optimum allocation in linear regression theory. *Ann. Math. Statist. 23*, 255-262 (1952).

14. R. A. Fisher, On the mathematical foundations of theoretical statistics. *Phil. Trans. Roy. Soc. (London) Ser. A 222*, 309-368 (1922).

15. P. R. Halmos and L. J. Savage, Applications of the Radon-Nikodym theorem to the theory of sufficient statistics. *Ann. Math. Statist. 20*, 225-241 (1949).

16. J. L. Hodges and E. L. Lehmann, The efficiency of some nonparametric competitors of the t-test. *Ann. Math. Statist. 27*, 324-335 (1956).

17. W. Hoeffding, Asymptotically optimal tests for multinomial distributions. *Ann. Math. Statist. 36*, 369-401 (1965).

18. J. Kiefer and J. Sacks, Asymptotically optimum sequential inference and design. *Ann. Math. Statist. 34*, 705-750 (1963).

19. J. Kiefer and J. Wolfowitz, Optimum designs in regression problems. *Ann. Math. Statist. 30*, 271-294 (1959).

20. S. Kullback and R. A. Leibler, On information and sufficiency. *Ann. Math. Statist. 22*, 79-86 (1951).

21. L. Le Cam, On some asymptotic properties of maximum likelihood estimates and related Bayes' estimates. *Univ. California Publ. Stat. 1*, 277-330 (1953).

22. J. Neyman, Contribution to the theory of the χ^2 test. *Proc. 1st Berkeley Symp. Math. Statist. Prob.*, 239 (1949).

23. J. Neyman, Optimal asymptotic tests of composite statistical hypotheses. In *Probability and Statistics, The Harald Cramér Volume*, (U. Grenander, ed.), pp. 213-234, Wiley, New York, 1959.

24. J. Neyman and E. S. Pearson, On the use and interpretation of certain test criteria for purposes of statistical inference. *Biometrika 20A*, Part I:175-240; Part II:263-294 (1928).

25. J. Neyman and E. S. Pearson, On the problem of the most efficient tests of statistical hypotheses. *Phil. Trans. Roy. Soc. London Ser. A 231*, 289-337 (1933).

26. G. Noether, On a theorem of Pitman. *Ann. Math. Statist. 26*, 64-68 (1955).

27. K. Pearson, On the criterion that a given system of deviations from the probable in the case of a correlated system of variables is such that it can reasonably be supposed to have arisen in random sampling. *Philos. Mag. 50* (5), 157-175 (1900).

28. E. J. G. Pitman, Non-parametric statistical inference (mimeographed lecture notes). University of North Carolina Institute of Statistics, Chapel Hill, N. C. (1948).

29. C. R. Rao, Information and the accuracy attainable in the estimation of statistical parameters. *Bull. Calcutta Math. Soc. 37*, 81-91 (1945).

30. G. Schwarz, Asymptotic shapes of Bayes sequential testing regions. *Ann. Math. Statist. 33*, 224-236 (1962).

31. A. Wald, Tests of statistical hypotheses concerning several parameters when the number of observations is large. *Trans. Am. Math. Soc. 54*, 426-482 (1943).

32. S. S. Wilks, The large sample distribution of the likelihood ratio for testing composite hypotheses. *Ann. Math. Statist. 9*, 60-62 (1938).

DEALING WITH MANY
PROBLEMS SIMULTANEOUSLY

CHARLES ANTONIAK and *BRADLEY EFRON* graduated together from the
California Institute of Technology in 1960. They then went
their separate ways.

   Efron went to the Statistics Department, Stanford University where he received his M.S. degree in 1962 and his Ph.D.
degree in 1964, and where he is now Professor of Statistics and
Biostatistics. He is a Fellow of the American Statistical Association and the Institute of Mathematical Statistics. He was
Theory and Methods Editor of the *Journal of the American Statistical Association* from 1969 to 1972 and has published over 25
papers.

   Antoniak went to work as a Research Physicist in the
Solid State Physics Research Lab at the U.S. Navy Electronics
Center in San Diego and simultaneously entered the graduate
Mathematics Department at San Diego State University where he
received his M.S. degree in 1963. Antoniak continued his studies
at the University of California at Los Angeles where he received
his Ph.D. degree in 1969. Since 1970 Antoniak has been with the
Department of Statistics at the University of California at
Berkeley. He is a member of the Institute of Mathematical Sta-
tistics, the American Statistical Association, the Mathematical
Association of America, and the Institute of Electrical and
Electronic Engineers. He has published several papers in technical journals.

DEALING WITH MANY
PROBLEMS SIMULTANEOUSLY

Charles E. Antoniak

Department of Statistics
University of California
Berkeley, California

Bradley Efron

Department of Statistics
Stanford University
Stanford, California

We shall discuss just a small bit of statistical history, and very recent history at that. The two protagonists, Herbert Robbins and Charles Stein, are still among the most active of contemporary statisticians, so it may be somewhat insulting to call it history at all. However, the results are too important to leave out of any serious discussion of history of statistics and probability, so history it will have to be. Indeed, in some ways, which we hope will be made clear, we believe these results to be the finest fruit from the tree of mathematical statistics, and to have the greatest implications for the future of this subject, both in its theoretical and applied manifestations.

Let us begin with death from horse kicks in Prussia. There are good theoretical reasons for believing that the number of Prussian soldiers dying from horse kicks in the year 1870 is the realization of a Poisson random process. If $x = 6$ such deaths were observed in 1870, it seems reasonable to estimate the true Poisson occurrence rate θ by $\hat{\theta} = 6$ deaths per year. Therefore, a reasonable prediction for the number of such deaths in 1871 on the basis of the 1870 data would be 6. As a matter of fact, it seems hard to imagine using any estimator for θ other than x in this situation. Classical statistical theory supports this impression: x is the maximum likelihood estimator of θ; it is minimum variance unbiased, and so on. Any elementary statistics student could vouch for its good character.

Now let us expand the scenario. Suppose we are given the horse-kick death data for 1870 for every major country in the world, say, x_1 for France, x_2 for Germany, x_3 for Austria, and so forth, up to x_n, where we have numbered the armies 1, 2, 3, ..., n in some arbitrary way. These are statistically independent random numbers, each estimating the corresponding Poisson occurence rate θ_i in army$_i$. There is no reason to suppose that the θ_i are all the same. Actually, there are many good reasons for assuming that they are different, for the various armies are of different sizes, have different horse populations, different military traditions, and so forth.

Surely were we to estimate all of the Poisson parameters θ_1, θ_2, θ_3, ..., θ_n under these conditions, could we do any better than to estimate each θ by its own best estimator x? Yes! Let N(x) be the number of the x_i equal to some particular value x. Robbins' fundamental paper of 1955 (7) shows that if our goal is to minimize the expected total squared error of our estimators, $\sum_{i=1}^{n}(\hat{\theta}_i - \theta_i)^2$, we would do better, if n is very large, to estimate θ_i by $(x_i + 1)[N(x_i + 1)/N(x_i)]$ instead of $\hat{\theta}_i = x_i$.

Perhaps it would help to give an intuitive explanation of Robbins' estimator. If we use $\hat{\theta}_i = x_i$ as an estimator, our estimate will always be an integer, even if the true value of θ is, say, 6.5. In fact, if the true value of θ is anywhere between 6 and 7, then $x_i = 6$ is the most likely observation, so the process will have done the best it can in informing us about the true value of θ_i with only one observation. If $\theta_i = 7$, then $x_i = 6$ and $x_i = 7$ are equally likely. The effect of Robbins estimator is to compare the ratio of the number of other observations where $x_j = 7$ with the number where $x_j = 6$. If these turn out to be equal, that is evidence that $\theta_i = 7$, even though $x_i = 6$; if these numbers are in the ratio of 6 to 7, this is exactly what we would expect if θ_i = 6, and the Robbins estimator would yield $\hat{\theta}_i$ = 6. Extending this idea, we would make similar modifications on the estimates of θ for each of the other countries. Thus, we find that in dealing with many "similar" problems simultaneously, we may allow our estimate for a parameter to be influenced by observations on other independent random variables.

Stein's result is even more striking because it seems to say that the estimates we are concerned with can be completely unrelated physically. The only "similarity" that is required is that the errors in measurement have the same Gaussian distribution. We begin with some fairly innocent-looking assumptions. Let X_i be a standard Gaussian (bell-shaped) random number with unknown mean θ_i and variance 1. Suppose we have several such X_i, independent of each other, each with its own mean θ_i. For estimating any one θ_i, the obvious estimator $\hat{\theta}_i = X_i$ is best under several classical criteria. In 1955, Stein (8) showed that in terms of expected total squared error, $\sum_{i=1}^{n}(\hat{\theta}_i - \theta_i)^2$, there is a better set of estimators when n is sufficiently large. Working with James in 1961 (3), Stein gave a more explicit result: the estimator $\hat{\theta}_i = X_i[1 - (n - 2)/\Sigma X_i^2]$ has uniformly smaller expected total squared errors than $\hat{\theta}_i = X_i$ for n equal to or greater than 3!

We can see a similarity between Stein's estimator and that of Robbins. In each, the "natural" estimate is being modified by the observations on other, independent random variables. The intuition behind the Stein estimator is the recognition that there are two factors that will tend to produce dispersion in the X observations--the inherent error or noise characterized by the variance σ^2, and the dispersion of the supposedly different values of the θ_i. If, upon examining these observations, we see only the dispersion we would expect from σ^2, this is evidence that the unseen θ_i's must all be very close together. The effect of the Stein estimator is to shrink the estimates of the θ_i together toward zero by a factor that is proportional to the amount by which the observed sample variance can be explained by the known "noise" component σ^2. The Stein estimator is a kind of transition toward the extreme case in which you know that each X_i is an observation on the same normal random variable. In this case, you would use for your estimator of the mean θ_i of each X_i the value $\hat{\theta}_i = \bar{X}$. Thus, each estimate θ_i is being affected by other *independent* observations from the same distribution. The observations X_i are stochastically independent, but they are related by the *fact* that they all have the same mean. It is this fact that induces us to make estimates in which each observation affects "all" the estimates.

Note that if instead of n observations X_1, X_2, ..., X_n, all from the same normal distribution with mean θ and variance 1, we say X_1 is from $N(\theta_1, 1)$, $X_2 \in N(\theta_2, 1)$, ..., $X_n \in N(\theta_n, 1)$, and that $\theta_1 = \theta_2 = \cdots = \theta_n$, this second model is mathematically identical to the first. Now, if we do not know for a fact that all of the observations are from the same normal distribution, but we have a very strong suspicion that this is the case, then it is not unreasonable to let our estimate of θ_i be affected by the "independent" but related observation X_j. In the case of the Stein estimator, this suspicion that all the means are the same is created by the absence of the variation among the observations that we would expect were the values of θ_i to be dispersed. For example, suppose, in the Stein situation, that we have n = 100, and that when we plot the 100 x_i values the histogram looks very much like a Gaussian curve with mean 0 and variance 1. That is, about 16% of the x_i are greater than 1, about 16% of the x_i are less than -1, etc. This gives strong evidence that most of the θ_i are quite near 0. Observing $x_1 = 0.92$, it is naive to estimate $\theta_1 = 0.92$ as does the maximum likelihood estimator. Stein's estimator will estimate θ_1 by (approximately) 0 in this case, recognizing that most of the variation of the x_i values from the central value 0 must be due to the Gaussian noise, and not to variation in the θ_i. In contrast, if the x_i values are not clustered near 0, Stein's estimator is nearly the same as the maximum likelihood estimator.

Lest you imagine Stein's scenario to be the product of perverse mathematical machinations, let us add that it is very easy to extend it to include what is customarily called "the linear model." This

includes all the familiar regression and analysis of variance models.
So any time you fit three or more parameters by least squares, remember that Stein's method will produce a different set of estimators
with guaranteed better estimation properties.   The difference is not
trivial.   Efron and Morris (2) give three examples in which the improvement runs from 25% to 75%.   Nor does the technique of dealing
with many problems simultaneously apply only to estimation.   In fact,
as early as 1950 Robbins (6) had proposed an equally controversial
approach toward simultaneous testing of many hypotheses, having
numerically illustrated the possible improvement in the expected
total frequency of errors.

These results seem paradoxical [so much so, that statistical literature is spotted with attacks on Stein's method as some sort of
mathematical trick, not the kind of thing a serious statistician
should dabble in--see discussions following Lindley and Smith (4)
and Efron and Morris (1), (4.1)].   What we have called the parameters
θ_i need not have any obvious connection with each other. θ_1 might
be the speed of light, θ_2 the tensile strength of bone, θ_3 the crop
yield for winter wheat in Montana, and so forth.   How can x_2 have
anything interesting to say about θ_1, under these circumstances?
Stein's method says it does, and so does Robbins'.   Briefly, much
too briefly, let us indicate the resolution of the paradox.

First, we must look closely at the estimator and note that it
will not differ appreciably from the "natural" estimator unless all
the θ_i are "close" together compared to σ^2.   One way in which a set
of observations could have this property in actual practice is if
all are observations on the same variable for a homogeneous group,
such as baseball players' batting averages.   However, if the effect
is to occur with "unrelated" observations, then the statistician
should arrange his measurement units and location so that all observations can be expected to have the same mean θ, say, $\theta = 0$, and the
same variance $\sigma^2 = 1$.   He could accomplish this, for example, by
first estimating the values he expects to observe for the speed of
light, the yield of winter wheat, and the tensile strength of bone,

and by then "observing" the standardized difference between his
estimates and the measured values, in each case dividing the observed
difference by the standard deviation appropriate for that measurement.
If the set of observations are all close to their original estimates,
the Stein estimator will move the estimate for each θ_i toward the
prior estimated values, in a kind of tribute to the statistician's
skillful choice of coordinates and scales. The greatest expected
improvement in estimation occurs when the statistician is able to
combine a large number of apparently unrelated problems in a way
that makes each observation a normal (0,1) random variable. If he
is successful in so organizing his work, it is only fair that the
Mathematical Muse reward him. It should also be emphasized that the
reward is his, and that it does not necessarily accrue separately
to each of his clients. Therefore, many of his clients, separately,
might be better off, in a single instance, if the statistician were
to use the naive estimate. (They can, however, reap their own rewards
by repeated patronage.)

Robbins coined the terms "empirical Bayes" and "compound Bayes"
to describe his methods. By this he meant that certain properties
of the set of θ_i values can be ascertained from the set of x_i values,
and that these collective properties can then be used to improve the
estimation of the individual θ_i. Both Robbins' and Stein's estimators
can be described as standard Bayes estimators when the exact form of
the estimator has been derived from the data. (It is one of the
ironies of the subject that these methods arose from the frequentist
rather than from the Bayesian school of statistical thought.)

We recognize that there has been very little historical reference
in these remarks. Unfortunately, this is because up to the last few
years the empirical Bayes breakthrough, as Neyman (5) justly characterized it in 1962, had surprisingly little impact on theoretical
statistics, and almost none at all on applications. Judging from a
recent flurry of papers, a few of which are referenced in our bibliography, the draught is over and the history of this subject should
now unfold along several lines: (1) a reassessment of older statistical

criteria such as unbiasedness and invariance; (2) the development of a subtler and more powerful statistical theory better able to deal with large data sets in complicated structures; (3) a reconciliation, at least partial, between the Bayesian and frequentist viewpoints; and (4) an expanded statistical methodology, which will hopefully allow empirical Bayes techniques to be brought to bear on applied problems.

REFERENCES

1. B. Efron and C. Morris, Combining possibly related estimation problems. (With discussion.) *J. Roy. Statist. Soc. Ser. B 35*, 379-421 (1973).

2. B. Efron and C. Morris, Data analysis using Stein's estimator and its generalization. Rand Corp. Report R-1394-OE0 (1974).

3. W. James and C. Stein, Estimation with quadratic loss. *Proc. 4th Berkeley Symp. Math. Statist. Prob. 1960 1*, 361-379 (1961).

4. D. V. Lindley and A. F. M. Smith, Bayes estimates for the linear model. *J. Roy. Statist. Soc. Ser. B 34*, 1-41 (1972).

5. J. Neyman, Two breakthroughs in the theory of statistical decision making. *Rev. Int. Statist. Inst. 30*(1) 11-27 (1962).

6. H. Robbins, Asymptotically subminimax solutions of compound statistical decision problems. *Proc. 2nd Berkeley Symp. Math. Statist. Prob. 1950 1*, 131-148 (1951).

7. H. Robbins, An empirical Bayes' approach to statistics. *Proc. 3rd Berkeley Symp. Math. Statist. Prob. 1955 1*, 157-164 (1956).

8. C. M. Stein, Inadmissibility of the usual estimator for the mean of a multivariate normal distribution. *Proc. 3rd Berkeley Symp. Math. Statist. Prob. 1955 1*, 197-206 (1956).

SOME REMARKS ON THE DEVELOPMENT
OF NONPARAMETRIC METHODS
AND ROBUST STATISTICAL INFERENCE

KJELL A. DOKSUM was born July 20, 1940, in Norway. He received his A.B. and M.S. degrees in Mathematics from San Diego State College in 1962 and 1963, respectively. He earned a Ph.D. degree from the University of California, Berkeley in 1965 and spent a year doing Postdoctoral Research at the Institut de Statistique de l'Université de Paris. He has been with the University of California, Berkeley since 1966. He is a Fellow of the Institute of Mathematical Statistics and was Associate Editor of the *Journal of the American Statistical Association* from 1969 to 1972.

11.

SOME REMARKS ON THE DEVELOPMENT
OF NONPARAMETRIC METHODS
AND ROBUST STATISTICAL INFERENCE*

Kjell A. Doksum

Department of Statistics
University of California
Berkeley, California

This paper concerns the development of some of the important aspects of nonparametric theory. It is not intended to be complete, as it excludes several important topics including robustness of normal theory tests, sequential rank tests, run tests, and the problem of ties, among others. In accordance with the wishes of the editors, the emphasis is intended to be on work in the United States, "recent" developments, brevity, and on being of general interest. Other review material is referred to in the text. For those who wish to learn more about the subject and its history, I would also like to mention Savage's (1962) bibliography of nonparametric statistics and the recent books by Hollander and Wolfe (1973) and Lehmann (1975).

The topics touched upon are: the nonparametric field; efficiency of rank tests--the two-sample case; other rank and related methods; the matched-pair problem, the k-sample problem, further problems in analysis of variance--aligned ranks, tests for independence, multivariate

---

*Dedicated to the memory of Jaroslav Hájek.

rank tests; confidence intervals and estimates based on rank statistics; robust estimation; optimality theory for nonparametric models;
permutation tests; goodness of fit; invariance; asymptotic theory; and
nonparametric Bayesian analysis.

THE NONPARAMETRIC FIELD

A nonparametric model makes relatively weaker assumptions about the
probability distribution of a population than does a parametric model.
Using a parametric model involves the assumption that this probability
distribution belongs to a family of distributions the form or shape of
which is known but for a few unknown real parameters. These parametric
families are appropriate when they can be derived from the physical
characteristics of the experiment, some example of which are the binomial, hypergeometric, and multinomial families of distributions. At
other times, parametric families can be decided on from past experience with similar observations or data. Continuous-type data such as
height, weight, volume, and so forth, are quite often assumed to be gen
erated by a member of the normal or Gaussian family of distributions;
however, this assumption often cannot be justified. Nonparametric models are appropriate when no parametric family of distributions can be
derived or determined from past experience.

    Nonparametric methods are statistical methods of analyzing data
that are designed to be applicable for nonparametric models. One of
the desirable properties of nonparametric tests is that of distribution
freeness, that is, the probability of falsely rejecting the hypothesis
is known exactly for a wide class of distributions rather than only
for a specific parametric family such as the normal family. One of
the most important developments in statistics was the derivation of
results which established that some of these nonparametric tests, namely rank tests, are often efficient when it comes to detecting alternatives to the hypothesis. We begin with a discussion of some of the
contributors to this development in the context of the two-sample problem, in which most of the important developments first took place.

EFFICIENCY OF RANK TESTS--THE TWO-SAMPLE CASE

The two-sample problem is the problem of deciding if two random samples were generated by the same probability distribution. An important application is to the question of determining whether or not a treatment has an effect on a certain characteristic of a population. In this situation, the first sample could be the responses of a control group, and the second sample could be the responses of the group receiving the treatment. Wilcoxon (1945) proposed ranking the two samples together and using the sum of the ranks of one of the groups as a test statistic. He was not the first* to propose this statistic, but he was instrumental in getting the development of nonparametric statistics on the right track, as were Mann and Whitney (1947).

Under the hypothesis that the two samples come from the same distribution, the distribution of the ranks is independent of this distribution, if it is only assumed that it is continuous. If the alternative we have in mind is that one of the two groups tends to have larger observations and thereby larger ranks than the other, we reject the hypothesis in favor of this alternative when the rank sum for this group exceeds some tabulated critical value.

Now, an important question that arises is how much is lost by using the Wilcoxon statistic when, in fact, the underlying distributions are normal. For the normal model, where both samples are normally distributed with a common variance, the optimal (uniformly most powerful unbiased) test is the two-sample t test. Thus, we can provide an answer to the above question by computing the ratio N_t/N_w of the samples sizes needed for the t and Wilcoxon tests to achieve the same preassigned probability β of detecting a given normal alternative. We do this assuming that the two tests have the same probability α of falsely rejecting the hypothesis. N_t/N_w is called the efficiency of the Wilcoxon test relative to the t test and depends on α, β and the alternative. The limit of N_t/N_w for sample sizes tending to infinity

---

*For an interesting history of the independent discoveries of this statistic, see Kruskal and Wallis (1952,1953) and Kruskal (1957).

was first obtained by Pitman (1948) in lecture notes at Columbia
University. He obtained the value $3/\pi = 0.955$, which is independent
of α, β, and the alternative considered. This result indicates a remarkable small loss in efficiency. In these notes Pitman also considered limits of ratios of required sample sizes in a more general
context and developed what today is commonly called Pitman efficiency
or asymptotic relative efficiency (ARE). This has turned out to be
a very useful tool for the comparison of nonparametric and parametric
tests. The first published proof of Pitman's result was by Mood (1954).
Noether (1955) carried Pitman's work on the concept of efficiency further. The efficiency N_t/N_w for finite sample sizes was considered by
Dixon (1954), who found that the efficiency is close to the asymptotic
value $3/\pi$.

Next, suppose that the normality assumption is dropped but that
we retain the shift model, that is, we assume that the distributions
of the two samples differ only in their means. The Pitman efficiency
of the Wilcoxon test relative to the t test can still be computed and
was obtained by Pitman (1948). Hodges and Lehmann (1956) found that
this efficiency has a lower bound of 0.864, whatever the distribution
of the samples, whereas there is no upper bound. Hence, in a situation wherein the underlying distribution is unknown and the sample
sizes are large, use of the Wilcoxon statistic never entails a serious
loss of efficiency, whereas the possible gain in efficiency is unbounded! The Wilcoxon test is preferable to the t test in terms of Pitman
efficiency for precisely the kinds of distributions for which one would
hope to increase the efficiency, namely "heavy tailed" distributions.
These are distributions likely to generate a few datum points far to
the left or right of the bulk of the data. This property of the Wilcoxon test is a consequence of the results of Hodges and Lehmann (1961)
and van Zwet (1964).

There are other rank statistics that consist of summing functions
of the ranks of one group. One such statistic is the normal scores
statistic of Hoeffding (1951) and Terry (1952). Hodges and Lehmann

(1956) conjectured that the Pitman efficiency of the normal scores test to the t test is 1 for the normal model and greater than 1 for all other shift models. This conjecture was established as Theorem 3 in the fundamental work by Chernoff and Savage (1958).

The Wilcoxon and normal scores tests were compared by Hodges and Lehmann (1961) and van Zwet (1964). They showed that the Wilcoxon test is preferable for distributions with "heavy tails" as described above.

There are numerous other rank tests for the two-sample problem, of which we mention a few of the more commonly used ones. For the alternative that the distributions of the two samples differ only in scale, a rank test that consists of summing the squared deviations of the ranks from their hypothesized expected values was introduced by Mood (1954) who also derived its Pitman efficiency. Other tests were proposed by Freund and Ansari (1957), David and Barton (1958), Siegel and Tukey (1960), and Ansari and Bradley (1960). These are all asymptotically equivalent. Their Pitman efficiencies relative to the studentized normal theory test were given by Klotz (1962).

Capon (1961) and Klotz (1962) proposed "normal scores" statistics with Pitman efficiency 1 in the normal case. The general Pitman efficiency properties of rank tests for scale are not as desirable as those of the Wilcoxon test for shift. Raghavachari (1965) showed that the Pitman efficiency of the normal scores test relative to the studentized normal theory test can take on any value between 0 and ∞.

When the two samples consist of "times until failure" of pieces of equipment, two extremely useful statistics are those of Lehmann (1953) and Savage (1956). For the case in which the distributions have increasing failure rates, desirable power properties have been established by Doksum (1967b; 1969).

OTHER RANK AND RELATED METHODS

The development of rank methods for other problems to a certain extent parallels that of the two-sample problem.

The Matched-Pair Problem

For matched-pair or paired-comparison experiments wherein subjects are
matched in pairs and one member of the pair is given a treatment while
the other serves as control, it is natural to base tests on the differences of the responses in each pair. Wilcoxon (1945) proposed computing the ranks of the positive differences among the absolute values
of all the differences and using the sum of these ranks as a test statistic. An absolute normal scores statistic based on the same ranks
was derived by Fraser (1957) and its performance for normal alternatives was given by Klotz (1963). The performance for other models was
studied by Thompson et al. (1967). The oldest and most famous statistic for this problem, which can also be used in a more general context,
is the sign statistic, that is, the number of positive differences.
The "Pitman efficiency" of the test based on this statistic relative
to the one-sample t test was already found by Cochran (1937) to equal
$\pi/2 = 0.64$ for the normal model. Its efficiency for finite sample
sizes was studied by Walsh (1946), Dixon (1953), and Hodges and
Lehmann (1956). The Pitman efficiency can take on any value between
0 and ∞.

The k-Sample Problem

For the problem of comparing k populations on the basis of k independent samples, one from each population, Kruskal and Wallis (1952,1953)
proposed to rank all of the samples together, form the average of the
ranks in each sample, and then use a weighted sum of the squared deviations of these averages from their hypothesized expected values. The
hypothesis considered is that the k populations have the same distribution. When the populations differ only in their means, the Pitman
efficiency of the Kruskal-Wallis test relative to the normal theory
F test was obtained by Andrews (1954) who showed that it is exactly
the same as the Pitman efficiency obtained for the Wilcoxon test relative to the t test. Normal scores and other tests were considered by
Puri (1964). Multiple comparison procedures based on ranks have been
developed by Steel (1960), Nemenyi (1963), and Dunn (1964). See Miller
(1966) for further work and references in this area.

Tests for whether the populations are stochastically ordered are those of Terpstra (1952), Jonckheere (1954), Chacko (1963), and Shorack (1967). For further work and references, see Barlow et al. (1972).

Further Problems in Analysis of Variance--Aligned Ranks

The contributors to rank tests for problems in analysis of variance other than the above k-sample or one-way layout problem are numerous. For the two-way layout with one observation per cell, Friedman (1937) proposed ranking the observations in each block and then essentially replacing the observations in the numerator of the normal theory F statistic with ranks. This statistic was extended to balanced incomplete blocks by Durbin (1951) and Benard and van Elteren (1953). The Pitman efficiency of the tests of Friedman and Durbin relative to the F test, as the number of observations per block tends to infinity, was obtained by van Elteren and Noether (1959). This efficiency favors the Friedman test only for distributions with very heavy tails. The Friedman test and its generalizations have been superceded by the much more efficient aligned block rank tests due to Hodges and Lehmann (1962). Their paper constitutes a major breakthrough in the problem of obtaining efficient rank tests for two and higher way layouts. In the two-way layout the idea is to align the blocks by subtracting the block mean or median from the observations in each block. Now, when *all* the aligned observations are ranked together, these aligned ranks will be conditionally distribution free when we condition on the configuration of aligned ranks in each block. A distribution-free test statistic can thus be obtained by replacing the observations in the numerator of the normal theory F statistic by the aligned ranks. The Pitman efficiency of the resulting test relative to the F test was given in a special case by Hodges and Lehmann (1962) and for a more general two-way layout by Mehra and Sarangi (1967). They found that for the normal linear model, the Pitman efficiency depends on the number k of levels of the main factor; however, the efficiency is bounded below by $3/\pi$! The values for k = 2 and 3 are $3/\pi = 0.955$ and 0.9662, respectively. Extensions to tests based on normal and more general scores were given

by Sen (1968a,b), and extensions to tests for interactions were given
by Mehra and Sen (1969). Other highly efficient tests for problems in
the two-way layout with one observation per cell are those of Hollander
(1967) and Doksum (1967a). These tests are related to rank tests but
are only asymptotically distribution free. Puri and Sen have also made
numerous contributions, mostly concerned with general scores statistics
(see, for instance, Puri (1972) for a description of this work).

Tests for Independence

For the problem of measuring the correlation between two variables on
the basis of a bivariate random sample, Spearman (1904) introduced a
rank correlation coefficient. As the name suggests, it consists of
ranking the two sets of variables separately and then computing the
correlation between the ranks. The Pitman efficiency of the correspond
ing test for independence in the bivariate normal model relative to
the test based on the usual correlation coefficient was shown by Hotelling and Pabst (1936) to equal $9/\pi^2 = 0.912$. The performance for finite sample sizes in the normal model was studied by Bhattacharyya
et al. (1970). The Pitman efficiency for general models was considered by Konijn (1956) and Gokhale (1966). A statistic closely related
to the Spearman correlation coefficient is Kendall's tau proposed by
him in 1938. For a history of the independent discoveries of this
statistic, see Kruskal (1958) and Kendall (1962). A "normal scores"
test with Pitman efficiency 1 in the bivariate normal case was developed by Fieller and Pearson (1961) and Bhuchongkul (1964). Among
other tests for independence, we mention the corner test of Olmstead
and Tukey (1947) and the quadrant test of Blomquist (1950). Unbiasedness and monotonicity of power functions for general alternatives have
been established for rank tests by Lehmann (1966) and Yanagimoto and
Okamoto (1969).

Multivariate Rank Tests

For testing that the median vector is the null vector in the bivariate
one-sample problem, bivariate sign tests have been introduced by Hodges
(1955), Blumen (1958), Bennett (1964), Chatterjee (1966), and by Puri and

Sen (1971). Joffe and Klotz (1962) tabulated the hypothesis distribution of the Hodges sign statistic, and Klotz (1964b) computed the power
of the Hodges and Blumen sign statistics for bivariate normal models.
He also computed the efficiency of the Hodges test relative to the
normal theory Hotelling T^2 and found that the efficiency ranged from
0.80 to 0.94 for bivariate normal alternatives. Rank tests for shift
in the p variate case and their Pitman efficiency relative to Hotelling's T^2 have been considered by Bickel (1965a). For the case of
two and more multivariate samples, rank tests have been introduced
by Chatterjee and Sen (1964).

CONFIDENCE INTERVALS AND ESTIMATES BASED ON RANK STATISTICS

A statistical theory and methodology that does not include confidence
intervals and point estimates of parameters would be regarded as incomplete. It is therefore important that these can be readily obtained
from rank statistics. Consider first the matched-pair experiments and
the model in which the sample variables (differences) are assumed to
be symmetrically distributed about a point θ. A confidence interval
for the point of symmetry can be obtained by considering the collection
of all θ_0 that will be accepted if we test the hypothesis that the
point of symmetry is θ_0 by means of the matched-pair Wilcoxon test. A
class of confidence intervals closely related to these was derived by
Walsh (1947,1949a,b) (with acknowledgments to Wilks and Tukey). Tukey
(1949) considered the Wilcoxon confidence intervals discussed above
and their relation to Walsh's intervals.

For the two-sample problem, Moses (1953) assumed a shift model
(see section on efficiency of rank tests) and derived a confidence
interval for the shift parameter based on the Wilcoxon statistic. The
efficiencies of these Wilcoxon-type confidence intervals were studied
by Lehmann (1963c) who showed that when compared to the normal-theory
t intervals, the efficiency is asymptotically the same as that of the
Wilcoxon test relative to the t test.

To obtain estimates of the point of symmetry θ and the shift
parameter Δ, one could use the midpoints of the above-described confidence intervals. If one lets the confidence coefficient tend to

zero, these intervals coincide with those considered by Hodges and
Lehmann (1963). In the case of θ and the Wilcoxon statistic, this
approach leads to the Hodges-Lehmann estimate $\hat{\theta}$, which is the median
of averages of two observations at a time. In the two-sample case,
the Hodges-Lehmann estimate $\hat{\Delta}$ similarly becomes the median of the differences when the observations in one group are subtracted from the
observations in the other group. Hodges and Lehmann established the
asymptotic properties of these estimates; they showed that if asymptotic efficiency of one estimate relative to another is defined as
the reciprocal of the ratio of the asymptotic variance of the first
estimate to the second, then the estimates $\hat{\theta}$ and $\hat{\Delta}$ have the same asymptotic efficiency relative to the sample mean estimates as the Wilcoxon
test has relative to the t test. The estimate $\hat{\Delta}$ was also proposed by
Sen (1963).

These estimates have been extended to the problem of estimating
contrasts in analysis of variance models by Lehmann (1963a,b;1964),
Greenberg (1966), and Spjøtvoll (1968). The properties of the Hodges-
Lehmann estimates for dependent observations have been considered by
Høyland (1968).

ROBUST ESTIMATION

Up to this point, statistical methods based on rank or closely related
statistics have been discussed. There is, however, a vast statistical
literature that addresses itself more directly to the problem of obtaining point estimates in a nonparametric context.

Several early workers in the field of estimation noticed that the
data they were handling often did not appear to have been generated by
a normal distribution. They noticed that there tended to be "outliers"
(i.e., observations far from the bulk of the data), which indicated
that a model with a heavier tail than the normal distribution would be
appropriate. For an interesting history of the early development of
robust estimation, see Stigler (1973). The more recent interest in
robust estimation was, in large part, inspired by Tukey (1960,1962).
An estimate is said to be robust, roughly speaking, if the distribution

of the estimate is insensitive to small changes in the distribution of
the population. Similarly, a parameter θ, which is a numerical characteristic of this distribution F, is called robust if small changes
in F results in small changes in θ. This latter concept was considered by Bahadur and Savage (1956). Both concepts were formalized by
Takeuchi (1967), and a detailed analysis was given by Hampel (1968,1971).

Huber (1964,1968,1972) considered the problem of estimating the
location parameter or point of symmetry of a symmetric distribution.
His fundamental work solved the problem of finding an asymptotically
optimal estimate (in the minimax sense) for a class of distributions
constituting a "neighborhood" of the normal distribution. Bickel
(1965b) showed that Huber's solution is asymptotically equivalent to
a trimmed mean, with the trimming proportion depending on F. Jaeckel
(1971b) showed that the trimming proportion can be estimated from
the data, and thereby obtained a truly asymptotically optimal estimate for all Huber-type neighborhoods of the normal distribution.

Three different classes of estimates of location have received
considerable attention: (1) those that are maximum-likelihood estimates for some strongly unimodal distribution (M estimates of Huber,
1964); (2) linear combinations of order statistics (see Stigler, 1973);
and (3) the estimates based on rank statistics (the Hodges-Lehmann
rank-based estimators discussed in the section on confidence intervals
and estimates based on rank statistics). For a given F symmetric about
θ, one can choose one member of each class which asymptotically minimizes the asymptotic variance among all translation invariant estimates.
One thereby obtains three estimates which are asymptotically optimal
and equivalent for F. A natural question is: How do these three estimates compare for some G ≠ F symmetric about θ? This question has been
answered by Scholz (1971,1974) who showed that the rank-based estimates
are always preferable to the linear combinations of order statistics,
whereas no such dominance holds between rank-based estimates and M estimates. Other relevant work is that by Gastwirth (1966).

Some recent results on the problem of estimating location in the
case of asymmetric F have been obtained by Jaeckel (1971a), Bickel and
Lehmann (1974,1975) and Doksum (1975).

Two excellent sources for further work are the paper by Huber
(1972) and the Princeton report by Andrews et al. (1972).

OPTIMALITY THEORY FOR NONPARAMETRIC MODELS

As several nonparametric methods are available for each nonparametric
model, it is desirable to formulate an optimality criterion to decide
which method makes the most efficient use of the data. One criterion
that is natural in the nonparametric context is the minimax criterion.
Consequently, for estimation problems, one tries to minimize the maximum variance, for which the maximum is taken over some nonparametric
class of plausible distributions; for testing problems, one tries to
minimize the maximum probability of falsely accepting the hypothesis,
or equivalently, maximize the minimum probability of correctly rejecting the hypothesis. For the problem of testing if the median is zero
against the alternative that it is positive, Hoeffding (1951) and Ruist
(1954) have shown that the sign test is minimax. Huber's work (1965,
1968) includes the model wherein, under the hypothesis, the distribution
F underlying the data is in the neighborhood of a normal distribution
with mean zero, whereas under the alternative it is in the neighborhood
of a normal distribution with mean greater than zero. His optimal test
consists of pulling the observations that are far away from the bulk of
the data toward the middle and of using the average of the new observations as a test statistic. This approach has the advantage that it can
also be adapted to optimality of confidence intervals and estimates
(Huber, 1964). Work relevant to this approach has since been done by
Huber and Strassen (1973).

For the two-sample problem, Doksum (1965,1966) considered the class
of alternatives in which the maximum difference between the distributions for the two groups is bounded below by a constant Δ, and he
showed that in the class of tests based on linear rank statistics (sums
of functions of ranks), the Wilcoxon test is asymptotically minimax.
Extensions have been obtained by Doksum (1966) (goodness-of-fit problem)
Bell and Doksum (1967) (independence problem), Doksum and Thompson
(1971) (matched-pair problem), Paik (1971) (independence problem),

Barlow and Doksum (1972) (testing for increasing failure rate), and Yuan (1973) (k sample trend problem). Borokov and Sycheva (1968) considered a class of Kolmogorov-Smirnov statistics with weight functions for the goodness-of-fit problem and obtained the asymptotically optimal weight function. Minimax tests for scale alternatives have been obtained by Doksum (1969). This last theory can readily be extended to confidence bounds and point estimates. D'Abrera (1973) obtained asymptotic minimax tests for the k-sample trend problem with shift parameters.

Optimality of rank tests for parametric models was considered by Hoeffding (1951), Terry (1952), Lehmann (1953), and Savage (1956), among others. Lehmann gave classes of alternatives, sometimes referred to as Lehmann alternatives, that are nonparametric in nature and for which it is possible to find a uniformly most powerful rank test.

Finally, we remark that optimality theory for some nonparametric models from the point of view of stringency has been developed by Schaafsma (1966).

PERMUTATION TESTS

There is a class of distribution-free tests that have not been considered up to this point, namely the permutation tests. They have been considered by Fisher (1935), Pitman (1937,1938), Scheffé (1943), Wald and Wolfowitz (1944), and others. Roughly, the best known of these consists of computing the values of the normal theory statistic for all possible relevant permutations of the data, and then rejecting the hypothesis if the observed value of the statistic exceeds $100(1 - \alpha)\%$ of these values. The resulting test is distribution free. Lehmann and Stein (1949) established the optimality theory for such tests within the class of distribution-free tests. The main result concerning permutation tests is due to Hoeffding (1952) who proved that the permutation tests as described above are asymptotically equivalent to their normal theory counterparts. Hence, the permutation tests do not have the efficiency advantages of rank tests. Moreover, they are at a disadvantage because of the difficulty in carrying them out except for small sample sizes.

GOODNESS-OF-FIT

For the problem of testing whether or not a sample comes from a specified distribution, Pearson (1900) proposed the χ^2 goodness-of-fit test.
Later, Cramér (1928) and von Mises (1931) suggested tests based on an
integral of the squared deviation of the hypothesized distribution function from the empirical distribution function. Kolmogorov (1933) proposed using the maximum absolute difference between these two distributions and derived the limiting distribution (when properly normalized)
of this test statistic. This statistic has the advantage that it leads
to readily computed distribution-free confidence bounds for an unknown
distribution function. Massey (1951) and Birnbaum (1952) tabulated
the distribution of the Kolmogorov statistic. Wald and Wolfowitz (1939)
and Smirnov (1939) gave a one-sided version of the Kolmogorov statistic
the distribution of which was derived by Birnbaum and Tingey (1951).

     Smirnov (1939) proposed as a two-sample statistic the maximum
absolute difference between the two empirical distribution functions.
Similar statistics for the k-sample problem have been studied by Kiefer
(1959), Birnbaum and Hall (1960), and Conover (1965).

     Multivariate Kolmogorov-Smirnov-type statistics have been investi-
gated by Kiefer and Wolfowitz (1958), Kiefer (1961), Bickel (1969),
and Straf (1972). Some of the early history of the numerous extensions
and variations of the above was given by Darling (1957).

INVARIANCE

Consider the two-sample problem with the alternative that the observations in one group are stochastically larger than the observations in
the other group. This problem is invariant under the group of continuous, strictly increasing transformations. It is known that the class
of tests invariant under this group of transformations is precisely the
class of rank tests.

     Bell (1964) showed that the class of statistics with distributions
that are invariant under the above groups of transformations is again
the class of rank statistics. Related results are due to Birnbaum and
Rubin (1954) and Bell (1960) (the goodness-of-fit problem); Bell and

Doksum (1967) (independence problem); and Berk and Bickel (1968) (general relationship between invariance and almost invariance).

ASYMPTOTIC THEORY

The distributions of rank statistics usually are so complicated that they cannot be computed except when the sample sizes are small. Therefore, approximations to these distributions are very important. Among the early work on this subject, we have already mentioned the results of Kolmogorov (1933) and Hotelling and Pabst (1936), who considered the asymptotic distributions of the Kolmogorov and Spearman statistics, respectively. Hotelling and Pabst considered not only the hypothesis distribution, but also bivariate normal alternatives.

For the two-sample case, contributions to the asymptotic theory of linear rank statistics under the hypothesis were made by Wald and Wolfowitz (1944), Noether (1949), Hoeffding (1951), Dwass (1955), Motoo (1957), among others. This work culminated with the results of Hájek (1961), who obtained necessary and sufficient conditions for asymptotic normality.

Two early papers that can be used to obtain asymptotic theory under general alternatives are those of von Mises (1947) and Hoeffding (1948). Lehmann (1951) used the results of Hoeffding to derive the asymptotic distribution of the Wilcoxon statistics under the alternative, and proved consistency of the test for general alternatives, as did van Dantzig (1951). Important contributions to the von Mises approach were made by Filippova (1962).

Chernoff and Savage (1958) established the asymptotic normality of general linear rank statistics. Their approach was to write the rank statistic as an integral which can be expanded into several terms, two of which are sums of independent variables, whereas the remainder of the terms converge appropriately to zero. Of the numerous extensions of their work, we mention Puri (1964) (k-sample problem); Bhuchongkul (1964) (independence problem); and Govindarajulu et al. (1967) (k-sample problem).

Another fruitful approach utilizes the concept of contiguity, attributable to Le Cam (1960). Hájek (1962) and Hájek and Šidák (1967) used this approach to develop a comprehensive asymptotic theory for rank statistics for general alternatives, the details of which may be found in the latter reference. Beran (1970) extended some of these results to vector parameters.

Yet another approach is based on the weak convergence of stochastic process. This approach is the one attributable to Doob (1949) and Donsker (1952) who applied it to the Kolmogorov statistic. It has been used by Bickel (1967) and Shorack (1969) to obtain the asymptotic distribution of linear combinations of order statistics, and by Pyke and Shorack (1968) to obtain the asymptotic theory of linear rank statistics. Further applications of this approach can be found in Billingsley's book (1968).

Finally, we mention the important projection approach of Hájek (1968), which consists of finding the best approximation (in the mean-square sense) to a linear rank statistic by sums of independent variables. This approach has been extended to the matched-pair case by Hušková (1970). Important results concerning the centering constants needed in this approach are those of Dupak (1970) and Hoeffding (1973).

Asymptotic expansions that yield approximations to a higher order than the normal approximation has recently been developed by Bickel and van Zwet (1974).

NONPARAMETRIC BAYESIAN ANALYSIS

Bayesian models are applicable when the distribution F underlying the data can itself be thought of as being generated by some random mechanism. In the context of nonparametric models, to make F random, we must define a probability P on a nonparametric family Ω of distributions such that the support of the probability is Ω. At the same time, P should be a plausible probability distribution for F, and it should be possible to compute posterior distributions given a sample. A breakthrough in the problem of finding such a P, a problem posed several times by Blackwell, was made by Ferguson (1973). He defined the Dirichlet prior P, the finite dimensional distributions of which are the familiar Dirichlet distributions, and in one of his many applications he obtained a statistic which is a linear function of the two-sample Wilcoxon statistic. Related work had been done earlier by Freedman (1963) and by Fabius (1964). Important extensions of Ferguson's Dirichlet prior were obtained by Antoniak (1974) and Doksum (1974). For a further review of the subject, see Ferguson (1974).

ACKNOWLEDGMENT

This paper was prepared with the partial support of National Science Foundation Grant GP-33697X1.

REFERENCES

Andrews, F. C. (1954). Asymptotic behavior of some rank tests for analysis of variance. *Ann. Math. Statist. 25*, 724-736.

Andrews, D. F., Bickel, P. J., Hampel, F. R., Huber, P. J., Rogers, W. H., and Tukey, J. W. (1972). *Robust Estimates of Location.* Princeton University Press, Princeton, N. J.

Ansari, A. R., and Bradley, R. A. (1960). Rank sum tests for dispersion. *Ann. Math. Statist. 31*, 1174-1189.

Antoniak, C. (1974). Mixtures of Dirichlet processes with applications to some nonparametric problems. *Ann. Statist. 2*, 1152-1174.

Bahadur, R. R., and Savage, L. J. (1956). The nonexistence of certain statistical procedures in nonparametric problems. *Ann. Math. Statist. 27*, 1115-1122.

Barlow, R. E., Bartholomew, D. J., Bremner, J. M., and Brunk, H. D. (1972). *Statistical Inference under Order Restrictions.* Wiley, New York.

Barlow, R. E., and Doksum, K. (1972). Isotonic tests for convex orderings. *Proc. 6th Berkeley Symp. Math. Statist. Prob. 1*, 293-323.

Bell, C. B. (1960). On the structure of distribution-free statistics. *Ann. Math. Statist. 31*, 703-709.

Bell, C. B. (1964). A characterization of multisample distribution-free statistics. *Ann. Math. Statist. 35*, 735-738.

Bell, C. B., and Doksum, K. (1967). Distribution-free tests of independence. *Ann. Math. Statist. 38*, 429-446.

Benard, A., and van Elteren, P. (1953). A generalization of the method of m rankings. *Indagationes Math. 15*, 358-369.

Bennett, B. M. (1964). A bivariate signed rank test. *J. Roy. Statist. Assoc., Ser. B 26*, 437-461.

Beran, R. J. (1970). Linear rank statistics under alternatives indexed by a vector parameter. *Ann. Math. Statist. 41*, 1896-1905.

Berk, R. H., and Bickel, P. J. (1968). On invariance and almost invariance. *Ann. Math. Statist. 39*, 1573-1576.

Bhattacharyya, G. K., Johnson, R. A., and Neave, H. R. (1970). Percentage points of non-parametric tests for independence and empirical power comparisons. *J. Amer. Statist. Assoc. 65*, 976-983.

Bhuchongkul, S. (1964). A class of nonparametric tests for independence in bivariate populations. *Ann. Math. Statist. 35*, 138-149.

Bickel, P. J. (1965a). On some asymptotically nonparametric competitors of Hotelling's T^2. *Ann. Math. Statist. 36*, 160-173.

Bickel, P. J. (1965b). On some robust estimates of location. *Ann. Math. Statist. 36*, 847-858.

Bickel, P. J. (1967). Some contributions to the theory of order statistics. *Proc. 5th Berkeley Symp. Math. Statist. Prob. 1*, 575-591.

Bickel, P. J. (1969). A distribution-free version of the Smirnov two sample test in the p-variate case. *Ann. Math. Statist. 40*, 1-23.

Bickel, P. J., and Lehmann, E. L. (1974). Measures of location and scale. *Proc. Prague Symp. Asymptotic Statist. 1*, 25-36.

Bickel, P. J., and Lehmann, E. L. (1975). Descriptive statistics for nonparametric models. *Ann. Statist. 3*, 1038-1069.

Bickel, P. J., and van Zwet, W. R. (1975). Asymptotic expansions for the power of distribution free tests in the two sample problem. *In preparation*.

Billingsley, P. (1968). *Convergence of Probability Measures*. Wiley, New York.

Birnbaum, Z. W. (1952). Numerical tabulation of the distribution of Kolmogorov's statistic for finite sample size. *J. Amer. Statist. Assoc. 47*, 425-441.

Birnbaum, Z. W., and Hall, R. A. (1960). Small sample distribution for multisample statistics of the Smirnov type. *Ann. Math. Statist. 31*, 710-720.

Birnbaum, Z. W., and Rubin, H. (1954). On distribution-free statistics. *Ann. Math. Statist. 25*, 593-598.

Birnbaum, Z. W., and Tingey, F. H. (1951). One-sided confidence contours for probability distribution functions. *Ann. Math. Statist. 22*, 592-596.

Blomquist, N. (1950). On a measure of dependence between two random variables. *Ann. Math. Statist. 21*, 593-600.

Blumen, I. (1958). A new bivariate sign test. *J. Amer. Statist. Assoc. 53*, 448-456.

Borokov, A. A., and Sycheva, N. M. (1968). On asymptotically optimal nonparametric criteria. *Theory Prob. Appl. 13*, 359-393.

Capon, J. (1961). Asymptotic efficiency of certain locally most powerful rank tests. *Ann. Math. Statist. 32,* 88-100.

Chacko, V. J. (1963). Testing homogeneity against ordered alternatives. *Ann. Math. Statist. 34,* 945-956.

Chatterjee, S. K. (1966). A bivariate sign test for location. *Ann. Math. Statist. 37,* 1771-1782.

Chatterjee, S. K., and Sen, P. K. (1964). Nonparametric tests for the bivariate two sample location problem. *Calcutta Statist. Assoc. Bull. 13,* 18-58.

Chernoff, H., and Savage, I. (1958). Asymptotic normality and efficiency of certain nonparametric test statistics. *Ann. Math. Statist. 29,* 972-994.

Cochran, W. G. (1937). The efficiencies of the binomial series test of significance of a mean and of a correlation coefficient. *J. Roy. Statist. Soc., Ser.* A 100, 69-73.

Conover, W. J. (1965). Several k sample Kolmogorov Smirnov tests. *Ann. Math. Statist. 36,* 1019-1026.

Cramér, H. (1928). On the composition of elementary errors. *Skand. Aktuartids 11,* 13-74; 141-180.

D'Abrera, H. (1973). Rank tests for ordered alternatives. Ph.D. thesis, University of California, Berkeley.

van Dantzig, D. (1951). On the consistency and the power of Wilcoxon's two sample test. *Indagationes Math. 13,* 1-8.

Darling, D. A. (1957). The Kolmogorov-Smirnov, Cramér-von Mises tests. *Ann. Math. Statist. 28,* 823-838.

David, F. N., and Barton, D. E. (1958). A test for birth order effects. *Ann. Eugen. 22,* 250-257.

Dixon, W. J. (1953). Power functions of the sign test and power efficiency for normal alternatives. *Ann. Math. Statist. 24,* 467-473.

Dixon, W. J. (1954). Power under normality of several nonparametric tests. *Ann. Math. Statist. 25,* 610-614.

Doksum, K. (1965). Asymptotically minimax distribution-free procedures. Ph.D. thesis, University of California, Berkeley.

Doksum, K. (1966). Asymptotically minimax distribution-free procedures. *Ann. Math. Statist. 37,* 619-628.

Doksum, K. (1967a). Robust procedures for some linear models with one observation per cell. *Ann. Math. Statist. 38,* 878-883.

Doksum, K. (1967b). Asymptotically optimal statistics in some models with increasing failure rate averages. *Ann. Math. Statist. 38,* 1731-1739.

Doksum, K. (1969). Minimax results for IFRA scale alternatives. *Ann. Math. Statist. 40,* 1778-1783.

Doksum, K. (1974). Tailfree and neutral random probabilities and their posterior distributions. *Ann. Prob. 2*, 183-201.

Doksum, K. (1975). Measures of location and asymmetry. *Scand. J. Statist. 2*, 11-12.

Doksum, K., and Thompson, R. (1971). Power bounds and asymptotic minimax results for one-sample rank tests. *Ann. Math. Statist. 41*, 12-34.

Donsker, M. D. (1952). Justification and extension of Doob's heuristic approach to the Kolmogorov-Smirnov theorems. *Ann. Math. Statist. 23*, 277-281.

Doob, J. L. (1949). Heuristic approach to the Kolmogorov-Smirnov theorems. *Ann. Math. Statist. 20*, 393-403.

Dunn, O. J. (1964). Multiple comparisons using rank sums. *Technometrics 6*, 241-257.

Dupak, V. (1970). A contribution to the asymptotic normality of simple linear rank statistics. In *Nonparametric Techniques in Statistical Inference, M. L. Puri (Ed.).* Cambridge University Press, Cambridge.

Durbin, J. (1951). Incomplete blocks in ranking experiments. *Brit. J. Psychol. (Statist. Sect.) 4*, 85-90.

Dwass, M. (1955). On the asymptotic normality of some statistics used in nonparametric tests. *Ann. Math. Statist. 26*, 334-339.

van Elteren, P., and Noether, G. E. (1959). The asymptotic efficiency of the χ^2-test for a balanced incomplete block design. *Biometrika 46*, 475-477.

Fabius, J., (1964). Asymptotic behavior of Bayes estimates. *Ann. Math. Statist. 35*, 846-856.

Ferguson, T. S. (1973). A Bayesian analysis of some nonparametric problems. *Ann. Statist. 1*, 209-230.

Ferguson, T. S. (1974). Prior distributions on spaces of probability measures. *Ann. Statist. 2*, 615-629.

Fieller, E. C., and Pearson, E. S. (1961). Tests for rank correlation coefficients. II. *Biometrika 48*, 29-40.

Filippova, A. A. (1962). Mise's theorem on the asymptotic behavior of functionals of empirical distribution functions and its statistical applications. *Theory Prob. Appl. 7*, 24-57.

Fisher, R. A. (1935). *The Design of Experiments.* Oliver & Boyd, Edinburgh.

Fraser, D. A. S. (1957). Most powerful rank-type tests. *Ann. Math. Statist. 28*, 1040-1043.

Freedman, D. A. (1963). On the asymptotic behavior of Bayes estimates in the discrete case. *Ann. Math. Statist. 34*, 1386-1403.

Freund, J. E., and Ansari, A. R. (1957). Two-way rank sum test for variances. *Tech. Rep.*, Virginia Polytechnic Institute.

Friedman, M. (1937). The use of ranks to avoid the assumption of normality implicit in the analysis of variance. *J. Amer. Statist. Assoc. 32*, 675-701.

Gastwirth, J. L. (1966). On robust procedures. *J. Amer. Statist. Assoc. 61*, 929-948.

Gokhale, D. V. (1966). Some problems in independence and dependence. Ph.D. thesis, University of California, Berkeley.

Govindarajulu, Z., Le Cam, L., and Raghavachari, M. (1967). Generalizations of theorems of Chernoff and Savage on the asymptotic normality of test statistics. *Proc. 5th Berkeley Symp. Math. Statist. Prob. 1*, 609-638.

Greenberg, V. L. (1966). Robust estimation in incomplete block designs. *Ann. Math. Statist. 37*, 1331-1337.

Hájek, J. (1961). Some extensions of the Wald-Wolfowitz-Noether theorem. *Ann. Math. Statist. 32*, 506-523.

Hájek, J. (1962). Asymptotically most powerful rank-order tests. *Ann. Math. Statist. 33*, 1124-1147.

Hájek, J. (1968). Asymptotic normality of simple linear rank statistics under alternatives. *Ann. Math. Statist. 39*, 325-346.

Hájek, J., and Šidák, Z. (1967). *Theory of Rank Tests*. Academic Press, New York.

Hampel, F. R. (1968). Contributions to the theory of robust estimation. Unpublished doctoral dissertation, University of California, Berkeley.

Hampel, F. R. (1971). A general qualitative definition of robustness. *Ann. Math. Statist. 42*, 1887-1896.

Hodges, J. L., Jr. (1955). A bivariate sign test. *Ann. Math. Statist. 26*, 523-527.

Hodges, J. L., Jr., and Lehmann, E. L. (1956). The efficiency of some nonparametric competitors of the t-test. *Ann. Math. Statist. 27*, 324-335.

Hodges, J. L., Jr., and Lehmann, E. L. (1961). Comparison of the normal scores and Wilcoxon tests. *Proc. 4th Berkeley Symp. Math. Statist. Prob. 1*, 307-319.

Hodges, J. L., Jr., and Lehmann, E. L. (1962). Rank methods for combination of independent experiments in the analysis of variance. *Ann. Math. Statist. 33*, 482-497.

Hodges, J. L., Jr., and Lehmann, E. L. (1963). Estimates of location based on ranks. *Ann. Math. Statist. 34*, 598-611.

Hoeffding, W. (1948). A class of statistics with asymptotically normal distribution. *Ann. Math. Statist. 19*, 293-325.

Hoeffding, W. (1951). A combinatorial central limit theorem. *Ann. Math. Statist. 22*, 558-566.

Hoeffding, W. (1951). 'Optimum' nonparametric tests. *Proc. 2nd Berkeley Symp. Prob. Statist.*, 83-92.

Hoeffding, W. (1952). The large sample power of tests based on permutation of observations. *Ann. Math. Statist. 23*, 169-192.

Hoeffding, W. (1973). On the centering of a simple linear rank statistic. *Ann. Statist. 1*, 54-66.

Hollander, M. (1967). Rank tests for randomized blocks when the alternatives have an a priori ordering. *Ann. Math. Statist. 38*, 867-877.

Hollander, M., and Wolfe, D. A. (1973). *Nonparametric Statistical Methods*, Wiley, New York.

Hotelling, H., and Pabst, M. (1936). Rank correlation and tests of significance involving no assumption of normality. *Ann. Math. Statist. 7*, 29-43.

Høyland, A. (1968). Robustness of the Wilcoxon estimate of location against a certain dependence. *Ann. Math. Statist. 39*, 1196-1201.

Huber, P. J. (1964). Robust estimation of a location parameter. *Ann. Math. Statist. 35*, 73-101.

Huber, P. J. (1965). A robust version of the probability ratio test. *Ann. Math. Statist. 36*, 1753-1758.

Huber, P. J. (1968). Robust confidence limits. *Zeit. Wahrsch. Verw. Geb. 10*, 269-278.

Huber, P. J. (1972). Robust statistics: A review. *Ann. Math. Statist. 43*, 1041-1067.

Huber, P. J., and Strassen, V. (1973). Minimax tests and the Neyman-Pearson lemma for capacities. *Ann. Statist. 1*, 251-263.

Hušková, M. (1970). Asymptotic distribution of simple linear rank statistics for testing symmetry. *Zeit. Wahrsch. Verw. Geb. 12*, 308-322.

Jaeckel, L. A. (1971a). Robust estimates of location: Symmetry and asymmetric contamination. *Ann. Math. Statist. 42*, 1020-1034.

Jaeckel, L. A. (1971b). Some flexible estimates of location. *Ann. Math. Statist. 42*, 1540-1552.

Joffe, A., and Klotz, J. (1962). Null distribution and Bahadur efficiency of the Hodges bivariate sign test. *Ann. Math. Statist. 33*, 803-807.

Jonckheere, A. R. (1954). A distribution free k-sample test against ordered alternatives. *Biometrika 41*, 133-145.

Kendall, M. G. (1938). A new measure of rank correlation. *Biometrika* *30*, 81-93.

Kendall, M. G. (1962). *Rank Correlation Methods*, 3rd ed. (1st ed. 1948). Griffin, London.

Kiefer, J. (1959). K-sample analogues of the Kolmogorov-Smirnov and Cramer-von Mises tests. *Ann. Math. Statist. 30*, 420-447.

Kiefer, J. (1961). On large deviations of the empiric distribution function of vector chance variables and a law of the iterated logarithm. *Pacific J. Math. 11*, 649-660.

Kiefer, J., and Wolfowitz, J. (1958). On the deviations of the empiric distribution function of vector chance variables. *Trans. Amer. Math. Soc. 87*, 173-186.

Klotz, J. (1962). Nonparametric tests for scale. *Ann. Math. Statist. 33*, 498-512.

Klotz, J. (1963). Small sample power and efficiency for the one-sample Wilcoxon and Normal scores tests. *Ann. Math. Statist. 34*, 624-692.

Klotz, J. H. (1964a). On the normal scores two-sample rank test. *J. Amer. Statist. Assoc. 59*, 652-664.

Klotz, J. H. (1964b). Small sample power of the bivariate sign tests of Blumen and Hodges. *Ann. Math. Statist. 35*, 1576-1562.

Kolmogorov, A. N. (1933). Sulla determinazione empirica di una legge di distribuzione. *Giorn. Inst. Ital. Attuari 4*, 83-91.

Konijn, H. S. (1956). On the power of certain tests for independence in bivariate populations. *Ann. Math. Statist. 27*, 300-323. [Errata: Ibid 29 (1958) 935.]

Kruskal, W. H. (1957). Historical notes on the Wilcoxon unpaired two-sample test. *J. Amer. Statist. Assoc. 52*, 356-360.

Kruskal, W. H. (1958). Ordinal measures of association. *J. Amer. Statist. Assoc. 53*, 814-861.

Kruskal, W. H., and Wallis, W. A. (1952,1953). Use of ranks in one-criterion variance analysis. *J. Amer. Statist. Assoc. 47*, 583-612; *48*, 907-911.

Le Cam, L. (1960). Locally asymptotically normal families of distributions. *Univ. Calif. Publ. Statist. 3*, 37-98.

Lehmann, E. L. (1951). Consistency and unbiasedness of certain nonparametric tests. *Ann. Math. Statist. 22*, 165-179.

Lehmann, E. L. (1953). The power of rank tests. *Ann. Math. Statist. 24*, 28-43.

Lehmann, E. L. (1963a). Robust estimation in analysis of variance. *Ann. Math. Statist. 34*, 957-966.

Lehmann, E. L. (1963b). Asymptotically nonparametric inference: an alternative approach to linear models. *Ann. Math. Statist. 34*, 1494-1506.

Lehmann, E. L. (1963c). Nonparametric confidence intervals for a shift parameter. *Ann. Math. Statist. 34,* 1507-1512.

Lehmann, E. L. (1964). Asymptotically nonparametric inference in some linear models with one observation per cell. *Ann. Math. Statist. 35,* 726-734.

Lehmann, E. L. (1966). Some concepts of dependence. *Ann. Math. Statist. 37,* 1137-1153.

Lehmann, E. L. (1975). *Statistical Methods Based on Ranks,* Holden-Day, San Francisco.

Lehmann, E. L., and Stein, C. (1949). On the theory of some nonparametric hypotheses. *Ann. Math. Statist. 20,* 28-45.

Mann, H. B., and Whitney, D. R. (1947). On a test of whether one of two random variables is stochastically larger than the other. *Ann. Math. Statist. 18,* 50-60.

Massey, F. J. (1951). The Kolmogorov-Smirnov test of goodness-of-fit. *J. Amer. Statist. Assoc. 46,* 68-78.

Mehra, K. L., and Sarangi, J. (1967). Asymptotic efficiency of certain rank tests for comparative experiments. *Ann. Math. Statist. 38,* 90-107.

Mehra, K. L., and Sen, P. K. (1969). On a class of conditionally distribution-free tests for interactions in factorial experiments. *Ann. Math. Statist. 40,* 658-664.

Miller, R. G., Jr. (1966). *Simultaneous Statistical Inference,* McGraw-Hill, New York.

von Mises, R. (1931). *Wahrscheinlichkeitsrechnung,* Leipzig.

von Mises, R. (1947). On the asymptotic distributions of differentiable statistical functions. *Ann. Math. Statist. 18,* 309-348.

Mood, A. (1954). On the asymptotic efficiency of certain nonparametric two sample tests. *Ann. Math. Statist. 25,* 514-522.

Moses, L. E. (1953). Nonparametric methods. In *Statistical Inference,* Walker and Lev, (Eds.), pp. 426-450, Holt, New York.

Motoo, M. (1957). On the Hoeffding's combinatorial central limit theorem. *Ann. Inst. Statist. Math. 8,* 145-154.

Nemenyi, P. (1963). Distribution-free multiple comparisons. Unpublished doctoral thesis, Princeton University, Princeton, N. J.

Noether, G. E. (1949). On a theorem of Wald and Wolfowitz. *Ann. Math. Statist. 20,* 455-458.

Noether, G. E. (1955). On a theorem of Pitman. *Ann. Math. Statist. 26,* 64-68.

Olmstead, P. S., and Tukey, J. W. (1947). A corner test for association. *Ann. Math. Statist. 18,* 495-513.

Paik, M. K. (1971). Asymptotically minimax rank tests for the independence problem. Ph.D. thesis, University of California, Berkeley.

Pearson, K. (1900). On a criterion that a given system of deviations from the probable in the case of a correlated system of variables is such that it can be reasonably supposed to have arisen in random sampling. *Phil. Mag. Ser. 5 50*, 157-175.

Pitman, E. J. G. (1937). Significance tests which may be applied to samples from any populations. I and II. *Suppl. J. Roy. Statist. Soc. Ser. B 4*, 119-130; 225-232.

Pitman, E. J. G. (1938). Significance tests which may be applied to samples from any populations. III. The analysis of variance test. *Biometrika 29*, 322-335.

Pitman, E. J. G. (1948). Lecture notes on nonparametric statistics. Columbia University, New York.

Puri, M. L. (1964). Asymptotic efficiency of a class of c-sample tests. *Ann. Math. Statist. 35*, 102-121.

Puri, M. L. (1972). Some aspects of nonparametric inference. *Rev. Int. Statist. Inst. 40*, 229-327.

Puri, M. L., and Sen, P. K. (1971). *Nonparametric Methods in Multivariate Analysis,* Wiley, New York.

Pyke, R., and Shorack, G. (1968). Weak convergence of a two-sample empirical process and a new approach to Chernoff-Savage theorems. *Ann. Math. Statist. 39*, 755-771.

Ragavachari, M. (1965). On the efficiency of the normal scores test relative to the t-test. *Ann. Math. Statist. 36*, 1306-1307.

Ruist, E. (1954). Comparison of tests for nonparametric hypotheses. *Ark. Mat. 3*, 133-163.

Savage, I. R. (1956). Contributions to the theory of rank order statistics--the two-sample case. *Ann. Math. Statist. 27*, 590-615.

Savage, I. R. (1962). *Bibliography of Nonparametric Statistics,* Harvard University Press, Cambridge.

Schaafsma, W. (1966). *Hypothesis Testing Problems with the Alternative Restricted by a Number of Inequalities,* P. Noordhoff, Groningen, The Netherlands.

Scheffé, H. (1943). Statistical inference in the nonparametric case. *Ann. Math. Statist. 14*, 305-332.

Scholz, F. W. (1971). Comparison of optimal location estimators. Ph.D. thesis, University of California, Berkeley.

Scholz, F. W. (1974). A comparison of efficient location estimators. *Ann. Statist. 2*, 1323-1326.

Sen, P. K. (1963). On the estimation of relative potency in dilution (-direct) essays by distribution-free methods. *Biometrics 19*, 532-552.

Sen, P. K. (1968a). On a class of aligned rank order tests in two-way layouts. *Ann. Math. Statist. 39*, 1115-1124.

Sen, P. K. (1968b). Robustness of some nonparametric procedures in linear models. *Ann. Math. Statist. 39*, 1913-1933.

Shorack, G. R. (1967). Testing against ordered alternatives in model I analysis of variance; normal theory and nonparametric. *Ann. Math. Statist. 38*, 1740-1752.

Shorack, G. R. (1969). Asymptotic normality of linear functions of order statistics. *Ann. Math. Statist. 40*, 2041-2050.

Siegel, S., and Tukey, J. W. (1960). A nonparametric sum of ranks procedure for relative spread in unpaired samples. *J. Amer. Statist. Assoc. 55*, 429-445.

Smirnov, N. V. (1939). On the estimation of the discrepancy between empirical curves of distribution for two independent samples. *Bull. l'Univ. Moscov 2* (2), 3-14.

Spearman, C. (1904). The proof and measurement of association between two things. *Amer. J. Psychol. 15*, 72-101.

Spjøtvoll, E. (1968). A note on robust estimation in analysis of variance. *Ann. Math. Statist. 39*, 1486-1492.

Steel, R. G. D. (1960). A rank sum test for comparing all pairs of treatments. *Technometrics 2*, 197-207.

Stigler, S. M. (1973). The history of robust estimation 1885-1920. *J. Amer. Statist. Assoc. 68*, 872-879.

Straf, M. L. (1972). Weak convergence of stochastic processes with several parameters. *Proc. 6th Berkeley Symp. Math. Statist. Prob. 1*, 187-221.

Takeuchi, K. (1967). Robust estimation and robust parameter. Unpublished manuscript, presented at the annual meeting of the Institute of Mathematical Statistics, Dec. 1967.

Terpstra, T. J. (1952). The asymptotic normality and consistency of Kendall's test against trend, when ties are present in one ranking. *Indag. Math. 14*, 327-333.

Terry, M. E. (1952). Some rank order tests which are most powerful against specific parametric alternatives. *Ann. Math. Statist. 23*, 346-366.

Thompson, R., Govindarajulu, Z., and Doksum, K. (1967). Distribution and power of the absolute normal scores test. *J. Amer. Statist. Assoc. 62*, 966-975.

Tukey, J. W. (1949). The simplest signed-rank tests. Mimeogr. Rep. No. 17, *Statist. Res. Group,* Princeton University.

Tukey, J. W. (1960). A survey of sampling from contaminated distributions. *Contributions to Probability and Statistics: Essays in Honor of Harold Hotelling,* pp. 448-485, Stanford University Press, Stanford, Calif.

Tukey, J. W. (1962). The future of data analysis. *Ann. Math. Statist.* *33*, 1-67.

Wald, A., and Wolfowitz, J. (1939). Confidence limits for continuous distribution functions. *Ann. Math. Statist. 10*, 105-118.

Wald, A., and Wolfowitz, J. (1944). Statistical tests based on permutations of the observations. *Ann. Math. Statist. 15*, 358-372.

Walsh, J. E. (1946). On the power of the sign test for slippage of means. *Ann. Math. Statist. 17*, 358-362.

Walsh, J. E. (1947). Some significance tests for the median which are valid under very general conditions. Ph.D. thesis, Princeton University, Princeton.

Walsh, J. E. (1949a). Some significance tests for the median which are valid under very general conditions. *Ann. Math. Statist. 20*, 64-81.

Walsh, J. E. (1949b). Applications for some significance tests for the median which are valid under very general conditions. *J. Amer. Statist. Assoc. 44*, 342-353.

Wilcoxon, F. (1945). Individual comparisons by ranking methods. *Biometrics 1*, 80-83.

Yanagimoto, T., and Okamoto, M. (1969). Partial orderings of permutations and monotonicity of a rank correlation statistic. *Ann. Inst. Statist. Math. 21*, 489-506.

Yuan, J. C. (1973). Asymptotically minimax rank tests for ordered alternatives. Ph.D. thesis, University of California, Berkeley.

van Zwet, W. R. (1964). *Convex Transformations of Random Variables,* Mathematics Centre, Amsterdam.

12.

SOME HISTORY OF THE
DATA ANALYSIS OF TIME SERIES
IN THE UNITED STATES

DAVID R. BRILLINGER was born October 27, 1937, in Toronto, Canada.
He received his B.A. degree in Mathematics from the University
of Toronto in 1959 and his M.A. degree in Mathematics from
Princeton University in 1960. He earned his Ph.D. degree in
Mathematics in 1961 from Princeton University where he worked
with Professor John W. Tukey who was then a major developer of
the subject of Time Series. From 1961 to 1962 he was a Social
Science Research Council Post Doctoral Fellow at the London
School of Economics. From 1964 to 1969 he was at the London
School of Economics. Since 1969 he has been a Professor of Statistics at the University of California at Berkeley. He is a
Fellow of the American Statistical Association and The Institute
of Mathematical Statistics. He is a Member of the International
Statistical Institute and author of a book on Time Series anal-
sis as well as some 50 papers. He is a Guggenheim Fellow for
1975-1976.

12.

SOME HISTORY OF THE
DATA ANALYSIS OF TIME SERIES
IN THE UNITED STATES

David R. Brillinger

Department of Statistics
University of California
Berkeley, California

The history of the data analysis of time series in the United States
is long and varied. It covers many individuals, many computing devices, and many fields of application. We divide our discussion into
the four epochs delineated by the 1930 Wiener paper on generalized
harmonic analysis (16), the 1949 Tukey paper on the sampling theory
of power spectrum estimates (33), and the 1965 Cooley-Tukey paper
on the fast Fourier transform (65). We have attempted to provide
some coverage of the important work before 1965. So much has been
done since 1965 we are only able to provide a small selection.

PRE-1930

In 1891, the Bostonian actuary and astronomer S. C. Chandler carried
out a numerical analysis of the variation of latitude with time (1).
His analysis suggested that the motion of the Earth's pole of rotation
was a composite, consisting of a component of a 12-month period and
another component of a near-14-month period. The long period of the

second component was unexpected under the current Eulerian model of
the Earth. Subsequently, scientists of the time were forced to revise
substantially their ideas of the physical properties of the Earth's
interior.

Also in 1891, the physicist A. A. Michelson of Chicago described
the interferometer, a device allowing the superposition of a light
signal on top of itself with prescribed delay. In one series of experiments, Michelson first band-pass filtered a light signal by passing it through a prism. He then used the interferometer to measure
the visibility of the superposed signal as a function of delay. The
curve he obtained was the autocovariance function of the original
signal. Michelson then used a mechanical harmonic analyzer to compute
the Fourier transform of the visibility curve; that is, he estimated
the power spectrum of the signal. This was done in order to examine
the fine structure of spectral lines of light (2,3).

The 1906 San Francisco earthquake stimulated a large-scale data
analysis of time series involving the collection and reading of seis-
mograms of the event as recorded at some 96 stations around the world
(4). The times of arrival and amplitude of the P and S seismic waves
were read from the records and were used both to construct a table of
the velocity of travel of waves through the Earth as a function of
depth and to check a procedure that had been suggested for determining the distance of an event from the difference in arrival times of
the P and S waves.

A numerical harmonic analysis of a time series was carried out
by H. L. Moore in 1914 (5), when he computed the periodogram of rain-
fall in the Ohio valley. The periodogram was suggested by the English
worker Schuster (6), a numerical form of the spectrobolometer that
had been invented by the American physicist S. P. Langley in 1881.

Also in 1914, A. E. Douglas of Arizona described a photographic
means of carrying out a harmonic analysis, which he applied to long
series of the annual growth of tree rings (7). In later years,
Douglas performed further analyses and attempted to relate periodicities in the growth of tree rings to the sunspot cycle (8).

In 1922, Crandall and MacKenzie (9) used resonance tubes to measure the distribution of the energy of speech as a function of frequency. Such analyses became routinely possible for electric signals with the invention of devices such as Campbell's wave filter (10). These three workers were at the Bell Telephone Laboratories. Also there at that period, W. A. Shewhart was pioneering the quality control of processes; he proposed that this be done by the plotting of relevant variates, such as the proportion of defectives in lots against time, and by examining the production process when the plot moved outside indicated limits (11). There too, T. C. Fry and E. C. Molina were laying the foundations for the application of congestion theory and queueing theory to telephone exchanges and related systems (12,13).

Throughout this period, the *Journal of the American Statistical Association* was filled with lively papers on the merits and disadvantages of a wide variety of techniques for the seasonal adjustment, smoothing, and detrending of economic time series. It appears that most of the issues of concern today on these topics had been raised at that time. S. Kuznets published a number of papers in the *Journal*. One interesting paper (14), contained computed plots of the variation of correlation coefficients with time. During this same period, Irving Fisher introduced readers of the *Journal* to "distributed lags" and wrote a book on the construction of index numbers (15).

These years of the data analysis of time series were typical in many ways of what has happened since that time. Graphical, digital, mechanical, electrical, and optical procedures were employed. As the means of computing improved, more detailed analyses with longer series were carried out. In the case of analog procedures, the particular device employed depended very much on the frequency range of the phenomenon under study.

THE PERIOD 1930-1949

In 1930, Norbert Wiener's monumental paper "Generalized Harmonic Analysis" (16) appeared. He stated in this paper, as he had mentioned earlier (17), that he had been motivated by the work of researchers in

optics such as Rayleigh and Schuster. The domain of application of
his results is much, much broader however. Wiener's achievements
included setting down precise definitions of the covariance function
and the power spectrum; the development of spectral representations
for the covariance function and the time series itself; and the mathematical description of the operation of filtering and its effect.
Wiener adopted a constructive approach (involving a section of increasing length of a single function) rather than an existential
approach (involving, for example, an ensemble of functions). This
meant that an interested worker could construct an associated data
analysis procedure directly. Wiener's work also provided a standard
language that researchers in many fields could use to describe and
compare their results with. Detailed descriptions of the impact of
Wiener's work may be found in refs. 18-21.

The years 1930-1942 saw marginal improvements in wave analyzers
and numerical procedures; the computation of many correlograms and
some Fourier transforms of these; and the collection of new data in
time series form, such as electrocardiograms and electroencephalograms.
The theory of economic time series models was beginning to develop,
and a remarkable book, *The Analysis of Economic Series* by H. T. Davis
(22), appeared. This book was written from a historical and multi-
disciplinary outlook and must have educated many to the existing and
most recent techniques of time series analysis. It presented some
new procedures as well, such as the moving periodogram, a periodogram
computed by sliding an observation window of fixed length along a series

During the World War II effort, Wiener worked at MIT on the design
of fire-control apparatus for anti-aircraft guns. This led to the
writing and restricted circulation in 1942 of his memorandum *The Extrapolation, Interpolation and Smoothing of Stationary Time Series*
(23). Although Wiener's "Generalized Harmonic Analysis" did not have
immediate influence, the memorandum (23), which was written in a more
accessible style, did. Engineers were led to model many of the phenomena that they studied as statistical processes; to measure parameters such as the cross covariance and the power spectrum; and to

design circuits in the light of these measurements. Almost immediately upon this circulation of Wiener's work, a large-scale application of his methods to weather forecasting was undertaken (24). The success of the application of these techniques led to the Symposium on Applications of Autocorrelation Analysis to Physical Problems at Woods Hole, Massachusetts, in June 1949, sponsored by the Office of Naval Research.

A clever device for the running display of the power spectrum of an audible signal was constructed in 1946 (25). It was called the sound spectrograph and displayed a running estimate of the signal energy at all frequencies with the intensity of the image produced related to the amount of energy present. In 1953, a corresponding device was constructed for the running display of a correlogram (26).

Other work on time series analysis in the 1940s included (a) the development of sequential analysis, a procedure for the making of repeated measurements on a variate until a criterion is satisfied followed by the then making of a decision or estimate (27); (b) the fitting and study of linear econometric models (28,29); (c) the analytic study of the null distribution of the correlogram in a series of papers in the *Annals of Mathematical Statistics* by R. L. Anderson, T. W. Anderson, W. J. Dixon, J. Von Neumann among others; and (d) the empirical study of the correlogram by H. R. Seiwell and P. Rudnick (30-32).

We close our discussion of this period by mentioning that Wiener's memorandum (23) became generally available in 1949.

THE PERIOD 1949-1964

The third epoch of the empirical analysis of time series in the United States began with the Woods Hole Symposium on Applications of Autocorrelation Analysis. There, J. W. Tukey presented the first of three papers (33-35), which he had written in the early years on spectrum analysis. These papers introduced the numerical spectrum estimate that was used by most workers throughout the whole epoch and described an approximate distribution for the estimate. This distribution was needed for the proper design of experiments for the collection of time series data. Tukey (36) described the situation that led to his work and which led to R. W. Hamming's suggestion of a smoothing of the discrete Fourier transform of an empirical correlogram. That paper (36) also described some of the history of the data analysis of time series.

During these years and continuing to the present, Tukey introduced a multitude of techniques and terms that have become standard to the practice of the data analysis of time series. We mention "prewhitening," "alias," "smoothing and decimation," "taper," "bispectrum," "complex demodulation," and "cepstrum." Tukey's papers of this particular period include refs. 37-42. The literature of applied time series analysis contains numerous acknowledgments of his suggestions. Tukey has also written a number of papers placing the data analysis of time series into perspective with current research in the physical sciences, in statistics, and in computing (43-47).

During this period, groups of workers on applied spectrum analysis appeared around the country, of which we mention only a few. The group at MIT included G. P. Wadsworth, S. M. Simpson, and E. A. Robinson. They were especially concerned with the numerical prediction and smoothing of seismic records. Prior to their work, Wiener's procedures had generally been realized in analog form. Lists of many of their reports may be found in refs. 24 and 48. W. J. Pierson and L. J. Tick were at New York University at the time. They were concerned with the analysis of oceanographic time series records (49). The thesis by N. R. Goodman (50), extending the results of Tukey (33-35) to bivariate series, and written under his supervision, appeared in the NYU series of reports. A group at La Jolla, California consisted of W. H. Munk, P. R. Rudnick, and F. E. Snodgrass. W. H. Munk and G. J. F. McDonald wrote a remarkable book, *The Rotation of the Earth* (51), which described a variety of empirical time series analyses and suggested a novel method of fitting a finite parameter model to an observed periodogram. In part, these authors were concerned with the Chandler wobble (1).

Along with groups of workers, packages of computer programs for the digital analysis of time series were appearing and being circulated. At Bell Telephone Laboratories, a collection of FORTRAN programs written by M. J. Healy were in use from 1960 on (52). The BOMM collection of programs was developed at La Jolla (53). The programs written at MIT are described (54). Some of the programs used by E. Parzen at Stanford University are given in ref. 55 (ref. 21 therein). References 56 and 57 give other collections of FORTRAN programs.

C. W. J. Granger's book *Spectral Analysis of Economic Time Series* (58) appeared in 1964. It described applications of many of the techniques suggested by Tukey to the analysis of univariate and bivariate time series that were carried out at the Econometric Research Program, Princeton University. Perhaps the most successful application of spectral techniques to economic series is its use for the description of the multitude of procedures of seasonal adjustment (59,60). Also during these years, the beginnings were made of the data analysis of spatial series and, indeed, spatial point processes. For example, Neyman and Scott (61) carried out an analysis of two-dimensional data consisting of the positions of the images of galaxies on photographic plates.

On the analytic front, a book by U. Grenander and M. Rosenblatt (62) appeared in 1957, which formalized many of the data analysis procedures and approximations that had come into use. E. Parzen began a long series of papers on time series analysis (55). A remarkable string of Ph.D. theses were written under his supervision at Stanford University. N. Wiener remained active until his death in 1964. His book (63) on nonlinear models for time series appeared. Wiener was especially concerned with analyzing and modeling brain waves (63, 64). He and Masani also developed an approach to multidimensional prediction theory (20,21). Wiener's death clearly marked the end of a period in the analysis of time series.

AFTER 1964

The year 1965 marks the beginning of the present era of time series
analysis with the publication of the paper by Cooley and Tukey on
the fast Fourier transform (65). The effect that this paper has had
on scientific and engineering practice cannot be overstated. The
paper described an algorithm for the discrete Fourier transform of
$T = T_1 \cdots T_p$ values by means of $T(T_1 + \cdots + T_p)$ multiplications
instead of the naive number T^2. Although such algorithms existed
previously (66), they seem not to have been put to much use. G. Sande
developed and programmed one algorithm that was widely circulated.

The existence of such an algorithm meant, for example, that the
following things could be computed an order of magnitude more rapidly:
spectral estimates, correlograms (!), filtered versions of series,
complex demodulates, and Laplace transforms (see, for example, ref.
67). General discussions of the uses and importance of fast Fourier
transform algorithms may be found in refs. 68 and 69. The Fourier
transform of an observed stretch of series can now be taken as a basic
statistic and classical statistical analyses--such as multiple regression, analysis of variance, principal components, canonical analysis,
errors in variables, and discrimination--can be meaningfully applied
to its values, (70) and references cited therein. Higher-order spectra
may be computed practically (71). Inexpensive portable computers for
carrying out spectral analysis have appeared on the market and may be
found in many small laboratories.

The years since 1964 have been characterized by the knowledge that
there are fast Fourier transform algorithms. They have also been
characterized by the rapid spread of type of data analyzed. Previously,
the data analyzed consisted almost totally of discrete or continuous
real-valued time series. Now the joint analysis of many series, such
as the 625 recorded by the Large Aperture Seismic Array in Montana
(72), has become common. Spatial series are analyzed (72). The statistical analysis of point processes has grown into an entirely separate
field (73). The SASE IV computer program developed by Peter Lewis
(74), has furthered such analysis considerably. We note that transforms other than the Fourier are finding interest as well (75,76).

A variety of alternate estimates of the power spectrum have been receiving practical attention lately. We mention the high-resolution estimate (72) and the autoregressive estimate (55, ref. 21 therein) (or the maximum entropy estimate, ref. 77). The usual estimate of a power spectrum is a quadratic function of the observations (when the data have not been prefiltered). These alternate estimates are more complicated functions. Their use in practice still requires study. Another area that has been popular is the fitting of finite parameter models, especially when there is an understandable nonstationarity present (78) and not much data available.

Throughtout the present paper, numerous references have been made to work carried out on the data analysis of time series at Bell Telephone Laboratories (refs. 9-13, 25, 26, 33-47, and 52). I end by pointing out two recent examples (79) and (80), of the work being done there.

ACKNOWLEDGMENTS

This work was prepared while the author was a Miller Research Professor and with the support of National Science Foundation Grant GP-31411.

I would like to thank Professors B. A. Bolt and J. W. Tukey for some helpful comments.

REFERENCES

1. S. C. Chandler, On the variation of latitude. *Astron. J. 11*, 83, 109 (1891).

2. A. A. Michelson, *Light Waves and Their Uses*, Chicago University Press, Chicago, 1907.

3. A. A. Michelson and S. W. Stratton, A new harmonic analyzer. *Amer. J. Sci. 5*, 1 (1898).

4. *The California Earthquake of April, 1906*, (2 vol. and atlas), Carnegie Institution, Washington, D.C., 1908.

5. H. L. Moore, *Economic Cycles: Their Law and Cause*, Macmillan, New York, 1914.

6. A. Schuster, On the investigation of hidden periodicities with application to a supposed 26 day period of meteorological phenomena. *Terr. Magnet. 3*, 13-41 (1898).

7. A. E. Douglas, An optical periodograph. *Astrophys. J. 41,* 173-186 (1914).

8. A. E. Douglas, *Climatic Cycles and Tree Growth,* Carnegie Institution, Washington, D.C. (Vol. 1, 1919; Vol. 2, 1928; Vol. 3, 1936.)

9. I. B. Crandall and D. MacKenzie, Analysis of the energy distribution in speech. *Bell Syst. Tech. J. 1,* 116-128 (1922).

10. G. A. Campbell, Physical theory of the electric wave-filter. *Bell Syst. Tech. J. 1,* 1-32 (1922).

11. W. A. Shewhart, Quality control. *Bell Syst. Tech. J. 6,* 722-735 (1927).

12. T. C. Fry, *Probability and its Engineering Uses,* Van Nostrand, New York, 1928.

13. E. C. Molina, Application of the theory of probability to telephone trunking problems. *Bell Syst. Tech. J. 6,* 461-494 (1927).

14. S. Kuznets, On moving correlation of time series. *J. Amer. Statist. Assoc. 23,* 121-136 (1928).

15. I. Fisher, *The Making of Index Numbers,* Houghton-Mifflin, New York, 1922.

16. N. Wiener, Generalized harmonic analysis. *Acta Math. 55,* 117-258 (1930).

17. N. Wiener, Verallgemeinerte trigonometrische Entwicklungen. *Gesell. Wiss. Gott. Nach.* 151-158 (1925).

18. *Selected Papers of Norbert Wiener,* MIT Press, Cambridge, Mass., 1964.

19. E. Olson and J. P. Schadé, A tribute to Norbert Wiener. In *Cybernetics of the Nervous System* (N. Wiener and J. P. Schadé, eds.), pp. 1-8, Elsevier, Amsterdam, 1965.

20. *Norbert Wiener 1894-1964. Bull. Amer. Math. Soc. 72,* 1-145 (1966).

21. T. Kailath, A view of three decades of linear filtering theory. *IEEE Trans. Inf. Theory IT-20,* 146-180 (1974).

22. H. T. Davis, *The Analysis of Economic Time Series,* Principia Press, Bloomington, Ind., 1941.

23. N. Wiener, *The Extrapolation, Interpolation and Smoothing of Stationary Time Series,* Wiley, New York, 1949.

24. G. P. Wadsworth, E. A. Robinson, J. G. Bryan, and P. M. Hurley, Detection of reflections on seismic records by linear operators. *Geophysics 18,* 539-586 (1953).

25. W. Koenig et al., The sound spectrograph. *J. Acoust. Soc. Amer. 17,* 19-29 (1946).

26. W. R. Bennett, The correlatograph. *Bell Syst. Tech. J. 32,* 1173-1185 (1953).

27. A. Wald, *Sequential Analysis,* Wiley, New York, 1947.

28. H. B. Mann and A. Wald, On the statistical treatment of linear stochastic difference equations. *Econometrica 11,* 173-220 (1943).

29. *Statistical Inference in Dynamic Economic Models* (T. C. Koopmans, ed.), Wiley, New York, 1950.

30. H. R. Seiwell, The principles of time series analyses applied to ocean wave data. *Proc. Natl. Acad. Sci. USA 35,* 518-528 (1949).

31. H. R. Seiwell, Experimental correlogram analysis of artificial time series (with special reference to the analysis of oceanographic data). *Proc. 2nd Berkeley Symp. Math. Statist. Prob. 1,* 639-666 (1951).

32. P. R. Rudnick, Correlograms for Pacific ocean waves. *Proc. 2nd Berkeley Symp. Math. Statist. Prob. 1,* 627-638 (1951).

33. J. W. Tukey, The sampling theory of power spectrum estimates. *Symp. Appl. Autocorr. Anal. Phys. Prob. U.S. Off. Naval Res.* (NAVEXOS-P-735), 1949. Reprinted in *J. Cycle Res. 6,* 31-52 (1957).

34. J. W. Tukey and R. W. Hamming, Measuring noise color 1. *Bell Tel. Labs Memo* (1949).

35. J. W. Tukey, Measuring noise color. Unpublished manuscript prepared for distribution at the Institute of Radio Engineers Meeting, Nov. 1951.

36. J. W. Tukey, An introduction to the calculations of numerical spectrum analysis. In *Spectral Analysis of Time Series* (B. Harris, ed.), pp. 25-46, Wiley, New York, 1967.

37. H. Press and J. W. Tukey, Power spectral methods of analysis and their application to problems in airplane dynamics. *Bell Syst. Monogr. 2606* (1956).

38. R. B. Blackman and J. W. Tukey, The measurement of power spectra from the point of view of communications engineering. *Bell Syst. Tech. J. 33,* 185-282; 485-569 (1958); *also* Dover, New York, 1959.

39. J. W. Tukey, The estimation of power spectra and related quantities. In *On Numerical Approximation,* (R. E. Langer, ed.), pp. 389-411, University of Wisconsin Press, Madison, 1959.

40. J. W. Tukey, An introduction to the measurement of spectra. In *Probability and Statistics,* (U. Grenander, ed.), pp. 300-330, Wiley, New York, 1959.

41. J. W. Tukey, Equalization and pulse shaping techniques applied to the determination of the initial sense of Rayleigh waves. In *The Need of Fundamental Research in Seismology,* Appendix 9, pp. 60-129, Department of State, Washington, D.C. (1959).

42. B. P. Bogert, M. J. Healy, and J. W. Tukey, The quefrency alanysis of time series for echoes; cepstrum, pseudo-autoco-variance, cross-cepstrum and saphe-cracking. In *Time Series Analysis* (M. Rosenblatt, ed.), pp. 201-243, Wiley, New York, 1963.

43. J. W. Tukey, Discussion emphasizing the connection between analysis of variance and spectrum analysis. *Technometrics 3*, 1-29 (1961).

44. J. W. Tukey, The future of data analysis. *Ann. Math. Statist. 33*, 1-67 (1963).

45. J. W. Tukey, What can data analysis and statistics offer today? In *Ocean Wave Spectra*, National Academy of Sciences, Washington, D.C. and Prentice-Hall, Englewood Cliffs, N. J., 1963.

46. J. W. Tukey, Uses of numerical spectrum analysis in geophysics. *Bull. Int. Statist. Inst. 39*, 267-307 (1965).

47. J. W. Tukey, Data analysis and the frontiers of geophysics. *Science 148*, 1283-1289 (1965).

48. E. A. Robinson, *Statistical Communication and Detection*, Griffin, London, 1967.

49. W. J. Pierson and L. J. Tick, Stationary random processes in meteorology and oceanography. *Bull. Int. Statist. Inst. 35*, 271-281 (1957).

50. N. R. Goodman, On the joint estimation of the spectra, cospectrum and quadrature spectrum of a two dimensional stationary Gaussian process. Sci. Pap. No. 10, Engineering Statistics Laboratory, New York University, New York, 1957.

51. W. H. Munk and G. J. F. MacDonald, *The Rotation of the Earth*, Cambridge University Press, Cambridge, Mass., 1960.

52. M. J. Healy and B. P. Bogert, FORTRAN subroutines for time series analysis. *Commun. Soc. Computing Machines 6*, 32-34 (1963).

53. E. C. Bullard, F. E. Ogelbay, W. H. Munk, and G. R. Miller, *A User's Guide to BOMM*, Institute of Geophysics and Planetary Physics, University of California Press, San Diego, 1966.

54. S. M. Simpson, *Time Series Computations in FORTRAN and FAP*, Addison-Wesley, Reading, Penn., 1966.

55. E. Parzen, *Time Series Analysis Papers*, Holden-Day, San Francisco, 1967.

56. *BMD Biomedical Computer Programs* (W. J. Dixon, ed.), University of California Press, Berkeley, 1968 (and its X-suppl., 1972).

57. E. A. Robinson, *Multichannel Time Series Analysis with Digital Computer Programs*, Holden-Day, San Francisco, 1967.

58. C. W. J. Granger, *Spectral Analysis of Economic Time Series*, Princeton University Press, Princeton, 1964.

59. M. Nerlove, Spectral analysis of seasonal adjustment procedures. *Econometrica 32,* 241-286 (1964).

60. M. D. Godfrey and H. F. Karreman, A spectrum analysis of seasonal adjustment. In *Essays in Mathematical Economics* (M. Shubik, ed.), pp. 367-421, Princeton University Press, Princeton, 1967.

61. J. Neyman and E. L. Scott, Statistical approach to problems of cosmology. *J. Roy. Statist. Soc. Ser. B 20,* 1-43 (1958).

62. U. Grenander and M. Rosenblatt, *Statistical Analysis of Stationary Time Series,* Wiley, New York, 1957.

63. N. Wiener, *Nonlinear Problems in Random Theory,* MIT Press, Cambridge, Mass., 1958.

64. N. Wiener, Rhythm in physiology with particular reference to encephalography. *Proc. Roy. Virchow Med. Soc. NY 16,* 109-124 (1957).

65. J. W. Cooley and J. W. Tukey, An algorithm for the machine calculation of Fourier series. *Math. Comp. 19,* 297-301 (1965).

66. J. W. Cooley, P. A. W. Lewis, and P. D. Welch, Historical notes on the fast Fourier transform. *IEEE Trans. Audio Electracoust. AU-15,* 76-79 (1967).

67. C. Bingham, M. D. Godfrey, and J. W. Tukey, Modern techniques of power spectrum estimation. *IEEE Trans. Audio Electroacoust. AU-15,* 56-66 (1967).

68. E. O. Brigham and R. E. Morrow, The fast Fourier transform. *IEEE Spectrum,* 63-70 (1967).

69. *IEEE Trans. Audio Electracoust.,* Special fast Fourier transform numbers, *AU-15,* June 1967 and *AU-17,* June 1969.

70. D. R. Brillinger, *Time Series: Data Analysis and Theory,* Holt, New York, 1974.

71. D. R. Brillinger and M. Rosenblatt, Computation and interpretation of k-th order spectra. In *Spectral Analysis of Time Series* (B. Harris, ed.), pp. 189-232, Wiley, New York, 1967.

72. J. Capon, Applications of detection and estimation theory to large array seismology. *Proc. IEEE 58,* 760-770 (1970).

73. *Stochastic Point Processes* (P. A. W. Lewis, ed.) Wiley, New York, 1972.

74. P. A. W. Lewis, A. M. Katcher, and A. H. Weiss, SASE IV--An improved program for the statistical analysis of series of events. *IBM Res. Resp. RC2365* (1969).

75. *Proc. Symp. Walsh Functions* (1970-1973).

76. A. Cohen and R. H. Jones, Regression on a random field. *J. Amer. Statist. Assoc. 64,* 1172-1182 (1969).

77. R. T. Lacoss, Data adaptive spectral analysis methods. *Geophysics 36,* 661-675 (1971).

78. G. E. P. Box and G. M. Jenkins, *Time Series Analysis Forecasting and Control,* Holden-Day, San Francisco, 1970.

79. J. D. Gabbe, M. B. Wilk, and W. L. Brown, Statistical analysis and modeling of the high-energy proton data from the Telstar 1 satellite. *Bell Syst. Tech. J. 46,* 1301-1450 (1967).

80. P. D. Bricker, R. Gnanadesikan, and others, Statistical techniques for talker identification. *Bell Syst. Tech. J. 50,* 1427-1454 (1971).

A SNAG IN THE HISTORY
OF FACTORIAL EXPERIMENTS

ROBERT H. TRAXLER was born in 1935 in Childress, Texas. He received his Ph.D. degree in Physics from the University of California at Berkeley in 1963 and his Ph.D. degree in Biostatistics, also from the University of California at Berkeley, in 1974. He is presently Associate Professor in the Department of Mathematical Sciences at the New Mexico State University. He has interests in the applications of statistics in health and medicine.

13.

A SNAG IN THE HISTORY
OF FACTORIAL EXPERIMENTS

Robert H. Traxler*

Statistical Laboratory
University of California
Berkeley, California

R. A. Fisher's founding of the theory of experimental design represents one of the most fruitful developments in mathematical statistics
that occurred during the last half-century. Particularly the principle of randomization, introduced by Fisher, after it penetrated
the many domains of experimental work, was very helpful in achieving
the impressive advances in the various fields of science and technology. However, it does not happen often in the history of science
that *all* the practical uses of a novel fruitful idea are flawless.
Any such idea has its own limitations and, not infrequently, the early
enthusiasm of the users carries them too far. As a result, an unpleasantness occurs.

Something of this sort happened to a particular chapter of the
theory of experimentation, namely to factorial designs with their
now traditional method of analysis, in terms of "main effects" and

*Present affiliation: Department of Mathematical Sciences, New
Mexico State University, Las Cruces, New Mexico.

"interactions." As is well known, strict validity of the use of
"main effects" is limited to the case when the true values of
interactions are zero, which in some cases may be known a priori.
In such cases, the factorial design is very useful indeed in keeping
down the cost of experimentation. Also, it is very elegant. The
unpleasantness just mentioned occurred when the original enthusiasm
of experimenters, including apparently Fisher himself, caused them
to use the factorial design also in those numerous cases when the
true interactions are unknown and, indeed, when their values are of
independent particular interest. It so happens that the traditional
tests for interactions have rather low power. In consequence,
the nonzero interactions may go undetected quite frequently, even if
they are so large as to make the "main effects" unrepresentative of
the actual properties of the treatments studied.

The presence of the unpleasantness was discovered 40 years ago
and it may be questioned whether this historical incident is worth
recalling. However, there are important arguments in favor of bringing the matter up.

One part of the argument is that in a number of modern studies,
particularly in biology, medicine, and pharmacology, the principal
subjects of interest include the so-called "synergistic" effects.
For example, agents A and B applied singly to experimental animals
may cause no noticeable increase of cancer cases: they are not "car-
cinogens." However, if applied simultaneously or in succession, the
combination of the two, denoted by AB, may cause cancer with high
frequency. If so, then the two agents A and B are called "cocarci-
nogens," and their effect is described as "synergistic." Carcinogenesis is not the only field of "synergistic studies." Another
field is indicated by the special term "comutagens," etc. It will be
realized that in the statistical jargon, "synergistic effects" mean
interactions, and the studies of synergistic effects may be labeled
the interaction oriented studies.

The other part of the argument in favor of discussing here the
incident of 40 years ago is the following remarkable fact. As is
well known, there are now available many books and manuals on experimental design. These books contain many useful items of information,

numerous illustrative examples, convenient algorithms, etc. Indeed, many of these books discuss examples of distinctly interaction oriented studies. Yet, the present writer was not able to find a single book or research paper offering an interaction oriented methodology beyond the traditional one, found to be inefficient 40 years ago. Adaptation of the design and novel methodology, in some intelligible sense "optimal," remains a challenging problem of mathematical statistics.

This chapter is divided into three parts. First, basic concepts relating to factorial experiments are briefly described. Next, the unpleasantness discovered in 1935 is recounted. The third part deals with the recent extension of the traditional evaluation of factorial experiments to the domain of binary data.

TRADITIONAL FACTORIAL EXPERIMENTS

As quoted by Yates (1), Fisher's own motivation for using the factorial design was

> Nature...will best respond to a logical and carefully thought out questionnaire; indeed, if we ask her a single question, she will often refuse to answer until some other topic has been discussed.

Accordingly, in an agricultural experiment contemplated to investigate the effectiveness of three fertilizers [see Yates (1)], designated by n (nitrogen), k (potassium), and p (phosphorus), the subjects of interest are not only the so-called "simple effects," but something in addition. The simple effects mentioned are symbolized by the differences:

$$n - (\text{control})$$
$$k - (\text{control}) \hspace{4cm} [1]$$
$$p - (\text{control})$$

each representing the difference between the "true" mean yield per plot of the experimental field attainable with the indicated manuring and the "true" mean yield (control) from the "control" plots, receiving none of the three fertilizers studies. Let nk, kp, and pn designate the true mean yields attainable on plots of the same field given not

just one but two indicated fertilizers, in the same standard amounts
as in cases in which they are applied singly. Similarly, let nkp
stand for the true mean yield attainable from plots manured with all
three fertilizers, applied with same standard doses.

The questions addressed to Nature, additional to the above
simple effects [1], in a factorial experiment are the so-called first-
order interactions. The simple first-order interaction of n and k
is defined by

$$[nk - k] - [n - (\text{control})] = nk - n - k + (\text{control}) \qquad [2]$$

which is the difference between the effect of n in the presence of k
and that in the absence of k. The reader will have no difficulty in
visualizing interaction between n and p and between k and p, and,
also, the interaction of n with k and p combined. Finally, the traditional evaluation of factorial experiments involves the so-called
main effects of the three treatments. The main effect of n, symbolized by N is the arithmetic mean of four differences:

$$n - (\text{control})$$
$$nk - k$$
$$np - p \qquad\qquad [3]$$
$$nkp - kp$$

The main effects of k and p are defined similarly. The attractive feature of factorial experiments and main effects is that each
of the latter utilizes the data from all the plots of the experimental field. Of course, the same remark applies to experiments other
than in agriculture. For example, in a clinical trial the main
effects of each of the treatments studied will utilize the data from
all the patients involved in the experiment, etc.

Obviously, if a single factorial experiment could provide reliable information on all the different simple effects, on all the interactions and all the main effects, one could hope for nothing better.
But can all of these effects be estimated with satisfactory reliability? The answer to this question involves technicalities connected with the concept of "power" of a test introduced in 1933 by Neyman
and Pearson (2). Here, it must suffice that the standard error of a

main effect is one-half of that of a simple effect and that simple interactions of important magnitudes are more difficult to detect.

The traditionally recommended analysis of factorial experiments includes a rule, which will be called here the *main effect rule.* In the various sources it is formulated in general terms, not quite specifically. The following quotations, one from Yates in 1935 and the other from a book by Box and Draper published in 1969 (3), may help avoid misunderstandings.

According to Yates (ref. 1, p. 193), the appropriate way of interpreting the data is as follows:

> In general if there is no evidence of interaction, the mean responses of the two factors (i.e. their "main effects") may be taken as the appropriate measures of the responses to those factors, which may be regarded as additive.

According to Box and Draper (ref. 3, p. 70), the right way to proceed is as follows:

> Suppose we have a main effect A = 7 ± 2 where ± 2 refers to the 2SE limits. We immediately inspect all two-factor interaction effects which involve A....If all of the interaction effects involving A are negligible, then we can interpret the main effect directly. We can say that, for B, C., etc., within the ranges tested, the effect of changing A from a low level to a high level is 7 ± 2.

Both statements are somewhat unspecific. When, exactly, do we say that there is no evidence of interactions? Also, when, exactly, do we say that the interactions are negligible?

The interpretative parts of the two statements are, in a sense, permissive: "...may be taken..." and "We can say that...." Regretfully, the user of the methodology is not warned that, even "within the ranges tested," the main effect A = 7 ± 2 may have no relation to the true effect of the treatment A used singly.

INCIDENT OF 1935

The incident in question occurred at the meeting of the Industrial and Agricultural Section of the Royal Statistical Society (London) at which Yates delivered his paper (1) describing the use of factorial

designs. The following passage (ref. 1, p. 182) stresses the point
that, while in some industrial experimentation the interactions
may be assumed negligible,

> In biological experiments such assumptions cannot be regarded
> as satisfactory...I need only instance dietetic experiments on
> animals, and especially on human beings...Here the interactions
> between the various components of any diet are of vital impor-
> tance, and factorial designs would appear to be as necessary as
> they are in the very similar problems encountered in agronomic
> research.

The published discussion of Yates' paper was predominantly lau-
datory. The only really critical contribution was offered by Neyman
(4) who questioned the validity of the methodology, particularly with
small numbers of replications. In essence, Neyman's argument was
that "if Nature chooses to be frivolous" and endows the treatments
studied with interactions unexpected by the experimenters, then, with
high frequencies, the conclusions drawn from a factorial experiment
like one of those described by Yates may be disastrously wrong,
treatments with deleterious effects being acclaimed beneficial, or
vice versa.

Having noticed a degree of vagueness in the specification of the
conditions under which the main effects ought to be taken as representing the performance of the treatments studied, Neyman adopted the
following "main effect rule." *If all the interactions involving a
factor A are not significant at the two-tail 5% level, while the main
effect of A is significant at 1% two-tail level, then the effectiveness of A will be judged by its main effect alone.*

Having adopted this rule, against which Yates did not protest,
Neyman proceeded to investigate how frequently it might lead to very
wrong conclusions regarding the effectiveness of the factor A. This
would occur when A applied singly is detrimental, even though its
main effect indicates that A is beneficial.

After some theoretical discussion of the power of the statistical tests involved, particularly with reference to a small number of replications, namely three, Neyman produced the results of 30 Monte Carlo simulated experiments modeled on one of those described by Yates. Specifically, all the observable variables were assumed to be normally distributed with the same, although unknown variance, etc. However, the "true" means attainable under all the treatments and their combinations were adjusted to imitate the frivolously behaving Nature. Expressed as percentages of the grand mean obtained in Yates' experiment, they are as follows.

Treatment	(control)	a	b, c, bc	ac, ab, abc	
Percentage	102	98	91	109	[4]

These numbers imply that

1. All the three treatments, a, b, and c applied singly are deleterious, b and c being worse than a.

2. The highest yields can be obtained using either of the three combinations including a, either ab or ac or abc.

3. The main effects are not representative of the performance of the treatments used singly. The main effect of a is positive and large. Those of b and c are zero.

Supposing that the above is the true state of Nature, the question arises: How frequently can a factorial experiment designed as was done by Yates (with three replications) discover anything like this general picture? An even more important question is: How frequently would this experiment lead to the very wrong conclusion that any of the treatments when used alone is beneficial, while in actual fact it is deleterious?

Table 1, reproduced from ref. 4, is indicative of what might happen.

TABLE 1

Results of Sampling Experiment: 'Main' Effect of *a*
and Interactions *a×b* and *a×c* in 30 Hypothetical Experiments

'True' values No.	'Main' effect *a*	S.D. (*a*)	Interaction *a×b*	Interaction *a×c*	S.D. (interaction)	Experiments in which the only significant effect is that of *a*
'True' values	1.755	0.4082	1.545	1.545	0.8165	
No.			Sample values			
1	1.64 SS	0.38	0.44	2.44 SS	0.76	
2	2.15 SS	.34	1.49 S	0.69	.69	
3	1.50 SS	.43	2.36 S	2.13 S	.87	
4	1.61 SS	.36	1.86	2.38 SS	.73	
5	1.50 SS	.48	2.03	2.13 S	.96	
6	1.90 SS	.31	2.83 SS	0.56	.62	
7	0.89	.42	1.86 S	0.88	.83	
8	1.77 SS	.42	1.80	2.20 S	.83	
9	1.66 SS	.30	0.44	1.11	.60	*
10	1.34 SS	.43	1.94 S	0.94	.87	
11	1.87 SS	.36	3.29 SS	1.62 S	.73	
12	1.08	.52	1.20	0.86	1.04	
13	2.01 SS	.25	-0.09	2.05 SS	0.50	
14	2.34 SS	.38	0.72	1.12	.76	*
15	2.66 SS	.40	1.72	1.55	.80	*
16	2.62 SS	.49	1.99	0.75	.99	*
17	1.38 SS	.36	0.50	2.40 SS	.73	
18	1.58 SS	.49	0.51	2.26 S	.99	
19	1.82 SS	.38	0.75	1.61	.76	*
20	1.69 SS	.49	1.03	1.60	.99	*
21	1.93 SS	.38	1.57	2.32 S	.76	
22	2.16 SS	.28	2.18 SS	-0.52	.55	
23	0.45	.30	2.18 SS	1.04	.60	
24	1.68 SS	.38	1.40	-0.41	.76	*
25	1.68 SS	.38	1.08	1.60	.76	*
26	1.76 SS	.36	0.93	2.99 SS	.73	
27	1.45 SS	.40	1.62	2.04 S	.80	
28	1.47 SS	.28	2.00 SS	1.25 S	.55	
29	2.29 SS	.38	1.13	1.36	.76	*
30	2.44 SS	.45	0.98	2.86 SS	.90	

'S' means significant at 5% point
'SS' means significant at 1% point

While the Monte Carlo experiments reported in Table 1 were performed to reflect the state of nature just described, for some reason, the results presented are not in round figures like 102, 98, etc., but in percentages of the SD per plot estimated by Yates, which creates a degree of messiness. Still, the asterisks in the last column tend to clarify the picture. Specifically, they show that, in nine simulated experiments out of 30, the only significant effect found is the main effect of treatment a, which is positive and "highly significant" (at 1%). Here, then, in spite of the 1% significance level used, the factorial experiment led to the "very wrong conclusion" in 9 cases out of 30. (Incidentally, the present writer computed the exact probability of this happening and his result is 0.298.) Remembering the insistence of Yates (see last quotation) that in dietetic as well as in agronomic research the interactions are "of vital importance," it is interesting to count those simulated experiments in which the interesting nonzero interactions have been found significant at 1%. For $a \times b$ the number is 5 out of 30; that for $a \times c$ is 6, say about 20%.

This 20% frequency of finding what is considered to be "vital," combined with 30% frequency of very wrong conclusions, can hardly be considered satisfactory. The remedy that immediately comes to one's mind is an increase of the number of replications. The present writer found that by doubling and by quadrupling the three replications used by Yates, the probability of the "very wrong conclusion" about treatment a is reduced to 0.078 and 0.002, respectively. These results are not general. They apply specifically to the particular frivolous state of Nature visualized by Neyman and need not apply to others of similar frivolity. The reasonable way to proceed seems to be to determine in advance, no doubt subjectively, the limits of error which in any given experimental situation are important to avoid and also the acceptable limits of the chance of committing such errors. Next, the study of power of the contemplated tests would indicate the necessary size of the experiment. In other words, it seems that considerations of power must be an integral part of the experimental design.

EXTENSION OF FACTORIAL DESIGN TO THE DOMAIN OF BINARY DATA

Over the last few decades factorial designs began to be used in experiments with binary (or quantal) responses. These are symbolized
by the experimental responses: "success" or "failure," "survival" or
"death," etc. The domains of research in which the use of factorial
experiments is spreading or, at least, where it is advocated includes
medicine. Sporadic favorable references are exemplified by the recent
review article on gastric ulcers (5) mentioning the "well known" factorial experiments by R. Doll. There is also book literature. A
symposium volume *Controlled Clinical Trials* published in 1960 contains
an article by Doll (6) in which factorial experiments with binary
responses are specifically advocated. Also, a methodological monograph by D. R. Cox (7), published in 1970, contains a special section
on binary response data stemming from factorial experiments. The
statistical tests used by Doll and Cox differ. Doll counts successes
and uses the χ^2 techniques. Cox begins with applying a logistic
transform followed by performing a study akin to the analysis of
variance. In what follows, it will be convenient to designate the
two approaches by the labels "χ^2" and "logistic." The subject of
our discussion will be factorial experiments typified by those of Doll.

Among other things, the paper of Doll (6) describes one of a
series of factorial experiments performed to investigate the effectiveness of 13 different treatments of gastric ulcers. The factorial
experiment actually described was with three different treatments
compared to controls and involved the total of 64 patients, or eight
per treatment category. The evaluation described is entirely in terms
of main effects. It is accompanied by the following comments.

> It was said above that this technique of the simultaneous trial
> of, say, two different treatments on the same patient allows the
> same amount of information to be obtained as would be obtained
> by two separate trials involving double the number of patients.
> In fact, this is not quite true. For example, if both the
> special treatments were effective, but their effects were not
> additive, the total number of patients required to produce a
> statistically significant result would probably be somewhat
> more than half the number required by two separate trials. This,

however, is not a very likely event and it is not important.
Perhaps more important is the positive advantage that the
simultaneous testing of several treatments allows observations
to be made on the interactions between treatments; that is it
provides an opportunity for seeing whether one treatment, perhaps
ineffective by itself, can enhance the effect of another.

It is seen that the author is generally optimistic about the
possibility of detecting interactions and shows no concern about the
reliability of main effects.

The situation with binary trials is analogous but not identical
with that discussed in the preceding section. There, the basic assumptions regarding the distribution of the observable random variables
were that they are normally distributed with the same, though unknown
variance. With binary response one has to consider the probability
of "success," and the variance of a number of successes observable
for any given treatment must be dependent on its mean. Thus, the
Monte Carlo simulation performed by Neyman need not be indicative of
what might happen with main effects and interactions computed for
binary data.

With this in mind, the present writer performed several Monte
Carlo simulations using the following set of probabilities of success.

Treatment	(control)	a	b, c, bc	ab, ac, abc	
Success probability	0.6	0.4	0.2	0.8	[5]

The general picture is similar to that ascribed above to frivolous
Nature. All three treatments applied alone are assumed deleterious,
with *b* and *c* more deleterious than *a*. The largest probabilities of
success, namely 0.8, are assumed to correspond to combinations involving *a*. The main effect of *a* is large and positive.

Simulation experiments were performed first using Doll's number
of observations under each treatment combination, namely 8, and also
double that number and its quadruple. In each case, for each sample,
both testing procedures were applied, the χ^2 and the logistic. There
appeared to be little difference in outcomes. Table 2 summarizes the
results obtained.

TABLE 2

Summary of 100 Monte Carlo Simulations of Binary
Factorial Experiments with Success Probabilities as in Setup [5]

No. of patients per treatment category	Percent of simulated experiments which resulted in:					
	"Very wrong conclusion" regarding treatment a		Failure to find interaction significant at 5% level			
			a × b		a × c	
	χ^2 method	logistic	χ^2 method	logistic	χ^2 method	logistic
8	19	21	65	70	68	71
16	4	4	43	48	49	51
32	0	1	9	17	8	12

Here, as before, the description "very wrong conclusions" des-
ignates the case in which the only significant effect found in an
experiment is the main effect of treatment *a*, which is positive and
significant at 1%. Since all other effects are not significant even
at 5%, the conclusion would be that treatment *a* is beneficial, while
in actual fact it is deleterious, see setup [5]. The other two
double columns of Table 2 give percentages of simulated experiments
which failed to find significant the two interesting interactions.

In conclusion, one can only hope that the properties of the 13
treatments against gastric ulcers studied in experiments of Dr. Doll
were substantially less tricky than those envisaged in setup [5].
However, a novel mathematical-statistical effort on factorial experiments is clearly indicated.

ACKNOWLEDGMENTS

I would like to thank Professor J. Neyman for the very generous help
he has given me in writing this paper. This investigation was partially supported by Research Grant GM-10525, National Institutes of
Health, U.S. Public Health Service.

REFERENCES

1. F. Yates, Complex experiments. *J. Roy. Statist. Soc. Suppl. 2*, 181-247 (1935); *see also Experimental Design, Selected Papers of Frank Yates*, pp. 69-117, Griffin, London, 1970.

2. J. Neyman and E. S. Pearson, On the problem of the most efficient tests of statistical hypotheses. *Phil. Trans. Roy. Soc. Ser. A. 231*, 289-337 (1933); *see also Joint Statistical Papers of J. Neyman and E. S. Pearson*, pp. 140-185, University of California Press, Berkeley, 1967.

3. G. E. P. Box and N. Draper, *Evolutionary Operation*, Wiley, New York, 1969.

4. J. Neyman, Complex experiments. Contribution to the discussion of the paper by F. Yates. *J. Roy. Statist. Soc. Suppl. 2*, 235-242 (1935); *see also A Selection of Early Statistical Papers of J. Neyman*, pp. 225-232, University of California Press, Berkeley, 1967.

5. K. J. Ivey, Anticholinergics: do they work in peptic ulcer? *Gastroenterology 68*, 154-166 (1975).

6. R. Doll, The concurrent assessment of several treatments. In *Controlled Clinical Trials* (A. Bradford Hill, ed.), Thomas, Springfield, Ill., 1960.

7. D. R. Cox, *The Analysis of Binary Data*, Methuen, London, 1970.

STATISTICAL THEORIES
AND SAMPLING PRACTICE

WEI-CHING CHANG was born December 28, 1937, in Taipei, Taiwan. He
obtained his B.A. degree in Philosophy from the National Taiwan
University in 1960. He earned an M.A. degree in Philosophy from
the University of Minnesota in 1964, an M.Sc. degree in Mathematics from the University of Oregon in 1967, and a Ph.D. degree
in Mathematics from the University of Toronto in 1973. He wrote
his doctoral dissertation on the history of statistics under
Professor Kenneth O. May. In 1974 he became an Assistant Professor of Mathematics at the University of Minnesota-Morris. He is
now the Research Officer of Travel Alberta, Government of the
Province of Alberta, Canada.

14.

STATISTICAL THEORIES
AND SAMPLING PRACTICE

Wei-Ching Chang
Department of Mathematics
University of Toronto
Toronto, Canada

The purpose of this report is to trace the close link between general theories of statistics and those used specifically for sample surveys prior to about 1935. Until recently, sampling theory became somewhat divorced from the elaborate theories of statistics. Modern textbooks on the subject, as Professor Kempthorne (1969, p. 673) observed at the Chapel Hill Symposium on Survey Sampling in 1968, pay excessive attention to tedious but elementary calculations of the means and variances of estimators of population means, with only marginal references to inferential processes. In order to emphasize the inferential aspect of sampling theory, I propose to review its early history.

The Danish author van Dantzig divided the development of mathematical statistics into four stages, which he succinctly characterized by the use (in the case of univariate distributions) of one, two, many, and no parameters, respectively (van Dantzig and Hemelrijk, 1954). I find van Dantzig's view very interesting, and will therefore briefly explain his thesis, using it to supplement my account of the history of sample surveys.

According to van Dantzig, the first stage of development began
with the discovery of regularities of certain statistical ratios. In
the area of demography where sample surveys were first used, the English merchant John Graunt's famous tract, entitled *Natural and Political Observations upon the Bills of Mortality* (1662), set the tone
for what was to follow during this stage of development. Graunt sur-
veyed families in certain parishes within the walls of London where
the registries were well kept, and found that there were on an average
three burials per year in 11 families. The annual total of burials
in London being about 13,000, he concluded that the total number of
families was about 48,000. Putting the average family size at 8, he
estimated the population of London to be 384,000. Graunt was apparently aware of the fact that these statistical averages varied in space
and time, but he made no provisions for these variations. His belief
in the constancy of statistical ratios of many vital phenomena was
apparently responsible for his inferences on vital statistics of the
whole country from a list of christenings and burials in a single
parish. Other leading demographers of the seventeenth and eighteenth
centuries, including William Petty and Edmund Halley of England, Per
Wargentin of Sweden, and Johann Peter Susmilch of Germany, made similar rash inferences on certain characteristics of an entire region on
the basis of the data from one section of that region (Westergaard,
1932,1968). In modern terminology, this state of knowledge about demographic phenomena is characterized by one parameter, viz. the mean of
the population under investigation.

The second stage, according to van Dantzig, was marked by the
growing awareness of variability. Various laws of error were suggested by the eighteenth century astronomers, culminating in the works of
Laplace and Gauss on the normal law of error. Mathematically speaking,
the population was characterized at this stage by two parameters, viz.
the mean and the precision constant (standard deviation), and, in the
more general case of multivariate distributions, by the first two moments. Consequently, all statistical theories based on the normal law
of error belong to this stage, including the least-squares method,
theories of correlation and regression, and the analysis of variance

and covariance. In the field of sampling theory, Laplace's work (1786,1812) on the precision of his estimate of the French population (to be discussed shortly) definitely belongs to this stage.

The works of Laplace and Gauss generated excessive reliance on the normal law, especially by such writers as Quetelet, Airy, and Galton. However, toward the end of the nineteenth century, empirical investigations gradually demonstrated normality to be the exception rather than the rule. This led to the development of Karl Pearson's system of skew curves and to the development of the Gram-Charlier series, the theory of curve-fitting by the method of moments, and Pearson's χ^2 goodness-of-fit test. This stage is therefore characterized mathematically by the use of many parameters in frequency distributions. In sampling theory, A. L. Bowley's contribution (1926) represents this stage of evolution.

The large sample theory developed during the second and third stages gave way to a more realistic small sample theory in the twentieth century. The search for new foundations of statistics in the 1920s and 1930s led to the replacement of the inverse approach by those of Fisher and the Neyman-Pearson. This desire for logical rigor, according to van Dantzig, was responsible for the increased interest in the nonparametric or distribution-free approach, which characterizes the fourth stage of development. That a similar sentiment was also prevalent in sampling theory may be seen from the following quotation from Professor Cochran's book on *Sampling Techniques* (1953, 1963, section 6.1):

> The preference in sample survey theory has been to make only limited assumptions about this frequency distribution (that it is very skew or rather symmetrical) and to leave its specific functional form out of the discussion.

In spite of this, present-day sampling theory is based essentially on the normal theory--as indicated by the Fisherian method of the analysis of variance and Neyman's reliance on the Gauss-Markoff least-squares method. Textbooks on sampling still justify the reliance on the first two moments by means of the normal theory (Yates, 1949, section 7.3; Cochran, 1963, section 1.6; Sukhatme and Sukhatme, 1970, section 1.11).

I will further elaborate van Dantzig's thesis in the course of
this discussion. My emphasis, however, will be to review the early
history of sampling from three different points of view that are associated, respectively, with the names of Laplace, R. A. Fisher, and
Neyman. Although less novel than van Dantzig's, my emphasis on inferential procedures adhered to by different schools of thought will com-
plement van Dantzig's approach and will lead to a fuller understanding
of the historical evolution of sampling theory.

THE INVERSE APPROACH

Let me start with the method of inverse probabilities. This method
largely dominated statistical thinking throughout the second and third
stages of development as specified by van Dantzig, and provided early
sampling theorists with an elaborate mathematical scheme of inference.

Pierre Simon Laplace (1749-1827) was apparently the first to apply
this method, which he reestablished in 1774, to the analysis of survey
sampling. The problem that attracted him and contemporary demographers
toward the end of the eighteenth century was the estimation of the
French population. Their method of sampling and estimation was similar to John Graunt's, and did not go beyond the simple arithmetics of
computing population totals from records of births, deaths and so on
from a certain number of parishes chosen subjectively throughout the
country. Laplace (1786) alone saw the need for and found a statement
of precision for his estimate. His solution was later incorporated
into his *Théorie Analytique des Probabilités* (1812, pp. 391ff.; 1820, pp.
398-401), which also contains the following brief description of a
sampling experiment conducted by the French government in 1802 at
his request:

> In the 30 *departments* distributed over the area of France, in a
> way to compensate the effects of the difference in climate, we
> chose those *Communes* whose mayors, by their zeal and intelligence,
> could furnish the most precise information.

From the sample so obtained, Laplace estimated the population total,
and, by resorting to Bayes' postulate of constant a priori distribution, he computed the a posteriori distribution of the difference of

the population and estimated totals. He finally appealed to his previous memoir on the central limit theorem (Laplace, 1783), and approximated this a posteriori distribution by a normal curve.

Ingenious as it was, Laplace's analysis was based on faulty assumptions to suit the Bayesian framework. First, Laplace employed a
cluster sampling design that relied heavily on subjective judgment and
practical expediency. Consequently, contrary to his assumption, his
sample was not taken at random. Second, Laplace took his sample from
the French population which, in turn, was regarded as the second sample from an infinite superpopulation. Furthermore, he assumed that
the two "samples" were independent, which apparently was not the case.*
Third, Laplace in effect assumed the homogeneity of the French population with respect to the annual birth rate. He was, therefore, unable
to free himself from the dogma of homogeneity which characterized the
first stage of development. It may be observed that it was this sweep-
ing dogma which prevented Laplace and others, including Karl Pearson,
from extending mathematical analyses to the design aspect of survey
sampling.†

Laplace's sampling experiment, unfortunately, was ignored by subsequent writers when the method of total enumeration gradually gained
ground in the nineteenth century. In fact, population censuses supplied many statisticians with such a feeling of certainty, formerly
unknown to them, that it became their aim to collect, by means of
complete censuses, as many direct observations as possible. Consequently, the German statistician G. von Mayer (1841-1925) and his

---

*The correct formalization, as Karl Pearson (1928) pointed out,
was to use a hypergeometric instead of Laplace's binomial model.
Pearson's solution differed from Laplace's. The difference was especially great when the sampling fraction was not small.

†That populations of natural and socioeconomic phenomena were
heterogeneous in general with respect to space and time was recognized
by many 19th century statisticians such as Poisson, Bienaymé and
Cournot, but it was not until the German economist W. Lexis' (1877)
work on the analysis of dispersion that a mathematical theory was
advanced to test the homogeneity of a statistical series. As is well
known, Lexis' work anticipated R. A. Fisher's analysis of variance,
which became a useful tool for sampling designs in the 1920s and 1930s.

school objected to the use of *indirect* methods of sampling to gather
information. The advocates of the representative method of sampling
in the late nineteenth and early twentieth centuries, therefore, had
to fight against the dogma that only by complete enumerations could
one hope to attain reliable results.

The British statistician-economist Arthur Lyon Bowley (1869-1957)
was an early crusader against this dogma. He was elected to the International Statistical Institute in 1903, the year when the Institute
endorsed the use of the Norwegian statistician A. N. Kiaer's representative method of sampling. Like many other mathematical statisticians,
among whom were von Borkiewicz (1901), Edgeworth (1912), and Westergaard (1916), Bowley (1906) was attracted to this method and drew
attention to the need and virtue of probability sampling. His subsequent pioneering work on the practical and theoretical aspects of survey sampling naturally led in 1924 to his appointment as a member of
the committee set up by the International Statistical Institute to
study the representative method in statistics. In order to supplement
the committee's recommendation "that the investigation should be so
arranged whenever possible, as to allow of a mathematical statement
of the precision of the results, and that with these results should be
given an indication of the extent of the error to which they are liable"
(Jensen, 1926, p. 378), Bowley (1926) wrote a report entitled "Measurement of the Precision Attained in Sampling," which summarized the mathematical theory of sampling known to him at that date.

To appreciate the historical significance of this report, we note
first that it belongs to the third stage of the evolution of mathematical statistics. Like Laplace before him, Bowley was interested in
a posteriori distributions of population characteristics, and not just
their means and standard deviations. Bowley, however, took advantage
of the development of theoretical statistics and sampling practice
during the late nineteenth and early twentieth centuries. Using the
method of moments, he approximated the hypergeometric distribution by
a Pearson curve, and derived both direct and inverse probability distributions of the difference of the sample and population means under
both simple and stratified random sampling schemes. These probability

distributions became normal when terms of order $n^{-1/2}$ were ignored
(where n is the sample size). Furthermore, following the work of Yule
(1911, pp. 281,349), Bowley demonstrated the basic algebraic identity
in the analysis of variance of stratified populations, and obtained,
from both direct and inverse points of view, the variance of the sample mean from stratified random sampling in relation to that from simple random sampling (Bowley, 1920, pp. 332ff.;1926). As a consequence,
Bowley, like Yule, had a clear conception of the superiority of stratified random sampling over simple random sampling. It must be remarked here that Bowley only considered proportional allocations in
the case of stratified sampling.

An interesting feature of Bowley's inverse approach was his very
weak postulate concerning the a priori distribution. Bowley (1923, p.
494) was well aware of R. A. Fisher's (1922) earlier refutation of
Bayes' postulate. This postulate was adopted by Laplace and his followers for its approximate validity in the case of large samples (Molina, 1931). But in the case of small samples, Fisher's criticism
was both valid and forceful. Fisher charged that this postulate was
inconsistent in that the assumption of the constant a priori distribution for a population parameter did not necessarily lead to the same
distribution for a function of that parameter. Bowley (1923,1926)
tried to circumvent this objection by merely postulating the continuity (and positiveness) of the a priori distribution in a small neighborhood of the observed value. Needless to say, this approach, which
von Mises (1919) and Neyman (1929) also adopted in different contexts,
was intended only for large sample theory. In effect, this reliance
on large sample theory characterized the early Bayesian approach to
the problem of sampling.

Another interesting aspect of Bowley's work is that it was a
product of the period in which the utility of the method of purposive
selection in survey sampling was generally accepted. In spite of his
consistent reliance on the method of random (or, rather, systematic)
selection, Bowley attempted a mathematical theory of purposive selection in the 1926 report. A distinctive feature of this method, according to Bowley, was that it was a case of cluster sampling. It was

assumed that the quantity under investigation was correlated with a
number of characters, called *controls,* and that the regression of the
cluster means of the quantity on those of each control was linear.
Clusters were to be selected in such a way that the average of each
control computed from the chosen clusters should (approximately)
equal its population mean. It was hoped that, due to the assumed correlations between controls and the quantity under investigation, the
above method of selection would result in a representative sample with
respect to the quantity under investigation. The existence of controls
also provided Bowley with supplementary information, which he used to
modify the usual estimate, the weighted mean of the selected cluster
means.

Unfortunately, it soon became apparent that Bowley's method was
wanting on both experimental and theoretical grounds. By means of
Bowley's method, the Italian statisticians C. Gini and L. Galvani
estimated the average rate of natural increase of the population of
Italy, using a purposive selection of 29 out of 214 administrative
districts called circondari̇̂ (Gini, 1928). They found their result
unsatisfactory, and proposed an alternative method of estimation.
Neyman (1934,1938b, pp. 91ff.) examined both methods of estimation,
and concluded that they were generally neither consistent nor efficient. Moreover, in the method of purposive selection, the source of
randomness necessary for the application of the probability theory was
not specified. When it was later specified by Neyman (1934), it turned
out that this method was, after all, a special case of the method of stratified random selection by groups (clusters). Consequently,
the method of purposive selection was soon eliminated from theoretical
discussions of sampling.

THE FISHER APPROACH

The battle that Neyman waged in the 1930s against the method of purposive selection may be regarded as an extension of the larger battle
for randomization carried out in the 1920s and 1930s by Ronald Aylmer
Fisher (1890-1962) and his school, first in the area of agricultural

field trials, then extended to agricultural surveys and other fields of research. In fact, the Fisher approach to survey sampling was a by-product of the theory of the design and analysis of experiments. Consequently, it is necessary to review first how Fisher developed the latter theory from his theory of estimation.

When Fisher (1921,1922,1925b) first propounded his theory of estimation in the early 1920s, he followed Galton and Pearson in considering the statistical data as an integral part of the problem, giving no thought to the manner in which the sample was taken. His chief objective at that time was the reduction of data in a most efficient manner so as to preserve as much as possible of the information contained in the original sample. If the functional form of the distribution was known, Fisher observed in the likelihood function a summary of information that was contained in the data. The method of maximum likelihood then provided the most efficient estimate, which could not be improved according to his theory. The method of inverse probabilities was rejected because the a priori specification of the Bayes-Laplace postulate appeared to him to be both irrelevant (to the reduction of data) and inconsistent (in the sense that it is not transformation-invariant). Also rejected were other less efficient methods of estimation for which the loss of information due to statistical reduction would not be minimal. It was this special concern for minimizing the loss of information due to the reduction of data which characterized the early stage of Fisher's approach to statistics.

That this conception of the role of the statistician was too narrow soon became evident to Fisher through his association with biological and agricultural research workers at Rothamsted Experimental Station (1919 to 1933). In order to assess the value of an experiment, it was necessary to estimate the experimental error, which should be as small as possible. An obvious desideratum for a good design, therefore, was to maximize "the amount of information created by a well-planned expenditure of experimental resources" (Fisher 1947, p. 435). It was natural that Fisher should extend his work beyond the post-data analysis, and investigate the way in which the data were to be collected through a well-planned experimental design.

The introduction of the technique of analysis of variance furnished research workers with a convenient analytical procedure for the design and analysis of experiments (Fisher and Mackenzie, 1923; Fisher, 1925a). Fisher first put forward this technique in relation to intraclass correlation, but he soon realized that it provided an extremely powerful method for the separation of sources of variations in orthogonal designs (such as randomized block designs). Fisher's (1921) work on the sampling distribution of the intraclass correlation (i.e., the F distribution) and the clear conception of degrees of freedom led to tests of significance and estimates of experimental errors. Fisher's analysis of variance thus completed Lexis' analysis of dispersions put forward half a century earlier.

Although Lexis had been impressed by the fact that economic series were rarely homogeneous, Fisher (1925a, p. 224) was struck by the heterogeneity of experimental plots whose fertility varied in a systematic and often complicated manner. It was at this point that Fisher (1925a, p. 224; 1926) modified the conventional procedure and introduced the new principle of randomization. This principle was designed to avoid certain pitfalls of the systematic arrangement of experimental plots, e.g., that it might have features in common with the systematic variation of fertility and that it would not permit valid estimation of errors. The two desiderata of the reduction and valid estimation of error then led in the mid-1920s to the recognition of the rational principles of good experimental designs, including randomization, replication, local control (blocking), factorial design, and so on (Fisher 1926). These principles, except that of randomization, were known and used before Fisher's time. Nevertheless, a clear understanding and the systematic use of them started in the 1920s at Rothamsted Experimental Station.

The principles of experimental design and the analysis of variance provided an adequate framework for the design and analysis of sample surveys when agricultural surveys were conducted at Rothamsted in the late 1920s and early 1930s. It was found that the principle of randomization with suitable local control (stratification) for increasing precision and/or with complex design (subsampling) for saving labor, and the use of the analysis of variance techniques for estimating sampling error in a replicated design were also essential to the rational design and analysis of sample surveys. The loss of information due to sampling and the gain in precision due to subdivision of plots were investigated. The use of supplementary information by the analysis of covariance technique was developed for the ratio estimate. The existence of nonsampling error was again brought to attention, and the danger of excessive reliance on sampling error was emphasized. The two papers by Yates (1935) and Yates and Zacopanay (1935) summarized the experience of the Rothamsted School up to 1935.

Fisher himself never published a paper on survey sampling, but he exerted a profound influence on the course of its development—through his association with research workers in England, India, and the United States, and also as a member of the United Nations Sub-Commission on Statistical Sampling, which met annually from 1947 to 1951. It was largely because of his influence that the battle for randomization was won in experimental sciences, including sample surveys. His work on experimental design furnished sampling theorists with sound principles and useful techniques. His philosophy of statistics led to a new approach alternative to Bowley's, and in recent years stimulated further search for new foundations for survey sampling based on the Fisherian concepts of likelihood, sufficiency, and the fiducial argument (e.g., Basu, 1958,1969; Godambe, 1969; Godambe and Thompson, 1971; Hartley and Rao, 1968,1969; Kalbfleisch and Sprott, 1969; Pathak, 1964; Royall 1968). In short, Fisher was very much responsible for pushing sampling theory toward the fourth stage of development characterized by the distribution-free approach.

THE NEYMAN APPROACH

If Fisher's influence became predominant in the area of agricultural surveys, the new approach advanced by Dr. Jerzy Neyman in the early 1930s replaced Bowley's in the field of social surveys. In 1930, at the request of the Institute of Social Problems in Warsaw, Neyman helped design a sample survey of the conditions of the working class in Poland. This experience, combined with his new statistical outlook,

resulted in a unified theory of sampling, which became widely accepted
in the ensuing years.

Neyman approached the problem of sampling from the point of view
of *confidence intervals*. This approach enabled him to avoid any "su-
perfluous" appeal to Bayes' theorem, thereby constituting a "revolution" in the theory of statistics, according to Neyman (1934). Inter-
estingly, Neyman originally formulated his theory in a Bayesian framework, and justified his solution on the grounds that it is independent
of any a priori assumptions. In contrast to Bowley's Bayesian large
sample approach, however, the new theory was also applicable to small
samples. This tendency toward small sample theory, as I mentioned
earlier, was an important feature of this stage of evolution.

In addition to the theory of confidence intervals, which he first
introduced into the English literature in his fundamental paper entitled "On the Two Aspects of the Representative Method" (1934), Neyman
also advocated a specific method of estimation for survey sampling.
This method, the Gauss-Markoff method of least squares, prescribes a
procedure for obtaining the minimum-variance linear unbiased estimator
for a linear model. Applying this method to stratified random sampling,
Neyman established an optimum property for the usual estimator of the
population total, namely, the weighted sum of stratum means. Moreover,
by minimizing the variance of this best linear unbiased estimator with
respect to the numbers of sampling units selected from different strata
Neyman obtained the solution for the optimum allocation of the sample
of a fixed size among the given strata. This allocation differed from
Bowley's proportional allocation, and it led to the later development
of a more general design of sampling with unequal probabilities in order to increase precision.

The main purpose of Neyman's 1934 paper was to demonstrate that
stratified random sampling by groups is "the only method which can be
advised for general use" (Neyman, 1934, p. 588). Historically, therefore, Neyman continued Fisher's battle for randomization in the field
of social surveys. This was done, in part, by arguing that the method
of purposive selection, when properly formulated, was nothing more than

the method of stratified random sampling by groups. Once this was established, it remained only to investigate the properties of stratified random sampling. Consequently, purposive sampling was spurned by sampling theorists, although it survived in practice in the form of quota sampling.

Neyman's work opened up many new avenues in theoretical and practical research. For example, Neyman (1934,1938a) criticized Gini and Galvani's choice of a small number of large districts in their sample and concluded that a subdivision of strata into a larger number of smaller districts would give a better result. This conclusion was in agreement with the work done at Rothamsted and stimulated further research into the optimum size of sampling units in relation to both precision and cost. Similarly, Neyman used supplementary information in the form of a pilot survey to stratify the population, and thus to increase the efficiency of surveys. Other uses of supplementary information, including his later work on double sampling (Neyman, 1938b), was certainly a natural outgrowth of his early contributions. A useful reference in this respect is P. V. Sukhatme's review paper (1966).

An important factor that made Neyman's paper a landmark in the history of sampling theory, was the fact that he published it in a leading British journal at a time when the need for such a theory was generally acknowledged. A completely different fate awaited the earlier Russian contributions. The Russian authors studied the mathematical foundations of probability sampling in the late nineteenth century and in the first quarter of the present century; they discussed simple and stratified random sampling with or without replacement of sampling units, which might be either individuals or clusters (groups). They arrived at the formula for the Neyman allocation, and dealt with the use of systematic sampling and supplementary information. Hence, it may be said without exaggeration that Russia was the first center of the modern mathematical theory of sampling. Unfortunately, the subsequent ideological interference made it neither possible to discuss probability sampling in Russia nor likely for Russian works to influence the development of sampling theory beyond the

Russian border.   Consequently, Russian works on sampling theory re-mained a forgotten chapter in the history of statistics (Zarkovic, 1956,1962).

ACKNOWLEDGMENTS

It is a pleasure to acknowledge my indebtedness to Professor K. O. May, whose encouragement and help resulted in the final version of this paper. He supervised my doctoral thesis at the University of Toronto on the history of the χ^2 test, which provided useful background material for the present article. I am also grateful to Ms. L. Cheu and Mr. P. Enros, who read the preliminary draft and offered helpful comments.

REFERENCES

Basu, D. (1958).   On sampling with and without replacement. *Sankhyā 20*, 287-294.

Basu, D. (1969).   Role of the sufficiency and likelihood principles in sampling survey theory. *Sankhyā 31*, 441-454.

von Borkiewicz, L. (1901).   Discussion on "Sur les méthodes représentatives ou typologiques appliqués à la statistique, rapport de Kiaer." *Bull. Int. Statist. Inst. 13*, 71-72.

Bowley, A. L. (1906).   Presidential address to the economic science and statistics section of the British Association for the Advancement of Science, York, 1906. *J. Roy. Statist. Soc. 69*, 540-558.

Bowley, A. L. (1920).   *Elements of Statistics* (4th ed.), King, London.

Bowley, A. L. (1923).   The precision of measurements estimated from samples. *Metron 2*(3), 494-500.

Bowley, A. L. (1926).   Measurement of the precision attained in sampling. *Bull. Int. Statist. Inst. 22*, 1-62.

Cochran, W. G. (1953,1963). *Sampling Techniques,* Wiley, New York.

van Dantzig, D., and Hemelrijk, J. (1954).   Statistical methods based on a few assumptions. *Bull. Int. Statist. Inst. 342*, 239-267.

Edgeworth, F. Y. (1912).   On the use of the theory of probabilities in statistics relating to society. *J. Roy. Statist. Soc. 76*, 165-193.

Fisher, R. A. (1921).   On the "probable error" of a coefficient of correlation deduced from a small sample. *Metron 1*(4), 1-32.

Fisher, R. A. (1922). On the mathematical foundations of theoretical statistics. *Phil. Trans. Roy. Soc. Ser. A 222*, 309-368.

Fisher, R. A. (1925a). *Statistical Methods for Research Workers,* Oliver & Boyd, Edinburgh.

Fisher, R. A. (1925b). Theory of statistical estimation. *Proc. Camb. Phil. Soc. 22*, 700-725.

Fisher, R. A. (1926). The arrangement of field experiments. *J. Min. Agric. 33*, 503-513.

Fisher, R. A. (1947). Development of the theory of experimental design. *Bull. Int. Statist. Inst. 31*, 434-436.

Fisher, R. A., and Mackenzie, W. A. (1923). Study in crop variation. II. The manurial response of different varieties. *J. Agric. Sci. 13*, 311-320.

Gini, C. (1928). Une application de la méthode représentative aux matériaux du dernier récensement de la population italienne (1er décembre 1921). *Bull. Int. Statist. Inst. 23*, 198-215.

Godambe, V. P. (1969). Some aspects of the theoretical developments in survey-sampling. In *New Development in Survey Sampling,* N. L. Johnson and H. Smith (Eds.), pp. 27-58, Wiley (Interscience), New York.

Godambe, V. P., and Thompson, M. E. (1971). Bayes, fiducial and frequency aspects of statistical inference in regression analysis in survey-sampling (with discussion). *J. Roy. Statist. Soc. Ser. B 33*, 361-390.

Graunt, J. (1662,1939). *Natural and Political Observations Made upon the Bills of Mortality,* edited with Introduction by W. F. Wilcox, Johns Hopkins, Baltimore.

Hartley, H. O., and Rao, J. N. K. (1968). A new estimation theory for sample surveys. *Biometrika 55*, 547-557.

Hartley, H. O., and Rao, J. N. K. (1969). A new estimation theory for sample surveys. II. In *New Development in Survey Sampling,* N. L. Johnson and H. Smith (Eds.), pp. 147-169, Wiley (Interscience), New York.

Jensen, A. (1926). Report on the representative method. *Bull. Int. Statist. Inst. 22*, 359-377.

Kalbfleisch, J. D., and Sprott, D. A. (1969). Applications of likelihood and fiducial probability to sampling finite populations. *New Development in Survey Sampling,* N. L. Johnson and H. Smith (Eds.), pp. 358-389, Wiley (Interscience), New York.

Kempthorne, O. (1969). Some remarks on statistical inference in finite sampling. In *New Development in Survey Sampling,* N. L. Johnson and H. Smith (Eds.), pp. 671-692, Wiley (Interscience), New York.

Laplace, P. S. (1783). Mémoire sur les approximations des formules qui sont fonctions de très grands nombres. *Oeuvres de Complètes Laplace 10*, 209-338.

Laplace, P. S. (1786). Sur les naissances, les mariages et les morts à Paris depuis 1771 jusqu'en 1784, et dans toute l'étendue de la France, pendant· les années 1781 et 1782. *Mém. Acad. Sciences Paris*, 693-702.

Laplace, P. S. (1812). Théorie analytique des probabilités. *Oeuvres complètes*, Vol. 7, Gauthier-Villar, Paris, 1891.

Lexis, W. (1877). *Zur Theorie der Massenerscheinungen in der menschlichen Gesellschaft*, Wagner, Feiburg im Breisgau.

von Mises, R. (1919). Fundamentalsätze der Wahrscheinlichkeitsrechnung. *Math. Zeit. 4*, 1-97.

Molina, E. C. (1931). Bayes' theorem. *Ann. Math. Statist. 2*, 23-37.

Neyman, J. (1929). Contribution to the theory of certain test criteria. *Bull. Int. Statist. Inst. 24*, 3-48.

Neyman, J. (1934). On the two different aspects of the representative method. *J. Roy. Statist. Soc. Ser. A 97*, 558-625.

Neyman, J. (1938a). *Lectures and Conferences on Mathematical Statistics*, The Graduate School of the U.S. Department of Agriculture, Washington, D.C.

Neyman, J. (1938b). Contribution to the theory of sampling human population. *J. Amer. Statist. Assoc. 33*, 101-116.

Pathak, P. K. (1964). Sufficiency in sampling theory. *Ann. Math. Statist. 35*, 785-809.

Pearson, K. (1928). On a method of ascertaining limits to the actual number of marked members in a population of given size from a sample. *Biometrika 20*, 149-174.

Royall, R. (1968). An old approach to finite population sampling theory. *J. Amer. Statist. Assoc. 63*, 1269-1279.

Sukhatme, P. V. (1966). Major developments in sampling theory and practice. In *Research Papers in Statistics*, F. N. David (Ed.), pp. 367-409, Wiley, New York.

Sukhatme, P. V., and Sukhatme, B. V. (1970). *Sampling Theory of Surveys with Applications*, Asia Publishing House, New York.

Westergaard, H. (1916). Scope and method of statistics. *J. Amer. Statist. Assoc. 15*, 225-276.

Westergaard, H. (1968). *Contributions to the History of Statistics*, Agathon, New York.

Yates, F. (1935). Some examples of biased sampling. *Ann. Eugen. 6*, 202-213.

Yates, F., and Zacopanay, I. (1935). The estimation of the efficiency of sampling, with special reference to sampling error for yield in cereal experiments. *J. Agric. Sci. 25*, 545-577.

Yates, F. (1949). *Sampling Methods for Censuses and Surveys*, Griffin, London.

Yule, G. U. (1911). *An Introduction to the Theory of Statistics*, Griffin, London.

Zarkovic, S. S. (1956). Note on the history of sampling methods in Russia. *J. Roy. Statist. Soc. Ser. A 119*, 336-338.

Zarkovic, S. S. (1962). Note on the history of sampling method in Russia (suppl.). *J. Roy. Statist. Soc. Ser. A 125*, 580-582.

STATISTICS IN ASTRONOMY
IN THE UNITED STATES

ELIZABETH L. SCOTT was born November 23, 1917, in Fort Sill, Oklahoma. She received her B.A. degree in 1939 and her Ph.D. degree in 1949, both from the University of California at Berkeley. In 1951 she became an Assistant Professor of Statistics at the University of California at Berkeley and moved up the ranks to Professor in 1962 where she has remained. She was Departmental Chairman from 1968 to 1973 and Assistant Dean from 1965 to 1967. She is a member of the International Statistical Institute and a Fellow of the Institute of Mathematical Statistics. Scott's early work was in Statistical Astronomy under the supervision of both Robert J. Trumpler, an outstanding astronomer formerly at the Lick Observatory but at that time on the Berkeley Campus, and Jerzy Neyman. She has a wide interest in statistics in the biological and physical sciences, but retains a very strong interest in astronomy. She has published more than 20 research papers in astronomy and numerous papers in statistics.

STATISTICS IN ASTRONOMY
IN THE UNITED STATES

Elizabeth L. Scott

Statistical Laboratory
University of California
Berkeley, California

Astronomy is the oldest of the sciences, yet in many senses it is the newest of the sciences. Through the centuries, statistics has aided astronomy and astronomy has aided statistics. We have read that Gauss developed the theory of errors while working on improving the determination of time by star transits. Even earlier, Laplace developed the first test of a statistical hypothesis while trying to determine whether comets are members of the solar system.

The history of statistics in astronomy is too vast a subject to cover in one paper. We are concerned with this history in the United States, and so I shall focus attention on the determinations of distance and of distributions in space. The distances to the objects that astronomers study today are determined through a complex chain where most of the links are statistical. The chain starts with the trigonometric determination of the distances to the planets and nearby stars, that is, in exploiting their apparent motion as a reflection of the Earth's motion around the Sun. The trigonometric parallax method of estimating distance can be used for nearby objects only. To go farther, astronomers must employ statistical techniques.

Independently, Hertzsprung at Leyden and Henry Norris Russell at
Princeton used the new observations of the near stars, plotting the
absolute luminosity against the spectral type, which is a measure of
the surface temperature of the star and which was then being determined by Annie J. Cannon at Harvard. They found that most
stars were aligned on a "main sequence" with only a few deviating
along an arm referred to as the "giant arm" (because these stars were
unusually bright, and luminosity was assumed to be associated with
size) and a few as "supergiants." Russell developed a theory of stellar evolution to explain what the observations revealed; with more observations, this theory was expanded by Otto Struve and his colleagues
at the Yerkes and McDonald Observatories.* The Russell-Hertzsprung
diagram, as refined, is used to estimate distance, extending as far
as the spectral type can be observed or estimated by means of color.
Using the diagram, the spectral type provides an estimate of the
absolute luminosity. When this is compared with the observed apparent
luminosity, we have an estimate of the distance. Thus, there is another link in the distance determination chain, utilizing observations
of spectral type and apparent luminosity and the statistical relationship of the Russell-Hertzsprung diagram.

The basis for the next step involves a statistical relationship
discovered by H. Sophia Leavitt in 1913 at Harvard for the Cepheid
variables, a particularly luminous class of pulsating stars. In the
Magellenic Clouds she noted a simple statistical relationship
between the period of pulsation of a Cepheid variable and its
apparent luminosity. Because the Magellenic Clouds are external
galaxies (island galaxies outside our own Galaxy), we can assume all
the stars in a Cloud to be at essentially the same distance. Thus,
Leavitt's relation will also hold for period versus absolute luminosity, with a shift in the zero point. If we can determine the

---

*Much of Struve's book *Stellar Evolution* was dictated into a re-
corder while he was observing at the Cassegrain focus of the then new
82-inch telescope at McDonald Observatory. As George Herbig remarks,
probably not many realize that the subtitle *An Exploration from the
Observatory* is almost literally true.

zero point in the relationship between period and absolute luminosity
for Cepheids in our Galaxy, and if we assume that Cepheids in external galaxies are like those in our Galaxy, we will form another
link for distance determination. Harlow Shapley, at Harvard, did
this using nebulae in our Galaxy, which contained both Cepheid variables and bright stars on the Russell-Hertzsprung diagram.
There still is no completely satisfactory explanation of the Leavitt
relation; presumably the curve also represents an evolutionary track
and this interpretation is useful.

With the steps we have just described, astronomers can determine
distances up to 25 million light years. The next link is again statistical, using observations of galaxies. With the great telescopes
of the Mt. Wilson Observatory, Edwin Hubble was photographing galaxies,
classifying them, and estimating the apparent luminosity of the galaxy
as a whole. Milton Humason observed the spectrum, as did N.
U. Mayall with the 36-in. Lick telescopes; they found that the
spectral lines tended to be shifted to the red compared to their laboratory location. Hubble noticed that the fainter the galaxy, the
larger the redshift. Here, then, is another statistical relationship
between apparent luminosity and a property that is observable, albeit
with difficulty, of an astronomical object. We need again to find the
zero point so as to convert the relation to absolute luminosity versus
redshift. The nearby galaxies provide the needed information, since
astronomers can detect and observe the bright stars, especially blue
supergiants, and, as described above, Cepheid variables. The relation
between the absolute luminosity of a galaxy and its redshift provides
another method for determining distance, often called the Hubble
distance. The Hubble relation and its interpretation are of interest
in themselves. In order to discuss them, we need to develop some
notions of the statistical studies of the motion and the distribution
in space of stars and galaxies. It is just this rather narrow domain
of statistics in astronomy that astronomers call *statistical astronomy*.

It is of interest to start with the motions and distribution in
space of the planets, even though we have come to consider them as

completely determined with no statistical interpretations required.
Indeed, neither their motions nor their distribution are completely
determined. Astronomers of the nineteenth century were disconcerted
by an unexplained perturbation in the orbit of Mercury, and searched
for a still undiscovered planet between Mercury and the Sun to try
to explain the irregularity in the motion of this innermost planet.
A thorough discussion of the problem by the American astronomer Simon
Newcomb, at the Naval Observatory, showed that the discrepency is
almost surely real. It remained unexplained until Einstein's theory
of relativity predicted such a perturbation, whence it became a deci-
sive confirmation of special relativity theory. However, a remnant
of unexplained perturbation still persists in the orbit of Mercury,
and this induces further searches for an inter-Mercurial planet and
for an improved theory.

In 1772, Bode described a striking statistical relationship among
the distances of the planets from the Sun (see Table 1). Called Bode's
law, the relation actually is an interpolation formula. When the new
planet Uranus, discovered by Herschel in 1781, turned out to have

TABLE 1

Comparison of Distance Predicted by Bode's Law
and the Mean Actual Distance (in units of 10 times Earth's distance)

Planet	Predicted by Bode's Law	Actual distance
Mercury	4 + 0 = 4	3.9
Venus	4 + 3 = 7	7.2
Earth	4 + 6 = 10	10.0
Mars	4 + 12 = 16	15.2
(Ceres)	4 + 24 = 28	27.7
Jupiter	4 + 48 = 52	52.0
Saturn	4 + 96 = 100	95.4
(Uranus)	4 + 192 = 196	191.9
(Neptune)	4 + 384 = 388	300.7

nearly the distance predicted by Bode's law, followed by the discovery
that the asteroids were in the gap between Mars and Jupiter (the first,
and largest, asteroid Ceres was discovered by Piazzi in 1801), astronomers felt that Bode's law had been confirmed, even though it was not
understood. Therefore, it is not surprising that when perturbations
were found in the orbit of Uranus and the possibility of still another
planet was contemplated, both Adams and Leverrier (in 1846) used the
Bode distance 388 in their computations to predict its position. The
value 388 is quite different from the actual distance 300.7; luckily,
the direction in which to search for the new planet was little affected
at that time. Had they been working 75 years earlier or 75 years later,
this error in distance would have caused quite wrong predictions of
position. Percival Lowell also used Bode's law in his search for another new planet, beyond Neptune, and established the Flagstaff Observatory for planetary research. After Lowell's death in 1916, the
search continued. Pluto was discovered there by Clyde Tombaugh in
1930, but this little planet does not obey Bode's law and is very
different from the other planets. It may be a wanderer captured by
the solar system. In retrospect, it is not clear just how helpful
Bode's statistical relation--still an unexplained interpolation for-
mula--has been in astronomy.

Knowledge of the distribution of the stars and galaxies in space
has developed slowly and in a stepwise fashion. In a monumental exploration of stellar positions and distances, Sir William Herschel
and his sister Caroline noticed that the frequency of pairs of stars
close to each other in the sky was larger than could reasonably be
explained by chance. He suggested that the pairs are the physical
associations, with the two components revolving around each other,
like the Earth in its orbit around the Sun; he then set out to discover
such a double star and determine its orbit. Arising from this statistical clue, Herschel thus demonstrated that the laws of gravita-
tion also operate outside our solar system. Herschel observed that
the distribution of stars on the sky was concentrated toward the plane

of the Milky Way, especially for the fainter stars, but that the
fuzzy objects which he called nebulae were not found there.   The
plane of the Milky Way was a zone of concentration for the stars
and a zone of avoidance for nebulae.

There are many multiple stars and also many groups of stars.   We
have noted that this fact is useful in estimating distances; because
all members of a distant group have essentially the same distance,
knowledge of the distance of one member of the group provides the
distances for all.   In the late 1920s, Robert Trumpler at the Lick
Observatory was comparing groups in different directions, looking into
the distributions of brightness and of color.   He discovered that there
is a great deal of absorbing material, interstellar dust, in the plane
of the Milky Way, which limits our view in the plane but hardly affects
observing perpendicular to the plane.   By a statistical argument,
Trumpler suggested a lenticular shape for our galaxy and established
the Sun's location in the outer part.   Also in the 1920s,
Harlow Shapley at Harvard was studying the groups known as globular
star clusters, which are visible only away from the plane of the Milky
Way, and found that their distribution was concentrated about a center
some 50,000 light years away in the direction of the constellation
Sagittarius rather than symmetrically about the Sun as had been indicated by the limited observations of nearby stars.   These new astronomical studies, each statistical in nature, reinforced each other.
The Sun was displaced from its central location in the universe to
become just an ordinary star (Shapley, 1918).

Half a century ago there was argument about the size of our Galaxy
and the identity of the nebulae.   Harlow Shapley of Harvard and Heber
Curtis of the Lick Observatory debated the subject before the National
Academy of Sciences, with Curtis favoring a Galaxy of some 30,000 light
years and identifying nebulae as star systems similar to the Galaxy
and far outside of it.   Shapley had already shown, through his distance
determinations, that these nebulae were external island universes,
but he argued for a much larger Galaxy with a diameter of 300,000
light years and no other systems similar to it because he utilized

the observations, known later to be proved wrong, made by A. van Maanen at Mt. Wilson Observatory that purported to show motion along the arms of some spiral nebulae. Within a few years, Edwin Hubble at Mt. Wilson showed that the brightest external nebula Andromeda is very distant; he detected and observed Cepheid variables in Andromeda and used them to estimate its distance as 1 million light years. Modern determinations of the diameter of our galaxy, taking into account the absorbing material, make it about 100,000 light years and put the distance of the Andromeda galaxy at about 2 million light years. In 50 years, statistical refinements in distance determination, such as noticing that there is more than one kind of Cepheid variable with somewhat different period-luminosity relationships, have doubled the distance scale for the near external nebulae, now generally referred to simply as galaxies, and have more than doubled the distances of far galaxies. Our own Galaxy is no longer thought to be unusual either in size or location.

Twenty-five years ago astronomers considered galaxies to be uniformly distributed in space with an apparent zone of avoidance caused by the absorbing material near the plane of our Galaxy. This was in spite of the fact that several rich clusters of galaxies were known and studied--in Coma, in Virgo, and in Corona Borealis. Clusters merely added interest, like lace on a petticoat, in the words of H. P. Robertson at Cal Tech. In 1950, C. D. Shane, then director of the Lick Observatory, came to Neyman and me for statistical advice. As director, Shane had inherited a long and ambitious project: to use the new astrographic telescope to photograph systematically the entire sky visible from Lick twice, first during the 1950s and then again after 50 years. The purpose was to obtain better estimates of the distances and motions of the stars and nearby galaxies by measuring their apparent movement with respect to distant galaxies as a reflection of the Sun's motion along the Milky Way and their real motion (which astronomers call peculiar motion). Thus would be established a more reliable trignometric basis for the determination of distances to galaxies and of motions and distances within our Galaxy.

Is there something that an astronomer who is exposing the first
set of plates each for 2 hr, night after night, and not even expecting
to live 50 years more to the second set and fruition, can accomplish
with the first set alone? Shane decided to study the distribution
of galaxies in space. He had started to count the galaxies
in 10' × 10' squares, about 3,000 galaxies per 6° by 6° plate, and
proposed to analyze the counts. The earlier analyses of Bart J. Bok
and others at Harvard, using the surveys instituted by Shapley,
had tested for a Poisson distribution by comparing the mean and variance of the galaxy counts made by research workers at Harvard (they
were known as kilogal counts). Shane realized that the correlations
between the counts in adjacent squares are important; this is where
he sought statistical advice. We devised a stochastic model of kth-
order clustering, a four-dimensional point process projected onto the
two-dimensional photographic plate. We estimated the parameters in
the model from the observed quasi-correlations and other moments,
taking into consideration the uncertainties in the counts due to the
difficulties in distinguishing a faint galaxy from a star on the
astrographic plates. In order to build a realistic model of the difficulties in counting, Shane and Wirtanen took duplicate plates
and made duplicate counts separated by several years.

Our first-order clustering model fitted the observations from the
10' × 10' counts very well. We also fitted the counts from cells
combined into a 6° × 6° square very well, but unfortunately the numerical estimates of the parameters of the clustering were quite
different in the two cases. As the expected number of galaxies per
cluster cannot depend on the size of the cell in which the galaxies
are counted, we must conclude that the first-order clustering model
needs refinement, for example, by including the evidence for subclus-
tering and superclustering which point to higher order clustering.
But, by now astronomers are convinced that galaxies tend to occur
in clusters.

During the last decades, there has been a great deal of interest
in the determination of distances, especially in the Hubble relation

between the apparent luminosity and the redshift of galaxies and its
extension to the similar relation between the apparent luminosity and
redshift of the brightest member of a cluster of galaxies (or the
fifth brightest member to avoid possible confusion with overlapping
foreground clusters). How well determined is such a relation and how
precise is its scale? We must look at the component parts. Starting
with the fundamental first link, which has as unit the mean distance
from the Earth to the Sun--called the *astronomical unit*--we recall
the paper of W. J. Youden, listing the successive determinations of
the astronomical unit in the past 70 years, and pointing out
that *every* one of these 15 estimates lies *outside* of the probable
error range of the preceding estimate. This warning that the internal
consistency of the observations is not a good indicator of the precision of the estimator applies to all the links in the determination
of distances. The assumptions underlying the estimator may be unrealistic. The striking changes the distance scale has suffered in the
last two decades are due essentially to refinements in the assumptions.
These are of two kinds. One is additional precision in the predictor
variables, which reduces the scatter in the statistical relation. We
already noted the striking example of the establishment that there is
more than one type of Cepheid variable with different relations between
period and luminosity. The resulting change of scale alone increased
the estimated distances to the galaxies by a factor of two. Further
refinements in the other links, now being discussed in a series of
papers by Allan Sandage of the Hale Observatories (Mt. Wilson
and Palomar), are creating further changes in the scale that
will increase distances by another factor of approximately two.

The other kind of refinement comes from the fact that astronomers
select the objects they will observe: (1) these objects must be sufficiently bright to be observed with the astronomer's facilities; (2)
they must be identifiable as the type of object the astronomer wants
to study; and (3), which is rather vague, the objects must be interesting or important to the astronomer. Under the assumptions that bright-

ness is the only restriction upon selection and that the true distribution of the variable under observation is normal, K. G. Malmquist
(1922) of Lund showed that the empirical distribution will also be
normal, but with the mean displaced (the so-called Malmquist correction). J. Neyman and E. L. Scott of Berkeley generalized Malmquist's
result by removing the assumption of normality. Utilizing the catalog
(Humason et al., 1956) of redshift measurements by M. Humason and N.
U. Mayall as distance indicators with the apparent luminosity determinations of Allan Sandage for field galaxies and small groups
of galaxies, Neyman and Scott could then estimate the relative
abundance in space for each morphological type of galaxy
as well as the space distribution of absolute luminosity and the selection probability function for each type. When they assumed that
the galaxies in nearby and in distant clusters had the same relative
abundances and the same space distribution of absolute luminosity,
and also the same selection probability function for each type as
had been estimated from the field galaxies and galaxies in small
groups, Neyman and Scott could estimate from the space abundances
and space distribution of luminosities to the catalog distributions
for each morphological type. They found remarkable agreement between
their predictions and the actual catalog distribution for almost every
type and kind of cluster, indicating that the space abundances, the
space distribution of luminosity, and the selection probability functions vary markedly from one morphological type to another, but that
they are the same for each type for field galaxies and for cluster
galaxies, at least for those clusters within the Humason-Mayall-Sandage
catalog. The striking empirical differences in the catalog distributions, well known to astronomers, are all easily explained by the differing selection probabilities operating at increasing distances.

The Hubble relation involves several statistical domains. The
expected relation between apparent luminosity and redshift depends
on the cosmological assumptions made, especially for very distant
galaxies. It is therefore of special interest as a test of cosmological theories: Do the observations deviate from a straight line

and, if so, which curve do they follow? But the statistical refinements themselves will cause the observations to deviate from a straight line, as shown by E. L. Scott (1957). In order that a distant galaxy be observable at all, it must be unusually bright. Therefore, the expected Hubble relation is not a straight line; the amount of deviation will depend on the dispersion in the space distribution of luminosity. Also, a cluster will be more likely to have a galaxy bright enough to be observed if the cluster has an unusually large number of members. Both of these effects--the tendency for a distant galaxy to be unusually luminous and the tendency to belong to a very rich cluster--will curve the Hubble relation and thereby confuse the cosmological tests. Astronomers have tried to obviate the effect by searching out objects with small dispersion in luminosity. As astronomers are working here at the limit of their instruments, this task is not easy and it is probably more convenient and more accurate to calculate the statistical refinement in the Hubble relation.

The recently discovered exciting objects of extremely high redshift, such as quasars, provide the possibility of new distance determinations. However, as discussed (1968) by Maarten Schmidt of the Hale Observatories, no reliable relation has as yet been found for quasars; they have many surprising properties--very small diameters, very rapid oscillations--that are not understood and so are disturbing. There remain many unsolved problems in astronomy, some of them quite new and many of them statistical, that promise exciting discoveries and important insights into the history and future of our universe.

ACKNOWLEDGEMENT

This work was prepared with the partial support of Grant DA ARO D-31-124-G31, University of California, Berkeley.

REFERENCES

Bok, B. J. (1934). The apparent clustering of external galaxies. *Harvard Coll. Observ. Bull. No. 895*, 1-8.

Hubble, E. (1929a). A relation between distance and radial velocity among extra-galactic nebulae. *Proc. Nat. Acad. Sci. 15*, 168-173.

Hubble, E. (1929b). A spiral nebulae as a stellar system, Messier 31. *Mt. Wilson Contrib. No. 36: Astrophys. J. 69*, 103-157 (contains references to earlier papers from 1924).

Humason, M., Mayall, N. U., and Sandage, A. R. (1956). Redshifts and magnitudes of extra-galactic nebulae. *Ast. J. 61*, 97-162.

Marquis de Laplace, P.-S. (1776). Mémoire sur l'inclinaison moyenne des orbites des comites. *Mémoires de l'Académie Royale des Sciences de Paris, Savants étrangers, année 1773*, Vol. 7, 1776; *Oeuvres Complètes de Laplace*, Vol. 8, pp. 279-302, Gauthier-Villars, Paris 1841.

Leavitt, H. S. (1912). *Harvard College Observatory Circular, No. 173* (partial reference starting 1908).

Malmquist, K. G. (1922). On some relations in stellar statistics. *Lund Medd. I No. 100; Ark. Mat. Ast. Fys. 16*, 1-52.

Neyman, J. (1974). *The Heritage of Copernicus: Theories "Pleasing to the Mind,"* pp. 24-139, MIT Press, Cambridge, Mass. (contains four interesting papers on problems discussed here).

Neyman, J., and Scott, E. L. (1952). A theory of the spatial distribution of galaxies. *Astrophys. J. 116*, 144-163.

Neyman, J. and Scott, E. L. (1958). Statistical approach to problems of cosmology. *J. Roy. Statist. Soc. Ser. B 20*, 1-43.

Neyman, J., and Scott, E. L. (1961). Field galaxies: Luminosity, red-shift, and abundance of types. Part I. Theory. *Proc. 4th Berkeley Symp. Math. Statist. Prob. 3*, 261-276, University of California Press, Berkeley.

Neyman, J., and Scott, E. L. (1974). Field galaxies and cluster galaxies: Abundances of morphological types and corresponding luminosity functions. In *Confrontation of Cosmological Theories with Observational Data*, M. S. Longair (Ed.), pp. 129-140, International Astronomical Union Symp. No. 63, D. Reidel, Dordrecht, Holland.

Robertson, H. P. (1955). The theoretical aspects of the nebular redshift. *Publ. Ast. Soc. Pacific 67*, 82-98.

Sandage, A. Steps towards the Hubble constant, and The redshift-distance relation. A monumental series of papers ending with (1973) *Astrophys. J. 183*, 743-757.

Schmidt, M. (1968). Space distribution and luminosity functions of quasi-stellar radio sources. *Astrophys. J. 151*, 393-409.

Scott, E. L. (1957). The brightest galaxy in a cluster as a distance indicator. *Ast. J. 62*, 248-265.

Shane, C. D., and Wirtanen, C. D. (1967). The distribution of galaxies. *Publ. Lick Observ. 22,* 1-60.

Shapley, H. (1918). Remarks on the arrangement of the sidereal universe. *Contrib. Mt. Wilson Solar Observ. No. 157* (see also Nos. 151 and 161).

Shapley, H. and Curtis, H. D. (1921). The scale of the universe. *Nat. Res. Council Bull. No. 11, 2* (3), p. 182. ("Great Debate" was on April 26, 1920)

Slipher, V. M. (1917). Nebulae. *Proc. Amer. Phil. Soc. 56,* 403-409.

Trumpler, R. J. (1930). Preliminary results on the distances, dimensions and space distribution of open star clusters. *Lick Observ. Bull. 14,* 154-188.

Youden, W. J. (1962). *Experimentation and Measurement,* pp. 93-95, Scholastic Book Services, New York.

DEVELOPMENT OF
THE DECISION MODEL

THOMAS S. FERGUSON was born December 14, 1929, in Oakland, California. He received his A.B. degree in 1951 and his Ph.D. degree in 1956 from the University of California at Berkeley. Presently he is Professor in the Department of Mathematics at the University of California at Los Angeles. He is a Fellow of the Institute of Mathematical Statistics and author of a widely used, advanced text, *Mathematical Statistics: A Decision Theoretic Approach*. He has served on the Editorial Board of the *Annals of Mathematical Statistics*, and was elected to the Council of the Institute of Mathematical Statistics in 1974.

16.

DEVELOPMENT OF
THE DECISION MODEL

Thomas S. Ferguson
Department of Mathematics
University of California
Los Angeles, California

It seems appropriate at this celebration of the bicentennial of the
United States to trace the development of a model of statistical theory
that has become a characteristic of the mode of statistical thought in
this country in the last 25 years--the decision model of mathematical
statistics. This model provides a general framework for viewing statistical problems, for formulating them and for analyzing them. From
the decision model point of view, statistics may be defined briefly as
the science of making decisions in the face of uncertainty.

There are other important viewpoints and definitions of statistics.
No one model could suffice to contain all viewpoints of statistics. In
particular, the prevailing point of view in the first half of this century was that of the inference school, of which the dominant figure and
symbolic leader was R. A. Fisher. This school considers statistics as
a means of making inferences, as a means of reducing uncertainty through
experiment and observation, or merely as a means of the reduction of data (Fisher, 1922). Günter Menges, in an interesting article (1971) comparing the inference and decision viewpoints, notes that "the followers
of the English Fisher school reject decision theory, as did R. A.

Fisher, who in his book *Statistical Methods and Scientific Inference*
disdainfully qualified decision theory as nonscientific. The opposite
position is taken by a great number of modern, in particular American,
authors...." It is not surprising that decision theory is dominant in
this country with such statistical leaders and teachers here as Jerzy
Neyman, Abraham Wald, and L. J. Savage, all principals in the development of the theory.

The main beauty of the decision model is the contrast between the
simplicity of its basic notions and the broad areas of its usefulness.
Its applications reach into mathematical programming and operations
research, and in its most general form it provides a model for individual behavior in mundane problems of everyday life. There is always
value in considering problems from a decision-theoretic viewpoint.

THE MODEL

In describing the model, I shall refer to the decision maker as "the
statistician." There are just four basic elements of the model:

1. The space α of actions available to the statistician.

2. The space Θ of states of the world, or states of nature.
    One of these is the "true" state, but the statistician does
    not know which one. The space Θ is also called the parame-
    ter space.

3. The loss function $L(\theta,a)$, representing the numerical loss to
    the statistician if he takes action a ε α, when the true
    state of nature is θ ε Θ.

4. An experiment yielding observations X, the distribution of
    which depends on the true state of nature, and which hopeful-
    ly will help the statistician to reduce his loss.

This model encompasses such diverse problems as estimating the
bacterial density of a reservoir in order to decide how much chlorine
to add, or watching the stock market indicators in order to decide how
much to invest, or deciding whether to give up smoking on the basis of
the statistical data that has been amassed so far. Often an explicit
loss function is not available. The problem of deciding whether to
give up smoking is an extreme example. In a personal problem of this
sort, one must measure numerically the pleasure one gets from smoking

and compare on the same scale the distress due to the prospect of a shorter life.

A little terminology is useful in describing exactly the statistician's problem. *A decision function,* δ, for the statistician is a rule that associates an action with each possible outcome of the experiment, $\delta(x) \in \alpha$. For a given decision function, *the risk function* is the expected value of the loss, as a function of the true state of nature. The objective of the statistician is to choose a decision function whose associated risk function is small in some sense. He may, for example, seek to minimize the maximum risk (the minimax approach) or to minimize the average risk given some prior distribution on the states of nature (the Bayes approach).

EARLY DECISION THEORISTS

When I studied decision theory as a student in Berkeley in the early 1950s, I felt that the model itself was so simple that it must have been in use since Adam bit the apple. However, the statement of the model in the generality I have described seems to have been published for the first time by Wald in 1947. That does not mean that earlier statistical scholars did not think in decision-theoretic terms.

An early clear example is Daniel Bernoulli who, in 1730, introduced the notion of moral expectation and described a very important utility function. Bernoulli points out that the value of a small sum of money to a poor man is not the same as it is to a rich man. Judging the value of a small sum of money as being relative to one's fortune, Bernoulli arrives at the notion of the log utility function, in which the utility of a monetary fortune is measured as the logarithm of the amount of the fortune. He defines moral expectation as the expected value of the logarithm of the fortune, and explicitly suggests taking actions that yield a large moral expectation. As an example, Bernoulli considers the problem of sending a sum of money from one port to another. One has the option of sending it all on one ship, or of splitting the sum of money into two parts and sending it on two separate ships. With each ship there is a small probability that the ship

will be lost. Bernoulli recommends taking that action with the higher moral expectation, i.e., send the money on two ships. This analysis
is the forerunner of the advice stock market analysts give today--di-
versify! Sancho Panza's advice to Don Quixote not to venture all his
eggs in one basket has a similar justification.

Two aspects of the decision model are clearly visible in Bernoulli's example, the space of actions (send one or two ships), and the
loss function (the negative of the utility).

Although today much decision-theoretic work is done from a Bayesian viewpoint, it is interesting to note that Thomas Bayes himself was
apparently not decision oriented. In his basic paper of 1763, a careful solution is given to the problem of finding the posterior distribution of the probability p of the occurrence of an event, given the
number of times the event has occurred and failed to occur in n trials,
when the prior distribution of p is uniform on the interval [0,1].
However, no explicit use of this posterior distribution is suggested.
Not even the probability of the occurrence of the event on the next
trial is calculated. This was done by Laplace (1774).

Laplace should certainly be counted as one of the first statistical decision theorists, mainly in connection with the estimation of
orbits of planets and comets. Laplace found (1812) that for a given
distribution the expected distance to a point is minimized by choosing
the point to be the median of that distribution. As an estimate of a
location parameter with absolute error loss function, he suggested
using the median of the posterior distribution when the prior is taken
to be uniform. Here all four elements of the decision model are evident. The actions are the estimates, the states of nature may be taken
as the set of real numbers, the loss function is absolute difference,
and the observations are the sample from the distribution with the true
state of nature as a location parameter.

Gauss also used methods that today would be considered decision-
theoretic. Gauss saw (1821) that the mean of a distribution renders
the expected squared error a minimum. He stated clearly that the
choice of a loss function was somewhat arbitrary, that Laplace's choice

of absolute error was no less arbitrary than his own choice of squared error. By proving what is now known as the Gauss-Markov Theorem, he hoped "to satisfy mathematicians by demonstrating ... that the method of least squares furnishes the most advantageous combination of the observations, not merely approximately, but in an absolute sense, and this for an arbitrary distribution for the errors,...." This absolute sense is a decision-theoretic one very familiar to statisticians today, and Gauss felt that it was a much sounder criterion than just minimizing the sum of squares of the errors or that of being the maximum likelihood estimate based on the normal distribution for the errors.

TESTING HYPOTHESES

A few examples of a decision-theoretic treatment of statistical data does not make a model. It takes a lot of examples in addition to repetitively dealing with similar problems that can be treated in a unified manner before the need for a general model arises. And then it takes a stroke of insight.

An example of this procedure of building a model occurred in the 1930s with the development of the Neyman-Pearson theory of testing hypotheses. From the 1928 papers on the likelihood ratio test through the 1938 papers in the *Statistical Research Memoires,* J. Neyman and E. S. Pearson erected that mathematical structure for testing hypotheses that has had such an immense impact on the statistical world. It did not suddenly appear in one paper, but was built up, not always in the most convenient order, slowly and carefully, step by step. According to this theory, the statistician is to decide whether to accept a given hypothesis concerning the true state of nature, or to reject it and accept some alternative hypothesis. Thus, there are just two actions. There are two types of error--rejecting the given hypothesis when it is true and accepting it when it is false. These errors are not presumed to be of the same magnitude. This provides a basis for a loss function that could be set to two given positive values for the two types of error, and to zero if no error is committed.

DEVELOPMENT IN WALD'S PAPERS

Wald's first paper on decision theory, "Contributions to the Theory of Statistical Estimation and Testing Hypotheses" (1939), combines the problems of estimation and testing hypotheses in one model. Here for the first time the four basic elements of the model are *explicitly* displayed. Unfortunately, the full generality of the model is restricted by Wald's selection of his action space as some system S of subsets of the parameter space, and by his assumption that the loss function (called then a weight function) take on the value zero if the true value of the parameter is an element of the selected set, and is otherwise nonnegative.

When S consists of two disjoint subsets of the parameter space, this reduces to a hypothesis testing problem. When S is the collection of all singleton point sets, it is an estimation problem. It is most probable that Wald took his action space to be a system of subsets of the parameter space in order to attempt to include Neyman's (1937) theory of confidence intervals, which he hoped to subsume by letting S be some collection of intervals in the parameter space. He states "The problem in this formulation is very general. It contains the problems of testing hypotheses and of statistical estimation treated in the literature, (see, for instance, J. Neyman, 1937)." This is interesting because Neyman's justification of confidence intervals is not decision-theoretic.

A second interesting feature of Wald's first paper on decision theory is that his definition of the important notion of admissibility is not the one we use today. A different definition independent of the loss function, is given. The contemporary notion of an admissible decision function is there called "admissible relative to the weight function W."

In 1944, an event occurred that hastened the development of the decision model. That event was the publication of the important book of J. von Neumann and O. Morgenstern, *Theory of Games and Economic Behavior*. The theory of games concerns two or more decision makers with partially opposing interests. The decision makers simultaneously

choose an action each from his personal action space, and individual
payoff functions determine what each one wins. A model for two persons
of completely opposing interests goes back to E. Borel (1921), but game
theory really dates from 1928 when von Neumann proved the fundamental
minimax theorem for finite two-person zero sum-games.

In his 1945 paper, Wald pointed out the similarity between the
theory he was developing and the theory of two-person games. One of
the players takes the role of the statistician and the other Wald refers to as "nature." The negative of the payoff function is taken as
the loss function. However, his own theory still has the action space
restricted as a system of subsets of the parameter space, and the loss
zero if the true parameter point lies in the selected subset. But the
definition of admissibility is changed into the one we use today.

The action space was finally set free of the parameter space in
Wald's paper (1947a). This paper is notable for two features. It is
the first paper to use the decision model as described earlier in its
general form, and it contains a celebrated theorem characteristic of
decision theory and of Wald's unique ability--the complete class theorem. The notion of an essentially complete class of decision functions
was introduced earlier that year by Lehmann (1947) in a hypothesis
testing setting. Wald showed that under fairly general conditions the
class of Bayes decision functions forms an essentially complete class;
in other words, for any decision function that is not Bayesian, there
exists one that is Bayes and is at least as good no matter what the
true state of nature may be. Also in 1947, Wald published an extension of the decision theory model to contain his theory of sequential
analysis. Further work finally culminated in his book *Statistical
Decision Functions* in 1950. In December of that year, Wald met a
sudden and untimely death in an airplane accident.

The theory Wald created lives on. In the early 1950s a prodigious
amount of work on the general theory was carried out by such men and
in such papers as C. R. Blyth (1951), M. N. Ghosh (1952), J. Kiefer
(1953), M. Sobel (1953), L. Weiss (1953), J. Wolfowitz (1951), and C.
Stein (1955). Two specific influences in this avalanche of work deserve

special mention. The first is the book of Blackwell and Girshick
Theory of Games and Statistical Decisions (1954) that did much to popularize the theory and point out its vast applications. The second
is the paper of Le Cam (1955) in which Wald's theory is extended and
put into a modern mathematical framework. The influence of these two
studies is complementary: Blackwell and Girshick stimulate interest
in specific decision problems, while Le Cam stimulates the fine mathematical treatment of decision theory. But more important than the
general theory is the pervasive influence the model has brought to
statistical thinking. The simple idea of attempting to put statistical problems in a form with a clear criterion of optimality has been
extremely fruitful.

UTILITY AND PERSONAL PROBABILITY

There are two further developments of this theory that have made it
much more valuable. The first is the mathematical axiomatization of
utility theory. For the decision model, the loss function is considered to be known. Wald assumed the loss function known, but allowed
that it may depend on the statistician. In 1939, he wrote "The question as to how the form of the weight function $W(\theta,\omega)$ should be determined, is not a mathematical or statistical one. The statistician who
wants to test certain hypotheses must first determine the relative importance of all possible errors which will entirely depend on the special purposes of his investigation."

In a sense, the question of determining the weight function was
turned into a mathematical one by the axiomatization of utility theory
created in the book of von Neumann and Morgenstern. In this approach,
it is assumed that the statistician can express his preferences between random outcomes in a manner consistent with certain axioms. As
a consequence, there exists a numerical valued function, called a utility and defined on the set of outcomes, for which the statistician's
preference between two actions corresponds to a preference for that
action with the larger expected utility. Moreover, approximations to

the utility function can be obtained by determining some of the statistician's preferences.

It is to be emphasized that this is a personal theory. If the losses are measured in monetary units, Bernoulli would suggest using the negative of the logarithm of the statistician's resulting fortune as the loss. For a group effort such as putting a missile into orbit, the monetary consequences of any errors are so great as to swamp any individual's fortune. Utility theory is not well developed for group efforts.

The second development, also a theory of personal behavior, has had a strong influence on statistical thought in the past twenty years. This is the theory of personal, or subjective, probability. This theory, initiated in this century by E. Borel, F. Ramsey, B. de Finetti, and B. O. Koopman, and carried to completion by L. J. Savage in his book *The Foundations of Statistics* (1954), views probability as a personal attribute of the decision maker. In particular, each person whose behavior is sufficiently rational as to satisfy certain reasonable axioms behaves as if he viewed the unknown true state of nature as having been chosen from some distribution, his personal prior distribution. Such a person would choose as a solution to a statistical decision problem a Bayesian decision function with respect to his prior distribution. This theory has provided new impetus to the study of Bayes decision functions. In addition, it provides an elegant personal theory of decision making, in which new information, statistical or otherwise, may be evaluated and summarized in the form of a posterior distribution. Indeed, some statisticians, notably L. J. Savage, have attempted to consider statistics entirely from this point of view.

One of the satisfying features of the theory is that it puts the prior distribution on the same personal basis as the utility or loss function. To Wald, the loss function was given, but a prior distribution would either not exist or be too vague to be useful. Yet the existence of both may be given analogous justifications. However, the theory of personal probability is, as a result, even more strongly restricted than utility theory to be a theory of personal behavior.

CONCLUDING REMARKS

It is always interesting, although not always productive, to speculate
as to why a given theory appears at a particular time. In the case of
the statistical decision model, it could have been invented at the time
of Laplace. All the ingredients were there: acting on the basis of
experimental data to estimate unknown but fixed parameters, the use
of a loss function known to be arbitrary to a certain extent, a notion
of utility, at least as applied to monetary outcomes, and the use of
prior probabilities (although not necessarily personal probabilities).
Yet it took another 150 years before a complete statement of the model
was made. Why did it take so long? I believe it is because a general
theory to be useful must encompass a wide variety of problems. The
decision-theoretic problem scientists were concerned with at the time
of Laplace was estimation of parameters and so the need for a general
theory was not great. At the beginning of this century, a remarkable
growth in the theory and scope of statistical methods took place under
the leadership of such statisticians as Galton, K. Pearson, Edgeworth,
Gosset ("Student"), and R. A. Fisher. Yet, most of the theory was
inferential or descriptive rather than decision oriented. When testing hypotheses as developed by Neyman and Pearson provided another
solid example of a decision oriented statistical problem, the need and
value of combining this with estimation existed but lay dormant. The
catalyst in the development was Wald's admiration for the Neyman theory
of confidence intervals. The resulting model in Wald's hands benefited
from the ideas of von Neumann and Morgenstern and grew into a very general theory. In the hands of L. J. Savage, it developed into an elegant approach to personal behavior.

ACKNOWLEDGMENT

The preparation of this manuscript was partially supported by National
Science Foundation Grant GP-33431X.

REFERENCES

Bayes, T. (1763). An essay towards solving a problem in the doctrine of chances. Reprinted in *Biometrika 45*, 293-315 (1958), preceded by a biographical note on Bayes by G. A. Barnard.

Bernoulli, D. (1730). Specium Theoriae Novae de Mensura Sortis. (English transl. by L. Sommer) *Econometrica 22*, 23-36 (1952).

Blackwell, D., and Girshick, M. A. (1954). *Theory of Games and Statistical Decisions*, Wiley, New York.

Blyth, C. R. (1951). On minimax statistical decision procedures and their admissibility. *Ann. Math. Statist. 22*, 22-42.

Borel, É. (1921). The theory of play and integral equations with skew-symmetric kernels. Translated in *Econometrica 21*, 97-100 (1953).

Borel, É. (1924). Apropos of a treatise on probability. In *Studies in Subjective Probability*, H. E. Kyburg, Jr. and H. E. Smokler (Transl. Eds.), Wiley, New York, 1964.

Finetti, B. de (1937). Foresight: Its logical laws, its subjective sources. In *Studies in Subjective Probability*, H. E. Kyburg, Jr. and H. E. Smokler (Transl. Eds.), Wiley, New York, 1964.

Fisher, R. A. (1922). On the mathematical foundations of theoretical statistics. *Phil. Trans. Roy. Soc. Ser. A 222*, 309-368.

Fisher, R. A. (1959). *Statistical Methods and Scientific Inference* (2nd ed.), Hafner, London.

Gauss, K. F. (1821). Theory of the combination of observations which leads to the smallest errors. *Gauss Werke 4*, 1-93. (Translated by H. F. Trotter, Tech. Rep. No. 5, 1957, Statistical Techniques Research Group, Princeton University.)

Ghosh, M. N. (1952). An extension of Wald's decision theory to unbounded weight functions. *Sankhyā 12*, 8-26.

Kiefer, J. (1953). On Wald's complete class theorems. *Ann. Math. Statist. 24*, 70-75.

Koopman, B. O. (1940). The axioms and algebra of intuitive probability. *Ann. Math. 41*, 269-292.

Laplace, P. S. (1812). *Théorie analytique des probabilitiés*, Courcier, Paris.

Le Cam, L. (1955). An extension of Wald's theory of statistical decision functions. *Ann. Math. Statist. 26*, 69-81.

Lehmann, E. L. (1947). On families of admissible tests. *Ann. Math. Statist. 18*, 97-104.

Menges, G. (1971). Inference and decision. In *Inference and Decision,* G. Menges (Ed.), pp. 1-15, Wiley, New York.

Neumann, J. von (1928). Zur Theorie der Gesellschaftsspiele. *Math. Ann. 100,* 295-320.

Neumann, J. von, and Morgenstern. O. (1944). *Theory of Games and Economic Behavior,* Princeton University Press, Princeton.

Neyman, J. (1937). Outline of a theory of statistical estimation based on the classical theory of probability. *Phil. Trans. Roy. Soc. Ser. A 236,* 333-380.

Neyman, J., and Pearson, E. S. (1928). On the use and interpretation of certain test criteria for purposes of statistical inference. I, II. *Biometrika 20A,* 175-240;263-294.

Neyman, J., and Pearson, E. S. (1933). On the most efficient tests of statistical hypotheses. *Phil. Trans. Roy. Soc. Ser. A 231,* 289-337.

Neyman, J., and Pearson, E. S. (1936,1938). Contributions to the theory of testing statistical hypotheses. *Statist. Res. Mem. 1,* 1-37; *2,* 25-57.

Ramsey, F. P. (1926). Truth and probability. In *The Foundations of Mathematics and Other Logical Essays* (1950), R. B. Braithwaithe (Ed.), Humanities Press, New York.

Savage, L. J. (1954). *The Foundations of Statistics,* Wiley, New York.

Sobel, M. (1953). An essentially complete class of decision functions for certain standard sequential problems. *Ann. Math. Statist. 24,* 319-337.

Stein, C. (1955). A necessary and sufficient condition for admissibility. *Ann. Math. Statist. 26,* 518-522.

Wald, A. (1939). Contributions to the theory of statistical estimation and testing hypotheses. *Ann. Math. Statist. 10,* 299-326.

Wald, A. (1945). Statistical decision functions which minimize the maximum risk. *Ann. Math. 46,* 265-280.

Wald, A. (1947a). An essentially complete class of admissible decision functions. *Ann. Math. Statist. 18,* 549-555.

Wald, A. (1947b). Foundations of a general theory of sequential decision functions. *Econometrica 15,* 279-313.

Wald, A. (1950). *Statistical Decision Functions,* Wiley, New York.

Weiss, L. (1953). Testing one simple hypothesis against another. *Ann. Math. Statist. 24,* 273-281.

Wolfowitz, J. (1951). On ε-complete classes of decision functions. *Ann. Math. Statist. 22,* 461-465.

LARGE SAMPLE COMPARISON
OF TESTS AND
EMPIRICAL BAYES PROCEDURES

JACK C. KIEFER was born January 25, 1924, in Cincinnati, Ohio. His professional life has been spent in the Mathematics Department of Cornell University after receiving a Ph.D. degree from Columbia University in 1952. He is a member of the National Academy of Sciences and is also a Fellow of the Institute of Mathematical Statistics (President, 1969-1970).

DAVID S. MOORE received his A.B. degree from Princeton University in 1962 and his Ph.D. degree in Mathematics from Cornell University in 1967. Since that time he has been at Purdue University where he is now an Associate Professor in Statistics. He is Associate Editor of the *Journal of the American Statistical Association* and author of a number of papers on large sample theory of tests of fit and procedures using order statistics.

17.

LARGE SAMPLE COMPARISON
OF TESTS AND
EMPIRICAL BAYES PROCEDURES

Jack C. Kiefer

Mathematics Department
Cornell University
Ithaca, New York

David S. Moore

Department of Statistics
Purdue University
Lafayette, Indiana

The organizers of this conference have made a selection of recent
influential statistical ideas, and have asked us to present an ex-
position of the two topics of the title. The emphasis of the first
of these is to be on the use of limit theorems other than the central
limit theorem in large sample comparison of tests, in contrast with
the now more familiar "local" comparison treated by Pitman, Wilks,
Wald, LeCam, Neyman, Weiss, and Wolfowitz, among others. The non-
local comparison of tests was developed by Chernoff, Hodges and Lehmann,
and Bahadur, producing a striking result in a paper of Hoeffding
(1965). Empirical Bayes procedures were introduced by Robbins (1955).

LARGE SAMPLE COMPARISON OF TESTS

Introduction

We begin with a simple testing model: one observes independent and
identically distributed random variables $X_1, \ldots, X_n$. The probability
density function of X_i is unknown, but belongs to a known class

349

$\{f_\theta, \theta \in \Theta\}$ labeled in terms of an index set Θ. For example, Θ might be the upper half-plane and $f_{(\theta',\theta'')}$ the normal density with mean θ' and variance θ''. Or the class $\{f_\theta\}$ might consist of every symmetric density and $\theta = (\theta',g)$, where θ' is the median of f_θ and g the density of the "error" $X_1 - \theta'$, symmetric about 0.

It is desired to test the null hypothesis that $\theta \in \Theta_0$ for $\Theta_0 \subset \Theta$ against the alternative that $\theta \in \Theta - \Theta_0$. For simplicity, we shall assume throughout the first two sections of this paper that (1) the parameter space Θ is a subset of Euclidean k-dimensional space; (2) $\Theta_0 = \{\theta_0\}$, so that we are testing the simple null hypothesis H_0: $\theta = \theta_0$; (3) all critical regions considered are defined in terms of sums of iid random variables standardized to approach a normal distribution (under θ_0) by the central limit theorem. (The central limit theorem is not used in approaches that involve a computation like Eq. [2].) Given a sequence of critical regions $\{T_n\}$, we have two probabilities of error: the significance level or probability of erroneous rejection of H_0

$$\alpha_n = P_{\theta_0}[(X_1, \ldots, X_n) \in T_n]$$

and the probability of erroneous acceptance of H_0 when an alternative θ is true

$$\beta_n(\theta) = 1 - P_\theta[(X_1, \ldots, X_n) \in T_n] \qquad\qquad \theta \neq \theta_0$$

Suppose we have two competing families of critical regions for the same problem. (We say "family of critical regions" because the region actually used depends on the sample size n and the level α selected. Thus, one might compare--as Hoeffding did--the χ^2 and likelihood ratio families for a multinomial testing problem.) How shall we compare their performance? Given sequences $\{T_n\}$ and $\{T_n'\}$ of critical regions, if T_n has $\alpha_n = \alpha$ and $\beta_n(\theta) = \beta$ for a fixed alternative $\theta \neq \theta_0$, we may ask how many observations m are required for the critical region T'_m to attain $\alpha'_m = \alpha$ and $\beta'_m(\theta) = \beta$. The ratio n/m is the *efficiency* of $\{T'_n\}$ relative to $\{T_n\}$. Unfortunately, this relative efficiency usually depends on several of α, β, θ_0, θ, and n. It is therefore natural to seek some large sample simplification by investigating the behavior of the error probabilities as the sample size n increases. In addition to identifying good statistical procedures for large samples, such studies may suggest the form of good procedures for small n.

Any reasonable sequence of tests has the property that as n increases and the level α remains fixed, the probability $\beta_n(\theta)$ of erroneous acceptance approaches zero for any alternative θ. Thus, some quantity (α, β, or θ) in addition to n must change as n increases, and various approaches to large sample comparison of tests can be distinguished by the constraints placed on these quantities.

We wish to stress two themes in the development of this area: First, the use of tools other than the central limit theorem to compare tests. (This may be called the "mathematical front.") Second, establishment of the large sample optimality of procedures based on likelihood, in this case the likelihood ratio (LR) family of tests. (This is the "likelihood front," on which there have been significant advances in the theory of estimation as well as testing.) When the alternative as well as the hypothesis is simple, the Neyman-Pearson Fundamental Lemma, of course, states that any LR test is most powerful of its level for any sample size. Large sample optimality of LR tests (as of the analogous maximum likelihood estimators) has since been established with respect to a number of criteria. Hoeffding's contribution was to show that even families of tests that differ by little from LR tests and that are asymptotically equivalent to them in one sense (Pitman efficiency) may be inferior in another sense, in large samples.

Approaches to Large Sample Comparison

We will mention three approaches. The earliest of these was to study the relative efficiency of tests when α is fixed (or $\alpha_n \to \alpha$ and $0 < \alpha < 1$) and the alternative θ_n varies with n in such a way that $\beta(\theta_n) \to \beta$ for a fixed β, $0 < \beta < 1$. In the cases we are discussing, θ_n must approach θ_0 at rate $n^{-1/2}$ to obtain nontrivial α and β. This "local comparison" was systematized by Pitman in the one-dimensional case and bears his name. The central limit theorem is the essential mathematical tool in studying local alternatives. Early work in this setting is attributable to Wilks; Wald's definitive paper (1943) established several optimum properties of the LR and related families. Further work on local properties is contained in the work of LeCam, Neyman, and Weiss and Wolfowitz; this includes the more complex case of composite null hypotheses, various optimality criteria, and families of procedures other than LR tests. We omit details, as local comparisons are not our concern here.

[We remark that some of these last-mentioned developments are counterparts of the asymptotically efficient estimation results for Bayes and maximum likelihood estimators (LeCam, Wolfowitz), which are relevant to the discussion in the section on the standard Bayesian model (vide infra), and for Wolfowitz's maximum probability estimator.]

Other comparisons of tests leave the alternative θ fixed. Calculation of the probabilities of error can then no longer be handled by the central limit theorem, but requires results on probabilities of large deviations. (If t_n is a normalized sample mean, so that t_n converges in law to the standard normal distribution by the central limit theorem, $\{t_n \geq a_n\}$ is a *large deviation* of t_n if $n^{-1/2} a_n \to a$ for $0 < a < \infty$. In this case, the central limit theorem says only that $p_n = P[t_n \geq a_n]$ approaches 0, which is uninformative. We want to know the speed with which this probability approaches 0.) Crámer began the study of this probabilistic problem in 1938, and the literature now contains many sources for both order results (of the form $n^{-1} \log p_n \to c$) and asymptotic results (of the form $p_n/c_n \to 1$) for probabilities of large deviations. Only order results are required for the large sample comparisons of tests done to date.

Chernoff (1952) first used probabilities of large deviations to compare tests. We will mention two contrasting "fixed alternative" approaches. Hodges and Lehmann (1956) fixed θ and α $(0 < \alpha < 1)$ and studied the rate of convergence to 0 of $\beta_n(\theta)$. In the usual cases

$$\beta_n(\theta) = e^{-nc(\theta)[1+o(1)]} \quad \text{for all } \alpha \tag{1}$$

so that an asymptotic relative efficiency can be defined as the ratio of the indices $c(\theta)$ for competing families of tests.

Beginning in 1960, R. R. Bahadur produced an extensive theory of large sample properties of statistical procedures, which he recently summarized in a monograph, Bahadur (1971). His approach to tests can be stated as follows: fix θ and β and study the rate of convergence of α_n to 0. Again, one usually obtains

$$\alpha_n = e^{-nb(\theta)[1+o(1)]} \quad \text{for all } \beta \tag{2}$$

so that an asymptotic relative efficiency can again be defined. Bahadur's approach has borne more fruit than has that of Hodges and Lehmann for two reasons. First, it is easier; Eq. [2] requires an order result for probabilities of large deviations under θ_0, whereas Eq. [1] requires a similar result under the alternative θ. The Bahadur index $b(\theta)$ has therefore been computed for many more families of tests than has the Hodges-Lehmann index $c(\theta)$. Second, we have done Bahadur an injustice to have described his work in this framework. His basic idea was to study the behavior of the actually attained level of the test as a random variable. This natural "stochastic comparison" of tests turns out to be equivalent to the nonstochastic comparison based on Eq. [2]. Bahadur has shown in some generality that LR tests have maximum $b(\theta)$ and are therefore asymptotically optimal by his criterion.

Hoeffdings' Contribution

Hoeffding (1965) considered several testing problems involving the multinomial distribution with k cells and unknown vector $\theta = (\theta_1, \ldots, \theta_k)$ of cell probabilities. We will discuss only the problem of

testing the simple null hypothesis $\theta = \theta_0$ for a fixed probability
vector θ_0. Hoeffding made an advance on both the mathematical
front and the LR front. Mathematically, he built on work of Sanov
to give an order result for probabilities of large deviations in
this k-dimensional multinomial case. Previous comparisons of tests
had used such results only for sums of univariate random variables.

On the LR front, Hoeffding succeeded in distinguishing the large
sample performance of the LR family of tests for $\theta = \theta_0$ from that of
the familiar Pearson χ^2 tests for this problem. If $\bar{X}_{in}$ for i =
1, ..., k is the proportion of n observations falling in the ith
cell, the LR test is based on the information-distance statistic

$$L_n = \Sigma_i \bar{X}_{in} \log \frac{\bar{X}_{in}}{\theta_{0i}}$$

The χ^2 statistic is of course

$$Q_n^2 = \Sigma_i \frac{(\bar{X}_{in} - \theta_{0i})^2}{\theta_{0i}}$$

These tests had long been treated as being asymptotically equivalent
because of their equivalence under local comparison. Q_n^2 is the
dominant term in the Taylor's series expansion of L_n about θ_0; $2nL_n$
and nQ_n^2 have the same χ^2 limiting distribution under the null hypothesis; the two families of tests have the same large sample performance
against local alternatives.

The spirit and nature of Hoeffding's comparison can be demonstrated
with minimal mathematics in the two-cell (k = 2) case. This we do
in the next section, which may be omitted without loss of continuity.
Here, we content ourselves with observing that although Hoeffding's
precise comparison was not one of those discussed above, he implicitly
showed that the LR and χ^2 families have the same Hodges-Lehmann performance for all alternatives θ, but that *the LR family has strictly
better Bahadur performance for "most" alternatives.* As Bahadur's theory
has become a standard tool in the decade since Hoeffding's work, the

latter work is now most easily understood in Bahadur's framework.   That
it could be so understood was shown in detail by J. C. Gupta (1972).
Fixed-alternative comparisons ask more of the χ^2 test than its
creators probably intended.   The coincidence of power results for
local alternatives is closer to the motivation for Q_n^2, which involves
the relevance of the expected value of Q_n^2 and hence of normal theory.
Nevertheless, this common test has been discredited for large samples
and fixed alternatives by Hoeffding's result.

Progress on the LR front has, of course, continued.   Brown (1971)
has shown in considerable generality that appropriate tests of LR
type (actually LR tests of possibly larger hypotheses) are at least
as good as any given sequence of tests in both the Hodges-Lehmann
and Bahadur senses.   The more difficult task of analyzing what makes
an apparently equivalent test strictly inferior to a LR test for large
samples in the generality of Brown's setting awaits another advance
on the mathematical front--more general large deviation results for
multivariate problems.   Herr (1967) has done this in certain multivariate normal cases, but much work remains.   It would also be valuable to investigate the sample size required for LR tests to be close
to optimal (or, alternatively, to be superior to χ^2 tests in the multinomial case).

Hoeffding's Result for Two Cells

To illuminate Hoeffding's discovery that Q_n^2 is inferior to L_n, we will
consider the special case $k = 2$.   This amounts to observing n independent Bernoulli random variables $X_1, \ldots, X_n$ with $\theta = P[X_i = 1]$ unknown.   For $0 < p < 1$, the test statistics for the hypothesis $\theta = p$ are
$L_n = I(\bar{X},p)$ and $Q_n^2 = Q^2(\bar{X},p)$, where $\bar{X}$ is the sample mean of $X_1, \ldots, X_n$
and

$$I(\theta,p) = \theta \log \frac{\theta}{p} + (1 - \theta)\log \frac{1 - \theta}{1 - p}$$

$$Q^2(\theta,p) = \frac{(\theta - p)^2}{p(1 - p)}$$

The information distance $I(\theta,p)$ between two probability vectors [here
between $(\theta, 1 - \theta)$ and $(p, 1 - p)$] plays a central role in all large
deviation comparisons of tests.

Analysis of this special case requires only one mathematical tool,
an order result for probabilities of large deviations of binomial
random variables, obtainable from Cramér's inequality. Specifically,
if B_n is a binomial random variable with mean np, $q = 1 - p$ and

$$p_n = P\left[\frac{B_n - np}{(npq)^{1/2}} \geq bn^{1/2}\right] \qquad\qquad b > 0$$

then, for $p + b(pq)^{1/2} < 1$,

$$\frac{1}{n} \log p_n \to - I[p + b(pq)^{1/2}, p] \qquad\qquad\qquad [3]$$

From Eq. [3], one may first calculate that the Hodges-Lehmann
index $c(\theta)$ defined in Eq. [1] is $I(p,\theta)$ for both L_n and Q_n. Thus,
the Hodges-Lehmann approach also fails to distinguish Q_n from L_n.

The Bahadur index $b(\theta)$ defined in Eq. [2] depends on whether
$p > 1/2$ or $p < 1/2$. (When $p = 1/2$, the tests based on L_n and Q_n are
identical.) For the remainder of this discussion, we assume that the
hypothesized p exceeds $1/2$. Another application of Eq. [3] then shows
that for Q_n

$$\begin{aligned} b_Q(\theta) &= I(\theta,p) & 0 \leq \theta \leq p \qquad [4] \\ &= I[\theta - 2(\theta - p), p] & p \leq \theta \leq 1 \end{aligned}$$

whereas it is known that for the LR statistic L_n

$$b_L(\theta) = I(\theta,p) \qquad\qquad\qquad 0 \leq \theta \leq 1 \qquad [5]$$

The situation is illustrated in Fig. 1, where $Q^2(\theta,p)$ and $I(\theta,p)$
are drawn for $p = 3/4$. Note that $I(\theta,p)$ is not symmetric about p,
but increases more slowly for $\theta < p$ when $p > 1/2$. Thus, Eqs. [4] and
[5] say that $b_Q(\theta) < b_L(\theta)$ for $\theta > p$, so that Q_n is inferior to L_n
against alternatives $\theta > p$.

Let us now look at this comparison as Hoeffding did. For sufficiently regular sets A, he showed (for general k) that

$$\frac{1}{n} \log P[\bar{X} \in A | \theta] \rightarrow - I(A,\theta) \qquad [6]$$

where

$$I(A,\theta) = \inf_{\omega \in A} I(\omega,\theta)$$

is the information distance of A from θ. Suppose, next, that for δ > 0

$$A(\delta) = \{\theta: Q^2(\theta,p) \geq \delta\}$$

is a χ^2 critical region and

$$B(\delta) = \{\theta: I(\theta,p) \geq I(A(\delta),p)\}$$

is a corresponding LR critical region. These regions are illustrated in Fig. 1.

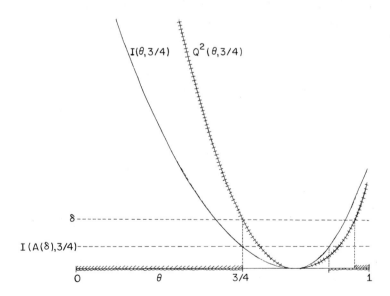

FIG. 1. $I(\theta,3/4)$ and $Q^2(\theta,3/4)$. The hatched region above the axis indicates the set A(δ) and the marked region below the axis indicates the set B(δ).

Applying Eq. [6] with $\theta = p$ and $A = A(\delta)$ or $B(\delta)$ shows that both critical regions have asymptotically the same $\log \alpha_n$, as both are the same information distance from p. The LR critical region includes the χ^2 region and is thus at least as powerful. More specifically, alternatives $\theta < p$ are the same information distance from both acceptance regions $A(\delta)^C$ and $B(\delta)^C$, and hence have asymptotically the same $\log \beta_n$. [In this heuristic sketch, we consider only alternatives in $A(\delta)$; this includes any given $\theta \neq p$ for sufficiently small δ.] But alternatives $\theta > p$ are strictly closer to the χ^2 acceptance region and therefore Q_n has larger $\log \beta_n$ than does L_n for these alternatives.

The geometry of Fig. 1 is indicative of the general case. Hoeffding showed that for any k the analogs of $A(\delta)$ and $B(\delta)$ have only finitely many common boundary points, one on each line segment joining $p = (p_1, \ldots, p_k)$ to the unit vector in the direction of smallest components p_i. L_n is superior to Q_n for all θ not lying on these line segments, by arguments indicated above. When $k > 2$, the exceptional line segments form a small portion of the parameter space, so that the superiority of L_n is more striking in these cases.

THE EMPIRICAL BAYES MODEL

This interesting model was introduced and first studied by Robbins (1955). We shall depart slightly from the usual development of background material by summarizing not only the standard Bayesian model, but also the notions of structural parameter models and adaptive estimators. All of these possess features reflected in some of the empirical Bayes concepts, as well as important differences from the latter.

For simplicity, we shall describe the ideas only for estimation problems in the absolutely continuous case. Regularity conditions that are required will not be listed in detail.

The simplest estimation model is that introduced at the beginning of the first section, except that the object is now to *estimate* some function ϕ of the unknown θ governing the probability law of the X_i. A common example is $\phi(\theta) = \theta'$ in either of the examples of the first section. An estimator t_n is a rule for guessing $\phi(\theta)$ on the basis of $X_1, \ldots, X_n$. As in the case of testing, it is often difficult to compute an estimator which is "optimal" in some prescribed sense for a given sample size n. It is again natural to study sequences $\{t_n\}$ of estimators as the sample size n increases in the hope of establishing desirable large sample properties.

The Standard Bayesian Model

The Bayesian model adds two assumptions to the estimation problem described above: (1) that the parameter θ can be regarded as a random variable, and (2) that the prior distribution G of this random variable is known. Note that the value of θ, once it is chosen according to G, remains the same in the density f_θ of each X_i.

Bayes' theorem combines G with the observed data to produce the posterior distribution of θ. Comparison of estimators in this model is based on the posterior expected loss $R(t_n, G)$ of an estimator t_n. A *Bayes estimator* $t^*_{G,n}$ of $\phi(\theta)$ is an estimator that minimizes this expected loss. For example, if (as in the examples of the first section) a real parameter $\phi(\theta)$ is to be estimated and loss is measured by squared error, then $t^*_{G,n}$ is the posterior expectation of $\phi(\theta)$.

A feature of interest to us in this Bayesian formulation is that the desired performance of the Bayes procedure is relatively insensitive to slight errors in the specification of G. More precisely

$$\frac{R(t^*_{G',n}, G)}{R(t^*_{G,n}, G)} \qquad [7]$$

is close to 1 when G is close to G' (under reasonable regularity conditions, as usual). Thus, using $t^*_{G',n}$ when the actual prior law is G (close to G') is almost as good as using the Bayes procedure $t^*_{G,n}$ relative to G. This is just the asymptotic optimality of Bayes estimators referred to in the section on large sample comparison of tests (vide supra).

If the Bayesian model as stated above is correct, there is no disagreement about using $t^*_{G,n}$. One source of controversy arises because doubt may be thrown on the simple-minded form assumed for $\{f_\theta\}$,

or for the assumed loss function, or on the stated aim of the inference [estimation of $\phi(\theta)$]. Another source of controversy lies in
the Bayesian assumptions (1) and (2). Bayesian statisticians feel
that a description of rational thought legitimizes the use of a subjective guess for G in the absence of knowledge of an actual G; others
disagree strongly, but we need not discuss this controversy in detail
here.

The Empirical Bayes Model

We now turn to Robbins' model. We are faced with a sequence of independent estimation problems, each of which must be acted on as it
arises. These problems are, however, related as follows: the observed
X_i in the ith problem has density f_{θ_i} once θ_i is given, but the θ_i
are themselves iid with unknown distribution G. So at the nth infer-
ence, we have available $X_1, \ldots, X_n$ and we can hope that if n is
large some information about the unknown prior law G can be wrung
from the past observations $X_1, \ldots, X_{n-1}$. If we knew G exactly, we
would estimate θ_n by $t^*_{G,1}(X_n)$, and in the absence of such exact knowledge it seems reasonable (as discussed below in the section on adaptive estimators) to use this estimator with G replaced by an estimator
of G; this is Robbins' proposal, which we now describe in further detail.

One can construct an empirical Bayes estimator of θ_n by (1) finding an estimator $\hat{G}_{n-1}(X_1, \ldots, X_{n-1})$ of G; and (2) using the Bayes
estimator $t'_n \overset{\text{def}}{=} t^*_{\hat{G}_{n-1},1}(X_n)$. One thus acts as if $\hat{G}_{n-1}$ were the known
prior law. If the estimator $\hat{G}_{n-1}$ of G is a good one, then on the
basis of Eq. [7] we expect that for large n

$$\frac{R(t'_n,G)}{R(t^*_{G,1},G)} \tag{8}$$

is close to 1. Robbins found appropriate $\hat{G}_{n-1}$ for several problems
and established Eq. [8] in these cases. Thus, we do as well asymp-

totically in estimating θ_n when G is unknown as we would if we knew
the prior law G.

The proof of Eq. [8] in various settings and the study of the
rate of convergence to 1 of the ratio in Eq. [8] has produced a body
of literature by Hannan, Johns, van Ryzin, Samuel, Gilliland, and
others. One can expect further research to yield reasonably efficient
procedures for small n.

Of interest to many observers will be the extent to which Bayesians
are able in practice to depart from the standard Bayesian model with
a subjective guess of G, and can instead imbed the problem at hand
as the nth one in the empirical Bayes model, to yield and use a more
formally described guess $\hat{G}_{n-1}$. To non-Bayesians, Robbins' model will
seem much more acceptable in many practical settings than the original
Bayesian formulation. For example, X_i might be an observation of some
biological characteristic of an organism at location i, governed by a
parameter θ_i of which distribution G is characteristic of the species
but is unknown. Or X_i might be the result of a diagnostic test on
individual i made at a preventive medicine clinic run for workers in
a large plant, and θ_i an index of the underlying condition having an
unknown distribution characteristic of this population of workers.

Structural Parameter Models

Robbins' model may be compared with one which had already been the
subject of a large body of literature by 1955, estimation of structural
parameters. This is a non-Bayesian framework.

Here again the observations X_1, ..., X_n are independent, but the
density of X_i is indexed by (α, θ_i). The structural parameter α is to
be estimated, while θ_i is an incidental parameter which varies from
observation to observation. In the most common example, the X_i are
points in the plane derived from an unknown line α by adding independent error vectors with zero means to points on α with abscissas θ_i.
This simplest line-fitting situation is illustrated in Fig. 2. One
approach to the structural parameter problem is to consider the θ_i to

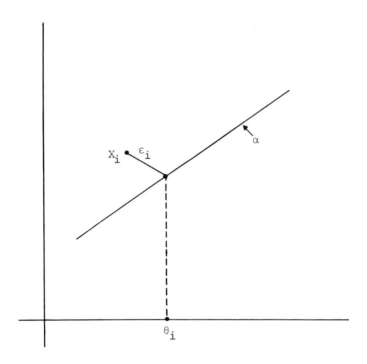

FIG. 2. The structural parameter model with α an unknown line, and observation X_i determined by adding a random error vector ε_i to the point on α with abscissa θ_i.

be independent random variables with the same law G. Indeed, the term "structural model" is sometimes reserved for this case. Such models have been studied by Geary, Reiersol, Wald, Neyman and Scott, Wolfowitz, and others. A survey of their work is given by Moran (1971).

The frameworks of the empirical Bayes model and the structural parameter model have been exhibited in Fig. 3 to point out their considerable similarities. The difference between the models lies primarily in the inference to be made.

EMPIRICAL BAYES MODEL

$X_1, \ldots, X_n$ independent observations

X_i has parameter θ_i

$\theta_i, \ldots, \theta_n$ iid with law G unknown

X_i has marginal density

$$f_G(X) = \int f_\theta(x) \, dG(\theta)$$

Use $X_1, \ldots, X_n$ to estimate θ_n

STRUCTURAL PARAMETER MODEL

$X_1, \ldots, X_n$ independent observations

X_i has parameter (α, θ_i)

$\theta_1, \ldots, \theta_n$ iid with law G unknown

X_i has marginal density

$$f_{\alpha, G}(x) = \int f(x | \alpha, \theta) \, dG(\theta)$$

Use $X_1, \ldots, X_n$ to estimate α

FIG. 3. Comparison of the empirical Bayes and structural parameter models for estimation.

Adaptive Estimators

The methodology of estimation used in the empirical Bayes setting is related to a methodology arising in the example $\theta = (\theta', g)$ of the first section and which can also be employed in the structural parameter model. If we knew g in the nonparametric problem given in the first section, we could use some known good estimator, e.g., Pitman's location parameter estimator $t_{g,n}$ (say), for estimating θ'. The form

of this estimator of course depends on g. Because we do not know g,
we find an appropriate estimator $\hat{g}_n$ of it, based on X_1, X_2, ..., X_n and
then use $t_{\hat{g}_n, n}$ to estimate θ'. This rough recipe requires care in
its execution, but such an approach has been carried out in various
settings by Weiss and Wolfowitz, LeCam, Hajek, and others. Under suitable assumptions on the unknown g, one can find an estimator of the
form $t_{\hat{g}_n, n}$ or something similar, whose accuracy is asymptotically the
same as that of the $t_{g,n}$ we would use if we *knew* g.

The spirit of this approach, of constructing procedures by *adapting* their form to what the data seems to say about the error law, is
also used in a number of small sample-size studies of "robust" estimators.

Empirical Bayes estimators make use of adaptive estimation in
the estimation of G by $\hat{G}_n$. It is also clear that a possible approach
to the construction of "good" estimators of α in the structural model
is to first estimate G by some $\hat{G}_n(X_1, ..., X_n)$, then substitute $\hat{G}_n$
into the estimator $t_{G,n}$ of α that we would use were G known. In some
of the work in this setting, G is only estimated implicity; in other
work, explicit estimates are given (e.g., Wolfowitz's minimum distance
estimator).

Thus, the empirical Bayes model is not only connected with the
Bayesian formulation of inference problems but is tied in spirit to
structural models and adaptive estimators. One may even ask if there
are some practical structural parameter problems in which the successive $θ_i$ are of enough interest to be estimated along with α. Empirical
Bayes methods can then be used.

ACKNOWLEDGMENT

The work of the first author was written under National Science
Foundation Grant 35816 GPX. The work of the second author was sponsored, in part, by the Air Force Office of Scientific Research, Air
Force Systems Command, USAF, under Grant No. AFOSR-72-2350. The U.S.
government is authorized to reproduce and distribute reprints for
governmental purposes notwithstanding any copyright notation hereon.

REFERENCES

Bahadur, R. R. (1971). *Some limit theorems in statistics. Reg. Conf. Ser. Appl. Math. 4,* Soc. Ind. Appl. Math., Philadelphia.

Brown, L. D. (1971). Non-local asymptotic optimality of appropriate likelihood ratio tests. *Ann. Math. Statist. 42,* 1206-1240.

Chernoff, H. (1952). A measure of asymptotic efficiency for tests of a hypothesis based on the sum of observations. *Ann. Math. Statist. 23,* 493-507.

Gupta, J. C. (1972). Probabilities of medium and large deviations with statistical applications. Ph.D. thesis, University of Chicago.

Herr, D. G. (1967). Asymptotically optimal tests for multivariate normal distributions. *Ann. Math. Statist. 38,* 1829-1844.

Hodges, Jr., J. L. and Lehmann, E. L. (1956). The efficiency of some nonparametric competitors of the t-test. *Ann. Math. Statist. 27,* 324-335.

Hoeffding, W. (1965). Asymptotically optimal tests for multinomial distributions. *Ann. Math. Statist. 36,* 369-408.

LeCam, L. (1956). On the asymptotic theory of estimation and testing hypotheses. *Proc. 3rd Berkeley Symp. Math. Statist. Prob. 1,* 129-156.

Moran, P. A. P. (1971). Estimating structural and functional relationships. *J. Multivariate Anal. 1,* 232-255.

Neyman, J. (1959). Optimal tests of composite statistical hypotheses. In *The Harold Cramér Volume,* pp. 213-234. Almquist & Wiksell, Stockholm.

Robbins, H (1955). An empirical Bayes approach to statistics. *Proc. 3rd Berkeley Symp. Math. Statist. Prob. 1,* 157-163

Wald, A. (1943). Tests of hypotheses concerning many parameters when the number of observations is large. *Trans. Amer. Math. Soc. 54,* 426-482.

Weiss, L., and Wolfowitz, J. (1969). Asymptotically minimax tests of composite hypotheses. *Z. Wahrschein. Verw. Geb. 14,* 161-168.

THE BIRTH, GROWTH, AND BLOSSOMING
OF SEQUENTIAL ANALYSIS

DONALD A. DARLING was born May 4, 1915, in Los Angeles, California. He received his A.B. degree from the University of California at Los Angeles in 1939 and his Ph.D. degree from the California Institute of Technology in 1947. He joined the faculty of the University of Michigan in 1949 and stayed there until 1968, when he went to the University of California at Irvine. He is a Fellow of the Institute of Mathematical Statistics and was a Fellow of the John Simon Guggenheim Memorial Foundation in 1958. He has given numerous invited addresses and is author of over 35 research papers.

THE BIRTH, GROWTH, AND BLOSSOMING
OF SEQUENTIAL ANALYSIS

Donald A. Darling

Department of Mathematics
University of California
Irvine, California

Sequential analysis is a manifestation of the general trend since the turn of the century in which the statistician has changed from the passive role of drawing charts, calculating averages, and so forth, once the data are in, to the active role of intervening in the design of experimentation and collection of data. In fact, the sequential analyst represents the extreme case of the statistician who decides after each datum is collected to stop taking data and to make a decision or to take another observation. He has thus changed from a static compiler to the dynamic organizer and analyzer.

The "classical method," by which I mean an experiment having a fixed number of observations, decided in advance, dominated the first three decades of this century. This was largely due, I think, to the nature of the experiments which were performed. If one is engaged in collecting, e.g., agricultural or biological data where large scale planning in advance is required, it may not be feasible to "take another observation." As statistics developed and was applied to other areas where one could easily augment the number of observations, the possibility of sequential analysis naturally presented itself to a number of

people--indeed, nowadays, the idea is so natural that it is hard to
conceive of statistical tests generally in other than sequential terms.

The advent of newer applications, particularly in acceptance sampling, led to sequential methods. Here, items are produced in large
quantities and both buyer and seller agree that screening defectives
by the seller would be too costly--the buyer agrees to accept a certain fraction defective to avoid prohibitive inspection costs, and he
controls what he buys by inspecting small samples and accepting or
rejecting large lots on the basis of the fraction defective found in
these samples. The method lends itself admirably in practice to the
theory of tests of hypothesis formulated by Neyman and Pearson, elaborate procedures for which were developed for the practitioners. It
soon became apparent that sampling could be curtailed in some cases,
since in the model for "randomness" of sample all permutations were
equally probable, and in some of them the defectives would appear early
and obviate the necessity to continue inspection.

A procedure to utilize this saving was formalized in 1929 by H. F.
Dodge and H. G. Romig (1929) of the Bell Telephone Laboratories in a
double sampling system in which the results of an initial smaller sample
sometimes eliminated the need for continuing with a second sample. This
was extended by W. Bertky (1943) to multiple sampling. The idea of
splitting an experiment up into stages and making action taken at any
stage dependent on the results of earlier stages was investigated earlier, and constituted a precursor of true sequential analysis--examples
were presented by P. C. Mahalanobis in 1940 and H. Hotelling in 1941.
Also, the classical methods were found to be frustrating in many cases,
inasmuch as sometimes if there were "just a few more data exhibiting
the same characteristics as the data in hand" a nonsignificant result
would become significant. But adjoining new data thus is not permitted
under the classical analysis, although there are published papers in
which it was done--this is the so-called "optional stopping" paradox
and it is well known that in the classical tests the null hypothesis
can always be rejected when it is true if one uses optional stopping.
Finally, one can mention that the method invented by R. A. Fisher

for combining tests of significance is, in effect, a kind of sequential procedure.

There were a number of workers who independently considered sequential methods, and who could lay claim to be the discoverer. This
was because, in addition to the reasons just mentioned, the period
during and immediately following World War II, there was a massive
treatment of military production by statistical methodology in the
United States and in England, and sequential methods lent themselves
to many of the problems raised. In England, in particular, G. A.
Barnard may be mentioned as an early leader.

However there is no dispute possible that the discovery by Abraham Wald in 1943 of the sequential probability ratio test (SPRT) made
all other sequential procedures obsolete. In this brilliant research,
initially classified as Restricted by the military agencies, Wald also was the first to carefully define and study a very important notion
of a *stopping time* of a sequence of random variables. This notion of
stopping times has become one of the most useful and fruitful notions
in the study of modern probability, and it has been the cause of strides
in probability theory over the past decade. Roughly it is a rule to
tell one when to stop observing a sequence of random variables, the
rule utilizing only present data and data anterior to it. Although
the notion is intuitively simple, it takes a certain care to make it
precise.

The methods of Wald were initially applied to acceptance sampling
procedures by a group at Columbia University, called the Statistical
Research Group. Wald showed that typically a saving of 50% in the
number of observations could be achieved by the SPRT, still maintaining control over the probabilities of errors. This is a truly remarkable achievement. In subsequent work by Wald (1947), published in his
famous book *Sequential Analysis*, he derived the operating characteristic
(or power function) for his text, and estimates for the number of observations needed to terminate the test--these had been done earlier by
others for the binomial case, which occurs in acceptance sampling.

Finally in subsequent work with J. Wolfowitz (A. Wald and J. Wolfowitz, 1948) he proved, confirming an earlier conjecture, the optimality of the SPRT--of all tests with specified error control, the SPRT has the least expected sample size. This result definitively closed the book for the class of tests applicable to SPRT--simple hypothesis vs. simple alternative, and by extension also to some composite cases. The virtues of the SPRT did not stop at reducing the sample size Wald had built better than he knew. The advantages of Wald's method were that it made more symmetric the role of probabilities of errors of the first and second kind, and it obviated the earlier objection of stopping prematurely and "running out of data" before rejecting a hypothesis. Also, surprisingly enough, it is simpler, in some respects, than the classical tests in not requiring tables of special functions. There are fixed levels--not depending on the distribution of the data-- and one evaluates partial sums of functions (logarithm of the likelihood ratio) of the observations and waits until the partial sums go outside these levels.

At present, there exist sequential analogs for a great many of the classical tests--Student's, Behrens-Fisher, etc.--but for them the results concerning the operating characteristic and the optimality are not known. Some involve considerable numerical calculations, which previously were prohibitive but today may become possible because of improved calculator technology. For some of the classical tests it is difficult to see how to adapt them to sequential analysis-- e.g., for the χ^2 test of goodness-of-fit and some of the tests used in the analysis of variance.

Sequential methods can be found for estimation problems and are generally presented as sequentially determining confidence intervals or regions. This is not quite so straightforward as sequential tests, and optimality is considerably more complex. These estimation problems were considered by Wald in his book.

I shall not attempt a survey of all the work since Wald since it is very large and still undergoing rapid development. I shall only give some recently discovered connections between sequential analysis

and other areas of modern probability theory, notably the theory of
optimal stopping. An optimal (*not* optional, as earlier) stopping
time is a stopping time described earlier which optimizes some pay-
off--evidently of great interest to gamblers and stock market specu-
lators. H. Robbins and others (Chow et al., 1971; Shirgaev, 1969)
have shown that some optimum sequential tests under the Neyman-Pearson
system are equivalent to certain optimal stopping problems. These optimal stopping problems are closely related to the modern theory of
martingales--Wald, in particular, would have been fascinated with this
relationship, and to know that his famous "fundamental identity of
sequential analysis" was a special case of the general theory of
stopped martingales. It is known that additionally the problem of
optimal stopping is in some cases that of solving a problem in dynamic
programming of R. Bellman or of linear programming (not the same thing).
Finally the optimum stopping problem is in some cases that of finding
the solution to a problem in partial differential equations--the so-
called free surface, or free interface problem. In its most general
form, the problem of finding an optimal stopping time is equivalent
to that of finding the least superharmonic majorant to the payoff
function.

I find that these interrelationships between the various fields
and how they coalesce are fascinating--they illustrate what has been
said earlier in the conference concerning the unity and interdepen-
dence between mathematics in general, probability and statistics.
Some of the latest and most sophisticated developments in probability
theory can now be brought to bear on the statistical problem of sequential analysis--but it must be said that, for all this, the number
of concrete problems that thus can be solved is still quite small.

There have appeared lately a new class of sequential tests--
the "open end tests" which I have had the privilege of working on
with H. Robbins. These don't come under the purview of the Wald
tests inasmuch as they admit the possibility of never terminating
the sampling--a case which happens naturally, e.g., in weather observations, some clinical trials, etc., where observations are "free"

and potentially unlimited. As discussed above, sequential tests, besides reducing significantly the sampling, have the added bonanza of making more symmetrical the errors, and in having you stop when you should. In testing composite hypotheses either in the classical case or in Wald's SPRT it was often necessary to introduce a "zone of indifference"--for hypotheses "close to" the null hypothesis, it is not a mistake to accept the null hypothesis. It is possible, with open-end tests, to eliminate this awkward ad hoc specification of "close to," as well as to specify errors of the second kind. It is possible to find sequential tests with uniformly high power over all alternatives, and indeed tests with power 1. This was the subject of the recent Wald lectures by H. Robbins (1970).

Professor J. Neyman (1971) has kindly pointed out to me a striking graph which he prepared for an address, "Foundation of Behavioristic Statistics." In this article, there are three interesting graphs of the power function curve corresponding to stages in the development of sequential tests from the Wald tests to the uniform power tests.

There is an early and historically important example provided by C. Stein (1945) which also shows that freeing the sample size can result in uniformity of power--in Stein's case, he found a sequential method (not Wald's) for Student's test, which made its power independent of the unknown variance, something that could not be done with any fixed sample size, as proved by G. Dantzig.

I would like to close on a personal remark. When I first entered probability, my elders were asserting that statistics only utilized distribution theory and weak limit laws, like convergence in probability. There was never any use, for example, for the strong law of large numbers, and following the philosophy of the mathematician, G. H. Hardy, this made such things much more attractive--i.e., their uselessness was a virtue. How time has proven such statements in error! For example, in the open-end tests just discussed, such a strong result as the law of the iterated logarithm is essentially used, and indeed certain refinements of it need to be developed.

REFERENCES

Barnard, G. A. Economy in Sampling with Reference to Engineering Experimentation, (British) Ministry of Supply, Advisory Service on Statistical Method and Quality Control, Technical Report, Series R, No. Q.C./R/7.

Bertky, W. (1943). Multiple sampling with constant probability. *Ann. Math. Statist. 14*, 363-377.

Chow, Y., Robbins, H., and Siegmund, D. (1971). *Great Expectations: The Theory of Optimal Stopping*, Houghton Mifflin Company, Boston.

Dodge, H. F., and Romig, H. G. (1929). A method of sampling inspection. *Bell Syst. Tech. J. 8*, 613-631.

Hotelling, H. (1941). Experimental determination of the maximum of a function. *Ann. Math. Statist. 12*, 20-45.

Mahalanobis, P. C. (1940). A sample survey of the acreage under jute in Bengal, with discussion on planning of experiments. *Proc. 2nd Indian Statist. Conf., Calcutta.*

Neyman, J. (1971). Foundation of behavioristic statistics. In *Symposium on Foundations of Statistical Inference*, V. P. Godambe and D. A. Sprott (Eds.), Holt, Rinehart and Winston, Montreal.

Robbins, H. (1970). Statistical methods related to the law of the iterated logarithm, *Ann. Math. Statist. 41*, 1397-1408.

Shirgaev, A. (1969). *Statistical Sequential Analysis*, Izd. Nauka, Moscow (In Russian).

Stein, C. (1945). A two sample test for a linear hypothesis whose power is independent of the variance, *Ann. Math. Statist. 16*, 243-258.

Wald, A. (1943). Sequential Analysis of Statistical Data: Theory, Restricted report submitted September 1943.

Wald, A. (1947). *Sequential Analysis*, John Wiley, New York.

Wald, A., and Wolfowitz, J. (1948). Optimum character of the sequential probability ratio test, *Ann. Math. Statist. 19*, 326-339.

BIOMEDICAL APPLICATIONS
OF STOCHASTIC PROCESSES

PREM S. PURI was born April 15, 1936, in Montgomery, Pakistan. He is a citizen of India and a permanent resident of the United States. He received his B.S. degree in 1953 and M.Sc. degree in 1956 from AGRA University, India. He received an M.A. degree in 1963 and a Ph.D. degree in 1964 from the University of California at Berkeley. He joined Purdue University in 1966, where he is now a Professor of Statistics. He is a Fellow of the American Statistical Association and the Institute of Mathematical Statistics. At the invitation of the Soviet Academy of Sciences, he spent three months in 1973 at the Mathematical Institute of V. A. Steklov of the Academy of Sciences USSR. He is author of over 40 research papers.

BIOMEDICAL APPLICATIONS
OF STOCHASTIC PROCESSES

Prem S. Puri

Department of Statistics
Purdue University
Lafayette, Indiana

It is well recognized that in the biological and medical sciences,
variability among the observations is much larger, more fundamental,
and intrinsic than in some other disciplines such as physics, in which
it is often possible (although not always) to dispose of a major part
of the variability by controlling certain relevant factors in the laboratory. This variability, in turn, makes the stochastic models much
more appropriate in biology and medicine than it is for their deterministic analogs. Again, the basic evolutionary characteristics of
living things such as births and deaths, growth and decay, change and
transformation, lead us in biology and medicine to many dynamic processes of development in time and space. Thus, one is led to the use
of the so-called stochastic processes as a natural vehicle for stochastic model-building for the study of various biological phenomena.

In short, a random phenomenon that arises through a process that
is developing in time or in space in a manner controlled by probabilistic laws is called a stochastic process. The examples are many,
e.g., the growth of a population such as a bacterial colony; the
spatial distribution of plants and animal communities; spread of an

epidemic; spread of cancer growth within the body; and so on. Mathematically a stochastic process is defined as a collection $\{X(t), t \in T\}$
of random variables indexed by a parameter t, which takes values over
an index set T of the process. Typically, either the set T consists
of nonnegative integers, or T is the nonnegative half of the real line,
i.e. $[0, \infty)$. In the first case, the collection can be rewritten as
$\{X_n, n = 0,1,2, \ldots\}$, the so-called discrete (time) parameter process.
The second case, $\{X(t), t \geq 0\}$ is called the continuous (time) parameter process. In practice, the discrete time case arises, for example, when X_n represents the number of progeny in the nth generation of
the growth of a tribe. In the continuous time case, $X(t)$ may represent the number of patients in a hospital present at time t, and so on.

The following sections briefly describe some of the key stochastic
processes, which have been commonly used as stochastic models suitable
for many of the biological phenomena. In each case, a brief sketch is
followed by a few examples of the real-life situations in which these
processes arise in practice. However, it is only fair to add the following remarks at this point for the sake of the reader. The author
could have selected for the following presentation the various stochastic processes according to their classical properties such as Markovian
or non-Markovian, stationary or nonstationary, etc. However, such
classifications appeared too broad for our purpose. Rather, it was
thought appropriate to mention only the "special processes," which
have emerged in their own right as useful models in biology and medicine. Of course, even within this limited scope, the author does not
claim the present account to be either complete or exhaustive in touch-
ing the various milestones in this area of biomedical applications of
stochastic processes. A similar remark applies to the reference list
(given at the end of the paper) of the various contributors in this
area. Also, it should not be taken to mean that the broad classifications mentioned above had in any way less influence on the study of
various live phenomena. In fact, these classifications are natural,
when one embarks on a theoretical investigation of the stochastic processes. The reader may find an excellent account of such investigations in treatises such as by Doob (17), Dynkin (19), and Feller

(28,29). In particular, in the last two references by Feller, the reader may find a colorful and rich account of probability, both in theory and its applications. Another two-volume treatise, recently published, is by Iosifescu and Tautu (45), in which the authors devote the first volume to the theory of stochastic processes and the second volume entirely to its applications in biology and medicine.

BRANCHING PROCESSES

An important class of processes arising in many live situations is what is commonly known as "branching processes," a term that appears to have been introduced in 1947 by the Russian mathematicians A. N. Kolmogorov and N. A. Dmitriev (62). Historically these processes* go back one hundred years ago to Francis Galton, a British biometrician. Galton was interested in 1870s in the decay of families of men who occupied conspicuous positions in the past. The question raised was whether this extinction of family names was merely a chance phenomenon or, in fact, whether the physical comfort and intellectual capacity were necessarily accompanied by a decrease in fertility. In 1873, Galton (33) posed the following problem, which appeared as problem 4001 in *Educational Times,* a mathematical periodical then published in London.

Problem 4001

A large nation, of whom we will only concern ourselves with adult males, N in number, and who each bear separate surnames, colonize a district. Their law of population is such that, in each generation, a_0% of the adult males have no male children who reach adult life; a_1% have one such child; a_2% have two, and so on. Find (1) what

----

*In a personal communication, Professor M. Iosifescu has kindly drawn the author's attention to the recent discovery of Heyde and Seneta (103) about the fact that I. J. Bienaymé (102) considered these processes in 1845, some 28 years before Galton and Watson, and in fact a correct statement of the criticality theorem related to these processes, was also known to him. For this the author is grateful to Professor Iosifescu.

proportion of the surnames will have become extinct after r generations
and (2) how many instances there will be of the same surname being
held by m persons. (In Galton's case, the number of possible positive
a_i's went as far as five, but this is a minor point.)

After some persuasion by Galton, Henry William Watson, a clergy-
man and a mathematician, attacked this problem. The underlying
discrete time stochastic process as described here is now commonly
known as the Galton-Watson process (G-W process for short). Watson
(98) showed that if q is the probability of a family name ulimately
becoming extinct, then this probability must satisfy the equation
$x = \Sigma_{k=0} a_k x^k$, where, of course, $\Sigma_{i=0} a_i = 1$. And because $x = 1$ is
a solution of this equation, Watson thus concluded from this, that
the probability of ultimate extinction is always one, no matter what
the probabilities a_0, a_1, a_2, ... are. What he failed to notice
was that under certain conditions on a_i's, there are two solutions of
the above equation lying between 0 and 1, and that it was smaller of
the two that was the correct answer. Anyway, it took about another
60 years before the problem was completely solved by J. F. Steffenson
(see refs. 93,94). The solution he gave is described as follows.

If $a_1 \neq 100\%$, then the probability of ultimate extinction of a
given family name is one if the *average* number of sons of the type
mentioned, born per male parent, is no more than one; and this extinction probability is strictly less than one if this average number is
greater than one. As it turned out, the later investigations showed
that the detailed behavior of these processes varies according to
whether the average number of sons is less than one (the so-called
subcritical case); is equal to one (the critical case); or is greater
than one (the supercritical case).

There is, of course, much more history behind all this, and instead the reader is referred to a paper by Kendall (56), in which an
excellent chronological account of the early history of these processes is given.

After 1940, interest in the branching process model increased
along with the interest in the applications of probability theory and
also, of course, because of its analogy between the growth of families

and the nuclear chain reactions. Several other aspects of the Galton-Watson problem were studied during 1940-1950. Furthermore, the original process was generalized in many directions inspired by various real-life situations. Particular reference should be made here to the works of Hawkins and Ulam (41), Harris (37,38), Everett and Ulam (20, 21), Bellman and Harris (8,9), and Otter (75), all in the United States, and of Kolmogorov and Dmitriev (62), Yaglom (101), Kolmogorov and Sevast'yanov (63), and Sevast'yanov (85-88), in the Soviet Union. Again the G-W process as defined above is a *Markov process*. The *Markov processes*, in simple terms, are defined as those processes wherein the future probabilistic course of the process depends only on the present state the process is in and not on the past history of the process. Thus, a non-Markovian generalization of a continuous time analog of G-W processes was developed in 1948 by Bellman and Harris (8), the so-called Bellman-Harris age-dependent branching processes, although a special case of these was already studied as early as 1939 by Feller (22). In analogy with cell growth, in this case an individual lives for a random length of time and then at death is replaced by a random number of progeny, the basic feature of the branching processes as always being that each individual, independent of the other individuals, undergoes the same chance process, under the same probability laws as did its parent. Another generalization of the original G-W process was developed and studied by Everett and Ulam (20,21) in 1948, and is called the multi-type Galton-Watson process, which was later generalized further to the age-dependent multitype branching processes. Here an individual could give offspring not only of its own type but of other types as well. An example, for instance, is in cell populations, where some of the progeny at birth may undergo mutation, yielding mutant types, different from the normal "wild" type.

In 1963, a book on branching processes by Harris (40) appeared, giving a complete account of these processes until about that time. This book served as a great stimulus for further research both in theory and in applications of these processes. Meanwhile, the work became so voluminous, that since that time three more books have come

out on the subject, one by Mode (68), and another by Athreya and Ney
(1), both in the United States, and finally the third in Russian by
Sevast'yanov (89), who, along with his students, has contributed considerably on the subject.

The list of the applications of these processes is of course too
long to mention. However, we shall mention a few. One application
is in disease epidemics, where the battle is between the infectives,
the carriers of the disease-causative agent and the susceptibles. In
1964, Neyman and Scott (73) studied a stochastic model of the phenomenon underlying the disease epidemic, which takes into account the spa-
cial movement of the infectives in the habitat, an important factor
that was missing in most of the earlier models. However, another important feature of their model is the extensive use of branching processes. Here, the progeny of an infective are to be identified with
the susceptibles getting infected by this infective during the time
he remains infectious.

Another example wherein the branching processes are used is the
extensive work of Karlin and McGregor (51) on genetic models, particularly the fixed size or the so-called finite population models, similar to the ones originally introduced by Wright (100) in 1931 and
R. A. Fisher (31) in 1930. At first sight, it would appear inconven-
ient to study a fixed size population model while using branching
processes, because the population size in these processes varies from
generation to generation. However, it is possible to define a branching process conditioned on the presence of the same number of individuals in each generation. This resulted in what Karlin and McGregor
call a "direct product branching process." They used this approach
to formulate among others, a one-locus, two-allele model of a finite
population with the incorporation in the model of factors such as
selection, mutation, migration, and drift due to finite population
size, and so forth. The reader may refer to Karlin (47,48) for an
expository account of these and other stochastic models in population
genetics.

Among many situations, in which the branching processes have
been used as *tools* for theoretical investigations of other related

processes, one situation arises in the study of the distribution of
the length of a busy period in queueing theory. Here, a branching
process is observed as being embedded in an M/G/1 queue, as pointed
out by Kendall (53). The symbol M/G/1 stands here for a queueing
system with a single server, in which customers arrive according to
a homogeneous Poisson process and in which the service times are
independent and identically distributed with an arbitrary common distribution. The embedding of a branching process as explained by
Kendall (53) goes as follows: Let the customer whose arrival initi-
ates the busy period be called the "ancestor." This customer forms
the zero-order generation corresponding to a G-W process. During his
service time, the new customers arriving, say, X_1 in number, form the
first generation. During their *total* service time, the further (new)
customers arriving, say, X_2 in number, constitute the second generation corresponding to a G-W process, and so on. Here, the new customers arriving during the service time of a customer constitute the
progeny of this customer. As it turns out, in an M/G/1 queue, the
random variables denoting the numbers of progeny for various customers
(defined in this manner) are independent and identically distributed,
a condition essential for X_n to be a G-W process. The busy period,
of course, terminates as soon as this process or the "family" becomes
extinct. The length in time of a busy period is equal to the aggregate
of the service times of all the individuals in various generations (including the ancestor) until the family becomes extinct. Thus, the a-
bove interpretation leads to a simple way of studying the busy period
and some other properties of an M/G/1 queue, with the help of already
well-developed G-W processes. The reader may refer to Neuts (70) for
an investigation of M/G/1 queue along these lines.

Finally, the reader may also refer to Harris (39), Bharucha-Reid
(10,11), Bartoszyński (5), Bühler (12), and Puri (78), for other applications of branching processes.

BIRTH AND DEATH PROCESSES

Another class of processes commonly arising in applications is the
so-called "birth and death processes" (B-D processes). These are

nonnegative integer-valued continuous-time Markov processes. Some members of this class are also known to share properties of those of
continuous time branching processes. Here, the process after waiting
for a random length of time in a given state jumps up by one step if
there is a birth and down by one if there is a death. The birth and
death events are, of course, also random events. Feller (22) was the
first to study these processes extensively as early as 1939. These
and his later contributions made far reaching impacts on the use and
further investigations of these processes. Polya studied a pure-time
nonhomogeneous (i.e., the birth rates being time-dependent) birth
process, now widely known as the Polya process. Here, the birth rates
are both time as well as state dependent. Bates and Neyman (7) also
used the birth processes in connection with stochastic models on accident proneness, the history of which Professor Neyman himself has
already touched in his presentation at this symposium. Kendall (52)
gave a complete solution for the first time of a linear time nonhomogeneous birth and death process. Lederman and Reuter (64) and later
Karlin and McGregor (49,50) made extensive studies of the spectral
properties of the time-homogeneous B-D processes, which led to further
insight into the behavior of these processes.

Practical situations in which B-D processes arise are beyond
enumeration. One such situation wherein these processes commonly a-
rise is in the queueing theory problems. Here, a person leaving the
queue after service is considered a death, whereas a person arriving
for service is considered a birth. Again, these processes have also
been used extensively in developing stochastic models for carcinogenesis; see for instance, Neyman (71), Neyman and Scott (73), and Kendall
(55). Another situation, in which a B-D process has been used as a
model initially in the work of Steinberg and Stahl (95) and later by
Gani (34), is that of bacteriophage reproduction. The reader is referred to a detailed account of this work and of other workers in this
area, which appeared in an expository paper by Gani (35). Some later
work also appears in Puri (79). Briefly, the nonmathematical details
are as follows.

The bacteriophage, or "phage" for short, is a virus that feeds
and multiplies only in a bacterium. Gani was concerned with the so-
called T bacteriophages. Such a phage in its mature form consists
of a DNA strand enclosed in a protein head, attached to which is a
syringe-type mechanism, which helps the phage to insert the DNA strand
into the bacterium. After insertion, the DNA strand, called the vegetative phage, begins to multiply in the bacterium. Meanwhile, the
protein coatings and other parts necessary for the assembly of a mature phage are also undergoing production. The bacterial growth processes are, of course, considered to be halted, and they are considered
dead after infection. Soon some of the vegetative phages start turning into mature ones, after their assembly. The phages in their mature
form are, of course, no longer capable of multiplying. Finally, after
a random length of time from initial virus infection, the life of the
bacterium ends with a burst, the so-called *lysis*, at which point it
yields a random number of mature phages, which are then ready to in-
fect other bacteria in the suspension medium, and the parasitic cycle
of the phage is repeated. Here, as the mature phage is no longer
capable of multiplying, Gani treated the conversion of a vegatative
phage to a mature one as a death, in his B-D process model of the a-
bove phenomenon. This is, of course, one of the many examples in which
B-D processes have been used as part of the underlying mechanism in
developing appropriate stochastic models. Again, an extensive literature is available on the discrete time analogs of B-D processes, the
so-called the "random walk models" defined on integers. The reader
may refer to an excellent treatise on this subject by Spitzer (92).

EMIGRATION-IMMIGRATION PROCESSES

Another class of processes of considerable importance in biology goes
under the name emigration-immigration processes; these are extensively
used in demographic models on population dynamics as the name suggests.
However, they also appear elsewhere. One of the early papers that gave
an impetus to further research in this area is that of Fix and Neyman
(32), which appeared in 1951, a study of a stochastic model of recovery,
relapse, death, and loss of patients in connection with cancer. Since

that time, many of Neyman's students and other workers have contributed
to the study of such processes and in particular to their applications.
Specifically, the work of Chiang (13,14), in connection with competing
risks of illness and death is worth mentioning. Here, an individual
is considered to be migrating from one state of illness to another or
from life itself to death, and so forth.

SEMI-MARKOV PROCESSES

Most of the models mentioned in the previous two sections are so-
called Markovian in nature; in the time-homogeneous case, this means
that the random length of stay of the process in a given state before
it moves to another state has an exponential distribution. Because
of the well-known lack of memory property of such distributions, this
means, e.g., in the case of emigration-immigration processes, that
when you leave the present state you are in and where you go, does
not depend on how long you have already been in that state. In many
diseases, such an assumption appears unrealistic. For instance, in
the case of cancer, the change of state of a person will depend more
critically on how long he has had the growth. Weiss and Zelen (99)
were led by these considerations, to the use of somewhat more general
processes, the so-called "semi-Markov processes." Here, the length
of stay in a given state is assumed to have an arbitrary distribution
not necessarily exponential. Such processes were originally introduced independently in 1954 by Lévy (65,66) and Smith (90), and later
in 1961 were extensively studied by Pyke (82,83). These processes
are now finding more and more use in many live situations. For instance, Weiss and Zelen (99) as mentioned earlier, have considered a
semi-Markovian model and have applied it to the study of behavior of
patients with acute leukemia. This, I believe, is a step in the right
direction, and is most welcomed insofar as the applications of these
processes in biology is concerned.

RENEWAL PROCESSES

Another class of processes, which may be considered as a special case
of semi-Markov processes, is called the "renewal processes." These

processes have traditionally been used extensively in the areas of
life-testing and reliability theory. In fact, the term "renewal theory" comes from problems in these areas, in which one is concerned with
the study of successive replacements (renewals) of items subject to
failure. An item may be a machine, a lightbulb, a vacuum tube, etc.,
which is replaced at the end of its lifetime by an item of the same
kind. It is assumed that the lifetimes of items (all of the same
kind) are independent and identically distributed random variables.
Mathematically, the renewal process $X(t)$ is defined as the number
(counts) of renewals (failures) occurring during the interval $(0,t]$.
An important special case where the common lifetime distribution is
exponential, is called the *Poisson process*, which, in turn, is also a
special case of birth processes touched upon in the section on birth
and death processes (vide supra). These processes have been used as
models for the number of car accidents occurring at an intersection,
the number of telephone calls arriving at a telephone exchange, the
number of immigrants arriving in a town, etc. Here the successive
interarrival times of calls (or accidents) are to be identified with
the lifetimes of the items mentioned earlier, whereas a call itself
is identified with a failure of an item.

Again, if the time is measured in discrete units, one gets the
so-called discrete time renewal processes. These are also commonly
known as recurrent event process, a term originally introduced by
Feller, who is also credited with having recognized and studied these
processes rather extensively (23,24,28).

The list of contributors to the theoretical study of renewal
processes is long and it will suffice for the present to mention two
references, the first of which is an expository paper by Smith (91),
who among others has contributed to this area extensively, and the
other a monograph by Cox (16).

Again, the Poisson processes have been extensively used for modeling in biology and medicine; on the other hand, their generalized
version, namely, the renewal processes, has just begun to find its
proper places in their applications to these areas. A recent example
is of their use in building models related to *neuron firing* (see Coleman and Gastwirth, ref. 15, and Hochman and Feinberg, ref. 42).

DIFFUSION PROCESSES

Most of the processes mentioned so far, are such that the process
X(t) takes discrete values usually over nonnegative integers, being
a number or a count of something. However, there is another class of
processes called the "diffusion processes," where X(t) takes values
on a continuous scale. For instance, in practical situations X(t)
could be the amount of sugar or cholesterol in the blood at time t,
and so on. The difference between these processes and the ones in
which X(t) takes discrete values, is basically as follows.

In the discrete valued processes, the probability of a transition
in a small interval of time Δt is small, but the size of the transition, when it occurs, is appreciable. For example, in a simple birth
process, the probability of a birth in interval Δt in a population of
size n is a small quantity, but when a birth does occur, it adds a
whole unit to the population. In contrast, in diffusion processes
where X(t) is continuous-valued, it is certain that some change will
occur during interval Δt; however, for small Δt, this change will also be small.

Among others, Soviet mathematicians such as Dynkin (18) and
American mathematicians such as Feller (25-27) and Stone (96) and
also Ito and McKean (46), have contributed extensively to the studies
of these processes. One way, in which these processes arise often in
biology, is described below.

Many problems in biology involve relatively large populations,
subject to the transitions resulting from birth, death, mutation, infection (in epidemics), etc. When the size of the population is large,
the transitions are relatively speaking small in size. However, for
a suitably chosen time scale, these transitions may be relatively frequent. Under these conditions, it has been possible in many situations to use an approximate model of the diffusion type in which both
the variable X(t) and the time are continuous. This is analogous to
the normal distribution approximation used in statistics for the sum
of a large number of small random variables. Among many, Feller (25),
Kolmogorov (61), and Kimura (60) have used these processes extensively

as suitable approximations to many live situations, arising particularly in the field of genetics in connection with gene frequency studies.

QUANTAL RESPONSE PROCESSES

Another type of process, called the "quantal response processes" arises as follows. Consider the situations wherein X(t) denotes the variable, such as the number of disease causing organisms such as viruses
or bacteria in the body of the host at time t, the size of the tumor
at time t, the number of tumors present at time t, or the amount of
toxic drug present in the body of the host at time t. In many live
situations, associated with such a variable is a well-defined response
of the host called the quantal response, such as death of the host,
burst of the cell, development of local lesion, or some other detec-
table symptom. The host may be animal, egg membrane, tissue culture,
or even a bacterium. The length of time t the host takes to respond
starting from a convenient origin, is typically known as the response
time and is a random quantity. In this connection, it is usually convenient to introduce the so-called quantal response process denoted
by Y(t), where Y(t) = 1 if the host has not responded by time t;
Y(t) = 0 otherwise, so that the random variables Y(t) and T are related as P(T > t) = P(Y(t) = 1). Here, one is concerned with the fundamental question as to the nature of connection between the quantal
response process Y(t) and the process X(t) itself. Until about 1963,
in many of the models concerned with the above situation, it was assumed that there exists a fixed threshold, identical for each host,
so that as soon as the process X(t) touches this threshold, the host
responds; i.e., the quantal response process Y(t) changes its value
from 1 to 0, an absorption state for the process Y(t). There are a
few situations in which such a hypothesis may appear reasonable such
as the models for neuron-firing, but in many other live situations,
such a fixed threshold hypothesis does not appear to be strictly correct. Thus, in ref. 77, an alternative hypothesis originally suggested
by LeCam, was adopted, namely, that the connection between the process
X(t) and the host's response is indeterministic in character. In other words, it is assumed that the value of X(t), or of a random variable

the distribution of which is dependent on the process $\{X(t)\}$, determines not the presence or absence of response, but only the probability of response of the host. Here, unlike the model based on a threshold hypothesis, the state of the process $X(t)$ at the moment of the quantal response is a random quantity. For further details about the quantal response processes based on the nonthreshold hypothesis, the reader may refer to Puri (77,80). Later similar nonthreshold assumptions were used in studying quantal response processes that arise, in controlling a lethal growth process by Neuts (69), in bacteriophage reproduction by Puri (79), and more recently in developing a stochastic model for rabies by Bartoszyński (6). A third hypothesis, also of the threshold type, is often used for situations involving biological assays (30). Here, each subject is assumed to have a threshold known as its tolerance limit. Unlike the first threshold type model, this limit is assumed to vary from subject to subject in a random fashion over the population of subjects. However, in such cases, this (tolerance) distribution is picked up typically on a rather adhoc basis. This and other objections led Puri and Senturia (81) to the consideration of models based on a nonthreshold hypothesis suitable for quantal response assays.

COMPETITION PROCESSES

Most of the processes mentioned thus far, with the exception of the multitype branching processes, involve a single population. Yet, in many applications especially in biology, medicine, and ecology, one is confronted with processes involving two or more interacting populations. These problems are relatively more challenging mathematically. These processes are typically called the "competition processes." Some of the probabilistic aspects of these processes have been studied in some generality by Reuter (84) and Iglehart (43,44). More recently, Kesten has studied the limit behavior of somewhat related processes falling in the present category, in a series of interesting papers (57-59). His work was inspired in part by the asymptotic behavior of certain branching processes and also the direct product branching processes introduced by Karlin and McGregor (51).

Historically, models that involve competition between species, the so-called prey-predator models, date back to Lotká (67) and Volterra (97) for their deterministic theories of struggle for existence. Extensive experimental work of Park (76) with flour-beetle *Tribolium* inspired further stochastic modeling in this area. The important paper by Neyman et al. (7) ("Struggle for Existence: The *Tribolium* Model") needs a special mention. Here, the competition between the two species of beetles is partly because they eat each others' eggs and also their own, so-called *cannibalism*.

Another area, in which models of two or more interacting populations arise, is that of disease epidemics, which involves a continuing battle between the infectives (plus the carriers) and the susceptibles. Here, the British researchers such as Bailey (2), Bartlett (3,4), Kendall (54), and Gani (36) deserve special mention.

Another situation that involves competition processes in biology is the following--in certain disease processes initiated within the body of the host by the invading organisms, such as viruses or bacteria, the host is known to put up some kind of defense mechanism through entities such as antibodies. Here, the battle is between the infecting organisms and the antibodies, with the host's life hanging in the balance. Unfortunately, not many models have been considered in literature to cover such situations.

And finally, in essence, if I may add another example, our life itself is full of all kinds of processes involving competition and interaction, and perhaps this is what makes life even more interesting.

ACKNOWLEDGMENT

This research was supported by the Office of Naval Research under Contract N00014-67-A-0226-00014 and Air Force Contract AFOSR-72-2350 at Purdue University. Reproduction in whole or in part is permitted for any purpose of the U. S. government.

REFERENCES

1. K. B. Athreya and P. E. Ney, *Branching Processes*, Springer-Verlag, New York, 1972.

2. N. T. J. Bailey, *The Mathematical Theory of Epidemics*, Griffin, London, 1957.

3. M. S. Bartlett, Monte Carlo studies in ecology and epidemiology. *Proc. 4th Berkeley Symp. Math. Statist. Prob. 4*, 39-56 (1960a).

4. M. S. Bartlett, *Stochastic Population Models in Ecology and Epidemiology*, Methuen, London, 1960b.

5. R. Bartoszyński, Branching processes and the theory of epidemics. *Proc. 5th Berkeley Symp. Math. Statist. Prob. 4*, 259-270 (1967).

6. R. Bartoszyński, A stochastic model of development of Rabies. *Proc. IASPS Symp.* (held during April 3-10, 1974, in Warsaw, Poland, in honor of J. Neyman), 1974.

7. G. E. Bates and J. Neyman, Contribution to the theory of accident proneness. I. An optimistic model of the correlation between light and severe accidents. *Univ. Calif. Pub. Statist. 1*, 215-254 (1952).

8. R. Bellman and T. Harris, On the theory of age-dependent stochastic branching processes. *Proc. Natl. Acad. Sci. USA 34*, 601-604 (1948).

9. R. Bellman and T. Harris, On age-dependent binary branching processes. *Ann. Math. 55*, 280-295 (1952).

10. A. T. Bharucha-Reid, On the stochastic theory of epidemics. *Proc. 3rd Berkeley Symp. Math. Statist. Prob. 4*, 111-119 (1956).

11. A. T. Bharucha-Reid, *Elements of the Theory of Markov Processes and Their Applications*, McGraw-Hill, New York, 1960.

12. W. Bühler, Quasi competition of two birth and death processes. *Biomet. Zeit. 9*, 76-83 (1967).

13. C. L. Chiang, A stochastic model of competing risks of illness and competing risks of death. In *Stochastic Models in Medicine and Biology* (J. Gurland, ed.), pp. 323-354, University of Wisconsin Press, Madison, 1964.

14. C. L. Chiang, Introduction to *Stochastic Processes in Biostatistics*, Wiley, New York, 1968.

15. R. Coleman and J. L. Gastwirth, Some models for interaction of renewal processes related to neuron firing. *J. Appl. Prob. 6*, 38-58 (1969).

16. D. R. Cox, *Renewal Theory*, Methuen, London, 1962.

17. J. L. Doob, *Stochastic Processes*, Wiley, New York, 1953.

18. E. B. Dynkin, Continuous one-dimensional Markov processes. *Dokl. Akad. Nauk. SSSR 105*, 405-408 (1955) (Russian).

19. E. B. Dynkin, *Markov Processes,* Vols. 1 and 2, Springer-Verlag, Berlin, 1965 (English transl.).

20. C. J. Everett and S. Ulam, Multiplicative systems. I. *Proc. Natl. Acad. Sci. USA 34,* 403-405 (1948).

21. C. J. Everett and S. Ulam, Multiplicative systems in several variables. III. *Los Alamos Declassified Document 707,* 1948.

22. W. Feller, Die grundlagen der Volterraschen theorie des Kampfes ums dasein in wahrscheinlichkeits-theoretischer Behandlung. *Acta Biotheoret. 5,* 11-40 (1939).

23. W. Feller, On probability problems in the theory of counters. *Courant Anniversary Volume,* pp. 105-115, Interscience Publishers, New York, 1948.

24. W. Feller, Fluctuation theory of recurrent events. *Trans. Amer. Math. Soc. 67,* 98-119 (1949).

25. W. Feller, Diffusion processes in genetics. *Proc. 2nd Berkeley Symp. Math. Statist. Prob.* 227-46 (1951).

26. W. Feller, Diffusion processes in one dimension. *Trans. Amer. Math. Soc. 77,* 1-31 (1954).

27. W. Feller, The birth and death process as diffusion processes. *J. Math. Pures Appl. 38,* 301-345 (1959).

28. W. Feller, *An Introduction to Probability Theory and its Applications,* (3rd ed.), Vol. 1, Wiley, New York, 1967.

29. W. Feller, *An Introduction to Probability Theory and its Applications,* (2nd ed.), Vol. 2, Wiley, New York, 1970.

30. D. J. Finney, *Statistical Method in Biological Assay,* Griffin, London, 1952.

31. R. A. Fisher, *The Genetical Theory of Natural Selection,* Clarendon Press, Oxford, 1930.

32. E. Fix and J. Neyman, A simple stochastic model of recovery, relapse, death and loss of patients. *Human Biol. 23,* 205-241 (1951).

33. F. Galton, Problem 4001, *Educational Times April 1,* p. 17 (1873).

34. J. Gani, An approximate stochastic model for phage reproduction in a bacterium. *J. Aust. Math. Soc. 2,* 478-483 (1962).

35. J. Gani, Stochastic models for bacteriophage. *J. Appl. Prob. 2,* 225-268 (1965).

36. J. Gani, On general stochastic epidemic. *Proc. 5th Berkeley Symp. Math. Statist. Prob. 4,* 271-280 (1967).

37. T. Harris, Some theorems on Bernoullian multiplication process. Ph.D. thesis, Princeton University, Princeton, 1947.

38. T. Harris, Branching processes. *Ann. Math. Statist. 19,* 474-494 (1948).

39. T. Harris, Some mathematical models for branching processes. *Proc. 2nd Berkeley Symp. Math. Statist. Prob.* 305-328 (1951).

40. T. Harris, *The Theory of Branching Processes,* Springer-Verlag, Berlin, 1963.

41. D. Hawkins and S. Ulam, Theory of multiplicative processes. *Los Alamos Declassified Document 265* (1944).

42. H. G. Hochman and S. E. Fienberg, Some renewal models for single neuron discharge. *J. Appl. Prob. 8,* 802-808 (1971).

43. D. Iglehart, Multivariate competition processes. *Ann. Math. Statist. 35,* 350-361 (1964a).

44. D. Iglehart, Reversible competition processes. *Zeit. Wahrsch. Verw. Geb. 2,* 314-331 (1964b).

45. M. Iosifescu and P. Tăutu, *Stochastic Processes and Applications in Biology and Medicine,* Vols. 1 and 2, Springer-Verlag, New York, 1973.

46. K. Ito and H. P. McKean, Jr., *Diffusion Processes and their Sample Paths,* Academic Press, New York, 1965.

47. S. Karlin, Equilibrium behavior of population genetic models with non-random mating. I. Preliminaries and special mating systems. *J. Appl. Prob. 5,* 231-313 (1968a).

48. S. Karlin, Equilibrium behavior of population genetic models with non-random mating. II. Pedigrees, homozygosity and stochastic models. *J. Appl. Prob. 5,* 487-566 (1968b).

49. S. Karlin and J. L. McGregor, The differential equations of birth and death processes and the Stieltjes moment problem. *Trans. Amer. Math. Soc. 85,* 489-546 (1957a).

50. S. Karlin and J. L. McGregor, The classification of birth and death processes. *Trans. Amer. Math. Soc. 86,* 366-400 (1957b).

51. S. Karlin and J. L. McGregor, Direct product branching processes and related induced Markoff chains. I. Calculations of rates of approach to homozygosity. In *Bernoulli-Bayes-Laplace Anniversary Volume* (J. Neyman and L. M. LeCam, eds.), Springer-Verlag, New York, 1965.

52. D. G. Kendall, On the generalized birth-and-death process. *Ann. Math. Statist. 19,* 1-15 (1948).

53. D. G. Kendall, Some problems in the theory of queues. *J. Roy. Statist. Soc. Ser. B 13,* 151-185 (1951).

54. D. G. Kendall, Deterministic and stochastic epidemic in closed populations, *Proc. 3rd Berkeley Symp. Math. Statist. Prob. 4,* 149-165 (1956).

55. D. G. Kendall, Birth-and-death processes and the theory of carcinogenesis. *Biometrika 47,* 13-21 (1960).

56. D. G. Kendall, Branching processes since 1873. *J. Lond. Math. Soc. 41*, 385-406 (1966).

57. H. Kesten, Quadratic transformation: A model for population growth. I, II. *Adv. Appl. Prob. 2*, 1-82; 179-228 (1970).

58. H. Kesten, Some nonlinear stochastic growth models. *Bull. Amer. Math. Soc. 77*, 492-511 (1971).

59. H. Kesten, Limit theorems for stochastic growth processes. I, II. *Adv. Appl. Prob 4*, 193-232; 393-428 (1972).

60. M. Kimura, Diffusion models in population genetics. *J. Appl. Prob. 1*, 177-232 (1964).

61. A. N. Kolmogorov, The transition of branching processes into diffusion processes and associated problems of genetics. *Teor. Verojatnost. Prim. 4*, 233-236 (1959) (Russian).

62. A. N. Kolmogorov and N. A. Dmitriev, Branching stochastic processes. *Dokl. Akad. Nauk SSSR 56*, 5-8 (1947).

63. A. N. Kolmogorov and B. A. Sevast'yanov, The calculation of final probabilities for branching random processes. *Doklady 56*, 783-786 (1947) (Russian).

64. W. Ledermann and G. E. H. Reuter, Spectral theory for the differential equations of simple birth and death processes. *Phil. Trans. Roy. Soc. London 246*, 321-369 (1954).

65. P. Lévy, Processus semi-Markoviens. *Proc. Int. Congr. Math. Amst. 3*, 416-426 (1954a).

66. P. Lévy, Systèmes semi-Markoviens à au plus une infinité dénombrable d'états possibles. *Proc. Int. Congr. Math. Amst. 2*, 294 (1954b).

67. A. J. Lotká, *Elements of Physical Biology,* Williams & Wilkins, Baltimore, 1925.

68. C. Mode, *Multitype Branching Processes,* American Elsevier, New York, 1971.

69. M. F. Neuts, Controlling a lethal growth process. *Math. Biosci. 2*, 41-55 (1968).

70. M. F. Neuts, The queue with Poisson input and general service times treated as a branching process. *Duke Math. J. 36*, 215-232 (1969).

71. J. Neyman, A two-step mutation theory of carcinogenesis. *Bull. Inst. Int. Statist. 38*, 123-135 (1961).

72. J. Neyman, T. Park, and E. L. Scott, Struggle for existence. The Tribolium model: Biological and statistical aspects. *Proc. 3rd Berkeley Symp. Math. Statist. Prob. 4*, 41-79 (1956).

73. J. Neyman and E. L. Scott, A stochastic model of epidemics. In *Stochastic Models in Medicine and Biology* (J. Gurland, ed.), pp. 45-84, University of Wisconsin Press, Madison, 1964.

74. J. Neyman and E. L. Scott, Statistical aspect of the problem of carcinogenesis. *Proc. 5th Berkeley Symp. Math. Statist. Prob.* *4*, 745-776 (1965).

75. R. Otter, The multiplicative process. *Ann. Math. Statist. 20*, 206-224 (1949).

76. T. Park, Experimental studies of interspecies competition. I. Competition between populations of the flour beetles Tribolium Confusum Duval and Tribolium Castaneum Herbst. *Ecol. Monogr. 18*, 265-308 (1948).

77. P. S. Puri, A class of stochastic models of response after infection in the absence of defense mechanism. *Proc. 5th Berkeley Symp. Math. Statist. Prob 4*, 511-535 (1965).

78. P. S. Puri, Some limit theorems on branching processes related to development of biological populations. *Math. Biosci. 1*, 77-94 (1967).

79. P. S. Puri, Some new results in the mathematical theory of phage-reproduction. *J. Appl. Prob. 6*, 493-504 (1969).

80. P. S. Puri, On the distribution of the state of a process at the moment of a quantal response. *Math. Biosci. 18*, 301-309 (1973).

81. P. S. Puri and J. Senturia, On a mathematical theory of quantal response assays. *Proc. 6th Berkeley Symp. Math. Statist. Prob. 4*, 231-247 (1972).

82. R. Pyke, Markov renewal processes: Definitions and preliminary properties. *Ann. Math. Statist. 32*, 1231-1242 (1961a).

83. R. Pyke, Markov renewal processes with finitely many states. *Ann. Math. Statist. 32*, 1243-1259 (1961b).

84. G. E. H. Reuter, Competition processes. *Proc. 4th Berkeley Symp. Math. Statist. Prob. 2*, 421-430 (1961).

85. B. A. Sevast'yanov, Branching stochastic processes for particles diffusing in a bounded domain with absorbing boundaries. *Theory Prob. Appl. 3*, 111-126 (1958) (English transl.).

86. B. A. Sevast'yanov, Extinction conditions for branching stochastic processes with diffusion. *Theory Prob. Appl. 6*, 253-274 (1961) (English Transl.).

87. B. A. Sevast'yanov, Age-dependent branching processes. *Theory Prob. Appl. 9*, 521-537 (1964) (English transl.).

88. B. A. Sevast'yanov, Limit theorems for age-dependent branching processes. *Theory Prob. Appl. 13*, 237-259 (1968) (English transl.).

89. B. A. Sevast'yanov, *Branching Processes*, 'Nauk' Publishing House, Moscow, 1971 (Russian).

90. W. L. Smith, Regenerative stochastic processes. *Proc. Roy. Soc. London Ser. A 232*, 304-305 (1955); cf. the abstract of this paper in *Proc. Int. Congr. Math. Amst. 2*, 304-305 (1954).

91. W. L. Smith, Renewal theory and its ramifications. *J. Roy. Statist. Soc. Ser. B 20*, 243-302 (1958).

92. F. Spitzer, *Principles of Random Walk,* D. Van Nostrand, Princeton, 1964.

93. J. F. Steffenson, On sandsynligheden for at afkommet udder. *Matem. Tiddskr. B*, 19-23 (1930).

94. J. F. Steffenson, Deux problèmes du calcul des probabilités. *Ann. Inst. Henri Poincare 3*, 319-344 (1933).

95. C. Steinberg and F. Stahl, The clone size distribution of mutants arising from a steady-state pool of vegetative phage. *J. Theoret. Biol. 1*, 488-497 (1961).

96. C. J. Stone, Limit theorems for random walks, birth and death processes, and diffusion processes. *Illi. J. Math. 7*, 638-660 (1963).

97. V. Volterra, Variazioni e. fluttuazioni del numero d'individui in specie animali convivente. *Mem. Acad. Lincei Roma 2*, 31-112 (1926).

98. H. W. Watson, Solution to problem 4001. *Educ. Times, August 1*, 115-116 (1873).

99. G. H. Weiss and M. Zelen, A semi-Markov model for clinical trials. *J. Appl. Prob. 2*, 269-285 (1965).

100. S. Wright, Evolution in Mendelian populations. *Genetics 16*, 97-159 (1931).

101. A. M. Yaglom, Certain limit theorems of the theory of branching random processes. *Dokl. Akad. Nauk (USSR) 56*, 795-798 (1947) (Russian).

102. I.J. Bienaymé, De la loi de multiplication et de la durée des familles. *Soc. Philomath. Paris Extracts Ser. 5*, 37-39.

103. C. C. Heyde and E. Seneta, Studies in the history of probability and statistics. XXXI. The simple branching process, a turning point test and a fundamental inequality: A historical note on I. J. Bienaymé. *Biometrika 59*, 680-683 (1972).

PIONEERING IN FEDERAL SUPPORT
OF STATISTICS RESEARCH

FRED D. RIGBY was born September 11, 1914, at Missoula, Montana. He received his B.A. degree in Mathematics in 1935 from Reed College. He earned his M.S. degree in Mathematics at the State University of Iowa in 1938, followed by a Ph.D. degree in 1940. He was the first head of the Logistics Branch of the Office of Naval Research which was formed in 1949. He remained Branch Head until 1958. He was Director of the Mathematical Sciences Division from 1958 through 1962 and Deputy Research Director from 1962 to 1963. From 1963 to 1968 he was Dean of the Graduate School at Texas Tech University and is presently Director of Institutional Study and Research at Texas Tech. He was the founding Editor of the *Naval Research Logistics Quarterly*.

PIONEERING IN FEDERAL SUPPORT
OF STATISTICS RESEARCH

Fred D. Rigby

Institutional Study and Research
Texas Tech University
Lubbock, Texas

When I was invited to participate in this symposium I was not told right away that it would be a banquet affair. When I found out I didn't object, but I did note that there is a constraint involved. In the past, I have sometimes criticized banquet speakers and for the sake of consistency, I guess I had better try to be either entertaining or brief. Of these alternatives, only brevity seems to have any real probability of attainment, so I'll try for it. I should apologize in advance for talking about myself such a great deal, but I must, for the activities to be described were carried on by a very small group of people to which I belonged.

Nearly all of the speakers (and other participants) at this symposium are distinguished statisticians. I am not. Indeed, I don't very often pretend to be any kind of statistician at all, but the present topic calls for discussion of one of those times when I did, I suppose, implicitly, at least. It may have been a disillusionment for the organizers of the symposium to find that the "founder" of the Office of Naval Research program in mathematical statistics was so little of a professional in the field. Some of you will recall that the population of such professionals was not very large in 1946, but perhaps I should begin with some description of how the circumstances came about.

Official history has it that at the end of World War II a group of naval reserve officers who were scientists in their civilian professions concluded that it would be very important to increase substantially the rate of scientific discovery in the United States. The idea was that scientists had been diverted to other activities, bringing that rate down close to zero, whereas every feasible effort had been made to exploit known science in the creation of technology. Accordingly, there would be a prolonged dead period in development of new applications of science to the benefit of society unless extraordinary steps were taken. The story goes that these officers were able to convince influential executives and politicians with the result that the Office of Research and Inventions was created in the Department of the Navy. I know of no reason to doubt that version of history, except that it is obviously oversimplified.

In the spring and summer of 1946, I was the most junior Lieutenant Commander, USNR, that you can imagine, having been a senior lieutenant only 28 days before being spot promoted as an inducement to stay in uniform 6 months longer than was legally required to help "wind down" the military machine. There was no great distinction attached to those spot promotions. I thought I was going back to academic life and felt no yearning to regain my civilian status in the middle of the spring semester, so I was easy to persuade. I did resist a mild effort to talk me into changing over to regular navy standing, feeling that the eagerness with which the navy had waived its vision requirements to get me into uniform, in the first place, might not persist during future eye examinations.

I was not a member of that legendary group of officer scientists that I mentioned--far from it. I was Statistics and Student Flow Officer on the staff of the Chief of Naval Air Training at Pensacola, Florida. I had reached that position rather involuntarily because someone had looked over the qualification cards of the officers who were teaching air navigation and concluded that anyone with a doctorate

in mathematics ought to be able to use a desk calculator. (I hasten
to say that this person was a reserve officer too, a lawyer in civilian
life.) But notice that word "statistics" in my title at that time.
It was deserved, in a way. I did maintain statistics on the flow of
students, both cadets and air crewmen, through the training system.
I used them to make frequent projections and then reported the statistics and the projections in such a way that it was easier for the
writers of transfer orders to fulfill my projections than to make innovations. Consequently, I was controlling the student flow, although
doing that was not explicitly part of my job. I now regard that practice as a very dubious extension of an analytical assignment, but at
the time I was just doing what I thought was needed to make the system work.

While I was doing that--and writing the job description of a
civil service position for the incumbent who would take over when I
left--the word came down through regular information channels that
there had been established an Office of Research and Inventions and
there were positions in it for properly qualified people. I wrote
in for details and then promoted a trip to Washington on official
business but with an afternoon free for a visit to ORI. So in August
of 1946 I became a civilian member of the ORI, holding a civil service job in a grade then called P-5 that was later merged into GS-12.
I was in the Mathematics Section of the Planning Division. The Head
of that Section was Mina Rees, whom some of you know, but her date of
appointment and mine were only 2 days apart. I shall mention her often and shall call her by her first name, which will seem natural to
those who know her. The other members were Eugene Smith, a junior
physicist-engineer, Vladimir Morkovin, a strong applied mathematician
who worked in aerodynamics, and Ernest Ryavec, the naval officer-ord-
nance engineer who had recruited the rest of us. The section still
had some growing to do. We added C. V. L. Smith who became our computer research man, with assistance on hardware matters from Gene
Smith, the physicist-engineer I mentioned. Presently, we lost our
aerodynamicist and then added Joe Weyl. No statistician in the group!
Indeed, most of us were pure mathematicians by training, although our

wartime experiences had carried us into various applied contexts.
After awhile, ORI became ONR, sections became branches, and divisions
proliferated so that, with addition of the continuum mechanics programs, we became the Mathematical Sciences Division, one of seven
divisions that ran the contract research programs.

Mathematical statistics was clearly an important part of our
program area, so Mina decided that one of us ought to be "in charge"
of it, and asked me to take that assignment. I do not remember how
I went about accepting it, but I must have pointed out the limitations
of my qualifications and Mina probably emphasized that none of us was
well qualified for it, etc. My graduate degrees are from the University of Iowa and I knew H. L. Rietz. Indeed, he exchanged my small,
1-year graduate fellowship in 1936 for a teaching assistantship that
I held 4 years. Without that, I'd have been an awfully hungry doctoral student if I'd held out for more than 1 year at all. Professor
Rietz also gave me valuable advice from time to time, but I never took
any course work with him. My graduate statistics was Allen Craig's
course, and they didn't come any better than that. It enabled me, a
couple of years later, to teach a graduate statistics course of my own
and to advise graduate students in other fields on their uses of statistical methods.

To build and manage a contract research program, however, I needed
more than that, so I wangled an opportunity to do some "retreading".
That next summer (1947) I spent 6 weeks at Princeton, studying Cramér's
Mathematical Methods of Statistics, having requested, received, and
taken Sam Wilks' advice as to how to become a better statistician.
Sam was helpful in other ways too, of course, and so was John Tukey
in a particular fashion that I shall mention presently. There is a
point here that is worthy of substantial emphasis. When we in ONR
felt we needed help we asked for it from whomever we thought best qualified to give it. To my knowledge, we never failed to get it when we
did ask. In this particular instance, though, nobody bothered to tell
the department chairman, Professor Lefschetz, what I was doing there,
and he was simultaneously brusque and plaintive when he found out. He
was also friendly in all of my subsequent contacts with him.

Now I have to say something that sounds incredible in this day and age. We actually had to go looking for people to take our money--our contracts, that is--in those days. Our considered strategy for statistics, as for applied mathematics generally, was to strengthen existing centers of excellence and to help start and develop new ones. As I recall, we had just one statistics research contract in being when I began my program management in the field, but there were other prospects, mostly because Mina's contacts were very good, especially with members of the wartime statistics groups that Professor Harshbarger spoke of last evening. One early event that I remember well, however, was a trip I made to the Pacific Northwest in the fall of 1946, looking for opportunities to foster development of a center there. That was my home territory, so I borrowed my father's car for visits in Oregon but used air and rail transport in the state of Washington. When we could, we used military aircraft, and on that trip we landed at Seattle in very thick weather, after "going around" three times and then just bare-ly catching a glimpse of the ground at the extreme low altitude permit-ted by the then brand new instrument landing system. The trip was a success, not in starting up a full blown center, of course, but in get-ting us one long term, high quality statistics contract--with Z. W. Birnbaum--and some contacts in other fields that were valuable later on. By way of geographical contrast, I recall a visit to Columbia University to try to persuade Abraham Wald and Jack Wolfowitz to pro-pose research on sequential analysis. This was a success, too. It was, I think, the only time I met Wald. Wolfowitz became a long-time star of our programs in statistics and, later, logistics.

It may seem to some of you that the grant is the "natural" vehicle for federal support of research, but in those days, and for some years thereafter, we didn't have the authority to make grants and had to use contracts adapted from those used for procurement of materials. Our legal people did the adapting, of course, and very well, too. Then we scientists became highly skilled at writing flexible "tasks" to attach to the basic contracts and cover our individual projects. Indeed, when we did get the grant authority, some of us thought it didn't have much advantage. We couldn't *do* anything with it that contracts wouldn't

handle; the gain was in lower costs of administration, not in flexibility. However, what we knew about government funds, controls, contracts, etc., and what university scientists imagined about them were
two quite different matters, so we had to do a lot of explaining. For
example, we often had to persuade a prospective principal investigator
not to tie his own hands by describing his proposed research in too
great detail. The proposal became a part of the contract package and
would be used in determining the allowability of expenditures. Another
item I should mention that took a lot of explaining was overhead--it
still does, as I'm sure you all know. That one had to be explained to
us rather forcefully, too. We understood about indirect costs, but we
also knew that the services they pay for are less expensive for mathematical research than for nuclear physics, say. So we agitated for
our own overhead rate for our contracts. Finally, one day, Mina went
to the front office determined to "pound the table" about it--her
phrase, as I recall. After awhile she came back to tell us that the
battle was lost. We could have our own rates if we insisted, but if
we did, our budget would be cut back to compensate. We muttered a-
bout conspiracies of physicists but gave in, since such a budget cut
would have reduced our position to a mere nuisance. Incidentally, for
good and sufficient reasons, we paid much, much higher overhead rates
to commercial firms that had ONR contracts than to universities, and
this may have made us seem a bit insensitive on the topic when talking
with faculty people.

I remember an exchange that may have originated from the necessity
to explain contract operations, and I bring it up because it was Professor Hotelling with whom we were dealing. Our working system included efforts not to slow things down because of absences from our office
unless we absolutely had to. There were only a few of us and we did
travel quite a bit, but we could and did keep up with each other's
programs. (We didn't have much office space and sometimes we learned
more than we wanted to about each other's programs because we couldn't
separate conversations.) So one of us wrote to Professor Hotelling
about something and he replied in such a way as to inspire further
correspondence, but the original writer was away, so another of us

stepped in. Again, there was a reply needing response and both earlier writers were away; a third person signed for our office. I was not one of those three, but the next step was Professor Hotelling's visit to our office, on a day when I was there and the others were not. I had not met him before, but things worked out very well. With complete insincerity, he complained of his confusion at finding a different name signed to each letter he received from us and I responded with equally insincere apologies. I doubt that any textbook on such matters advocates initiating a working relationship by 10 minutes of total and obvious dishonesty on both sides, but both of us enjoyed it and our serious discussion was certainly facilitated by its light-minded introduction. I ought to elaborate on that matter of signing letters; it was one of the advantages of being embedded in a military organization. Those letters were fully official and formal communications and our signatures committed the Office of Naval Research in whatever way the text indicated. We all had the authority to sign "by direction" of the Chief of Naval Research, who was an admiral, after the earliest times when a captain held the post. There are various legends about the rigidity of military organizations but that widespread delegation of signature authority is a bit of flexibility matched in few civilian cases. In the Mathematics Section we did, once, have one member whose letters we subjected to special, careful scrutiny because he occasionally produced a strange one, but we took care of the problem ourselves. Mina's boss may have known about it--probably did--but he didn't interfere. There *was* a review of correspondence but its main function was as an internal communication device. In Mathematics we simply circulated a folder containing the correspondence of the last few days-- we called it the "daily life." It was my idea, but I had simply adapted a system I had seen in operation at Pensacola.

As we were pioneering a new kind of government operation, we did some things that subsequently underwent change, sometimes quite radically. For the first couple of years we issued a monthly report on the status of contracts--to the navy, although it was a printed publication with some breadth of distribution. This evolved through a number of stages, but in the early days we wrote abstracts of the papers

and reports prepared by our contract researchers, that were presumed
to be interpretations of their results for readers in other parts of
the navy. We wrote longer articles of our own sometimes, too, but it
is the abstracts that bear on the point I am working up to. I do not
say that we wrote *good* ones. What we did do was study the reports and
papers we received until we understood them thoroughly. If our abstracts were not good ones, it was because we had a difficult expository problem, not because we didn't know our subjects. I'm sure you
realize that we could not keep on doing that as our programs grew and
one of us might be responsible for contracts aggregating several hundred thousand dollars in annual support rate. But starting out that
way established a tradition. We *knew* what our research projects were
doing. If we felt this knowledge weakening we made some visits and
repaired the damage--without ever attempting to direct the conduct of
project research. I think we were always consistent in living up to
our promises not to try to control the processes of research. That
tradition of knowing what went on persisted long after our monthly report abstracting died out; so far as I know, it still continues in ONR.

This brings me to another "sea story," this one involving no statisticians. I once rode from Madison, Wisconsin to Ann Arbor, Michigan with T. H. Hildebrand. That trip was memorable in part because
there were two cars, one driven by Professor Hildebrand himself and
one by his son. Father and son worked as a team by means of private
signals. Some of the resulting driving tactics were calculated to
alarm unsuspecting passengers. However, I mention the trip because
of another passenger, a young pure mathematician who was then something
of a "hotshot." He was on one of our contracts and I had abstracted
a couple of his papers for the monthly report, but he didn't know that.
He was convinced that the navy was deceiving itself--why would it be
supporting research in pure mathematics, otherwise? This was a question which we had examined exhaustively and answered quite thoroughly,
not only to our own satisfaction, but to that of the masters of the
channels through which our funds reached us. I offered him a short
version of our conclusions but he didn't want to hear it--preferred

his illusions about the "navy gravy," the term he actually used, as
I suppose others did also. His convictions included an absolute certainty that nobody employed by the navy could possibly understand research in pure mathematics anyway, his own in particular. I remember
losing patience at about that stage. I told him about the monthly
report and related activities, and I guess I was rather emphatic.
Professor Hildebrand looked around and grinned at me in the middle
of my disquisition, which immediately lost all pretense of effectiveness because I couldn't continue with a straight face once I realized
that the situation was funny. For some reason, that "navy gravy" incident reminds me of another that took place much later when my daughter was attending Antioch College, which alternates academic studies
with off-campus work intervals. She got a job at the National Bureau
of Standards one quarter, through the college, with no help from me.
She was in Ed Cannon's division and he knew she was my daughter, but
none of the others in the group did until one morning when a brash
young Ph.D. pure mathematician who had been on an ONR contract burst
into the common work room and pointed at her dramatically, exclaiming
"Do you know who she is?! Her father is Daddy Warbucks!"

Statistics did not give us any justification trouble, of course,
with sampling inspection and quality control a recent triumph of naval
application. Even so, however, when Professors Neyman and Scott reported their plan for research on astronomical problems and presently,
their early results, we were delighted. Astronomy is a science with
a long tradition in the navy, and to relate our new-fangled research
support activity to it was often a positive step in communication.
(We had an astronomy program too, of course, but that could be taken
for granted in a way that did not apply to other fields.)

Everybody knows that supporting research is gambling, by definition. If one knew what the product would be, the process wouldn't be
research. We were quite sensitive on this point as it was taxpayers'
money we were using and we got it from the Congress, which, although
not as hostile to the administration then as it is these days, is always a potentially severe critic. Our judgments as to which proposed

projects would turn out well were not always good, but we did not repeat our mistakes very often. I was responsible for one "lemon"--
again, not in statistics--quite early, but I never missed badly after-
ward, having learned an important lesson. That was that the one sure
way to get superior return on investment in research support is to make
certain that the money goes to people who are known, in one way or
another but with very little doubt, to be *good* at research. This is
pretty obvious to each individual good researcher, who feels that he
knows his own ability rather well. He isn't quite so confident about
some of his colleagues, though, and he knows of some people in his
field that have false reputations for expertise. A research program
manager needs to know, with as little bias as possible, who are the
good researchers in his field of responsibility. When I started man-
aging the ONR mathematical statistics program I didn't have that knowledge to an adequate degree. I knew the top 5%, perhaps, but that
isn't enough. I mentioned some help I got from John Tukey in the summer of 1947. That was it. He gave me the assistance I needed to extend that 5% to 25% or so, a tremendous boost. Let me emphasize that
he and I both knew perfectly well that I would combine what he gave me
with information from a good many other sources. He built me a plat-
form from which to take off, but I flew the craft myself, making my
own mistakes and scoring my own successes thereafter.

Knowing who the good researchers are is a kind of necessary condition for maintaining quality in a contract or grant research program,
but it is not sufficient in itself. It enables one to play safe, but
playing safe doesn't assure a good program. There is need for a concept of what constitutes excellence in that context, and that is fairly
easy in an established field of research but quite difficult in a new
one. Mathematical statistics was in the former, well established category, of course. There is also an absolute necessity for first class
proposal evaluation, assuming that proposals are as critically important as they were to ONR. This, by the way, is not quite as obvious
as it may seem. Surprisingly, we often found ourselves telling university people that we wanted unsolicited proposals and refusing to

tell them what we wanted done in the way of research. Nowadays, "every-body knows" that the "OK" procedure for proposal evaluation is to have panels and committees of reviewers drawn from the same population that originates the proposals, thereby securing the very best judgment a-vailable, that of the professionals in the field. This technique also provides a valuable defense against certain types of criticism. We didn't know all that in the "old days," and I still don't, although I cannot speak for the ONR of the last 10 years. We thought the best function of committees of researchers was to provide advice on policy and that the way to evaluate proposals was to do it ourselves, making sure that we were competent to do it well in nearly all cases, and ask-ing for help in the few instances in which we were not. Nobody is born with the ability to evaluate research proposals, and the process of be-coming a good researcher does not bestow it at a high competence level either. You have to work at it. We did, intensively, and we got very good at it, in spite of not being particularly good researchers our-selves. Moreover, and I know few of you will believe it, this is a skill that can be extended well beyond the bounds of one's own fields of specialization, as can its companion ability for comparative eval-uation of research results. That last remark is heresy on a university campus, but I am a heretic--a generalist, that is--anyway, and am ex-pected to have some peculiar notions.

I have a sea story on this point, too, although it is not from the early days. A friend of mine, after some years in ONR, transferred to the National Science Foundation, where the committee review of pro-posals was mandatory. A few months later, I asked him if it gave him any trouble and he said it did not, that he made up his own mind which proposals should receive support and then sent them for primary review to those prominent persons in his field who he was privately sure would agree with him. Then the formal committee review would almost certain-ly confirm the primary evaluation and his own judgment. I knew that this was a statistical description, not a claim that things invariably went that way, but beyond that I didn't doubt it at all. He had been a successful operator in ONR and therefore had the knowledge and com-petence to do just what he said he did. There is one conditioning

fact here that I ought to point out. The days of proposal scarcity
were long gone by the time of this incident as we, that is ONR, NSF
and the other research supporting agencies, normally had on hand high-
quality proposals aggregating in annual rate some two and a half times
our available funds. This made a statistical description appropriate.

The thing that a committee review system does not get for a program manager is an imaginative program, unless he is superlatively
skillful at tactics like those claimed by my friend. The do-it-our-
selves process we used in ONR does foster the exercise of imagination
and innovation, once it is mastered. There are times when these are
rather important.

I'm sure it has come through my remarks that ONR program managers
get to thinking they are pretty good at their jobs. They think that
about each other, too, and I want to emphasize again that our early
group in the Mathematics Section worked not only as individuals but
as a unit. We were *all* good at our jobs and Mina Rees was absolutely
magnificent at hers, which was the hardest one, of course. There is
a tendency to refer to the kind of thing she did as "leadership." I
deplore this, in view of what I observe about the behavior of clearly
identifiable leaders. I think the competence she displayed is a higher one than that, but I cannot give it a catchy name. In 1958 I became
the Director of the Mathematical Sciences Division, then composed of
five branches. The heads of those branches were the same kind of people Joe Weyl and Charlie Smith and I were in 1947-1949. They knew
their jobs and they knew they knew them. They were strongly competi-
tive, each in his own way, but with no tendencies toward destructive
competition. To get them to work as a unit when appropriate without
impairing their highly effective regular performance as individual operators wasn't easy but it was extremely rewarding. I can't be sure,
of course, but it seems very likely that Mina has a similar apprecia-
tion of her experiences as section head and then as first division
director. I hope she thought that highly of us, anyway.

One more sea story. Being a research program manager, at whatever
level, now and then gets one into ceremonial situations--often banquets.

Three of those stick in my mind. Once I sat, at the head table, next to the wife of the Russian consul to a city in The Netherlands where an ONR sponsored meeting took place under my partial supervision; conversation was difficult. Another time, again at the head table, I was unexpectedly called upon to say a few words, by a prominent astronomer who had been plying me with very strong gin all during the meal. These have no relevance to mathematical statistics; the meetings dealt with hydrodynamics and celestial mechanics, respectively. The third case was much earlier, when I was less sophisticated, and the memorable feature was simply that I sat next to R. A. Fisher. I was so impressed by this proximity to so awesome a personage that my normally negligible conversational articulateness vanished entirely, and this must have been obvious to Fisher. But he concealed his awareness helpfully, and I recovered, ultimately finding the conversation very interesting and enjoyable.

There is still a requirement for the management of high quality contract or grant research programs that I have thus far left implicit. It is really something that the organization involved must provide through the activities of its members, to facilitate their operations. It is awfully easy for those individual members to feel sure that they are *the* sources of the ideas that make the system work. Of course they do contribute those ideas, in their formulation for application, but in very large part what happens is that they extract them from the flow of information that characterizes their working environment, synthesizing as appropriate and adapting the resulting concepts to the purposes at hand. This is what good administration is about! However, if the information flow is not copious, rich, and varied, idea extraction and synthesis doesn't work very well. Thus it is the organization's business to see to it that an adequate (or better) flow of information does take place. This observation seems to me to be rather significant, and I wish I could claim credit for making it on my own, but I cannot. It dawned on me one day when I was talking with a physicist whose position in ONR was somewhat higher than mine and whose administrative talents I didn't particularly admire. Not only did I revise my opinion of him;

I also picked up a bit of proper and becoming modesty regarding my own contribution to the organization.

ONR did build itself an environment in which information concerning research in all its aspects flowed freely. To a generalist like me--I like that word "generalist" but my son tends to substitute "dilettante"--such an information environment is delightful. The ready availability of all sorts of research information, including results which would not see publication for a couple of years and some that were anticipated but not yet in hand, was the one thing above all that I missed when I left Washington and became a dean. Maybe the creation of that information flow was inevitable, but I doubt it. I think it was a consequence of inspiration, good judgment and good practice on the part of those who set the pattern of ONR operations. I was one of those, but only in a small way. There were some very foresighted people around the Office of Research and Inventions before I turned up.

Throughout much of this talk I have been claiming, indirectly, that it does not take a real professional in a field to build and manage a good contract research program in that field. There is lots of help to be had when it is needed. I do think my experience with the mathematical statistics program supports this position; it was already a good program when it was mine. However, we never assumed or concluded that a nonprofessional could or would do better than a professional. When we reached a stage where we could do it, we hired a real statistician and I gave him the program with my blessing and went off on a new adventure of my own, the logistics research program. This was a new domain for research and *it* had *no* professionals, though there were plenty of practical people who could be hard to convince. Linear programming had been discovered by George Dantzig only months before and it, with the closely related theory of games, opened a very exciting vista. With plenty of help, my colleagues and I came very close to creating a new science during those next few years. *That* was creative administration, in spite of the opinion of some of my friends that there is no such thing. The program in mathematical statistics

flourished under the management of Herb Solomon, Gene Lukacs, Ed
Paulsen, Dorothy Gilford and Bob Lundegard as the time went on. There
were close contacts and sometimes common management between statistics
and logistics. In the latter program I had statistical assistance from
Jack Laderman and Harry Rosenblatt at different times, for it has a
strong statistical content.

I have an afterthought. I've been further out of line, profes-
sionally, than I've confessed so far. One year in the mid-fifties,
the Electronics Branch of ONR, in the Physical Sciences Division, met
with a disaster. It had been headed by an officer, traditionally, and
the naval officer assignment system made a calamitous choice--not of
an incompetent, the system doesn't make that kind of error--but of a
man who could not work effectively in a free-wheeling environment.
After he had been transferred out, the solution applied was to assign
me as Head, Electronics Branch for long enough to repair the damage,
about 8 months, as things worked out. This was interesting but none
too flattering. One of my major qualifications for the assignment was
that I would realize it, when I didn't know enough, and would yell for
help. But I never had to. There were plenty of competent people in
the branch and an excellent program which really hadn't suffered much
harm. What they needed was a professional branch head! and that's
what I was, certainly not any kind of electronic engineer or physicist.
Of course, as soon as I could, in good conscience, I recommended one
of the branch members as future head and went back to my own shop. I
did take with me the computer science program--not so called--which we
had allowed to escape from Mathematical Sciences a couple of years earlier, only to find that it languished without our care.

Perhaps I could have arranged this talk so as not to have emphasized my personal experiences so strongly, but they did tell me I could
use an anecdotal format and be consistent with the symposium pattern,
so I have done so as it suited me. It must be plain that I think ONR
was and probably still is about the finest example of government service to the needs of society that ever was, and I take a great deal of
pride in having been a part of it. Moreover, it was my natural working

environment and even its frustrations seem like pleasures in retrospect. Why did I leave? The handwriting was on the wall in 1963,
and it was quite clear to me that a change for the worse was inevitable. Social experimentation was the coming thing and the glamour and
hence much of the political support would fade from scientific research.
ONR would have to draw funds from other agencies at the cost of much of
its freedom of action. Besides, in the spring of 1963 I had been presented with a pin recognizing 20 years of service, military and civilian, to the navy. If there is anything calculated to make a person
feel like a piece of the furniture, that's it. The Washington area
was pretty hard to take as a place to live in, even then, for a family
that prefers a relatively low population density. So I left and came
back to Texas Tech after 20 years instead of the five or six I had
contemplated when I had first gone to Washington in 1946.

THE IMPACT OF
COMPUTERS ON STATISTICS

H. O. HARTLEY was born April 13, 1912. He earned a Ph.D. degree in
Mathematics at Berlin University in 1934. He worked under Professor J. Wishart at the School of Agriculture at Cambridge,
England, from 1934 to 1936, and took the Ph.D. degree in Statistics at Cambridge University in 1940. He received a D.Sc. degree
in Mathematical Statistics at the University of London in 1953.
He went to Iowa State College (now University) in 1953 and stayed
there until 1963 when he moved to Texas A & M University, where
he remains as the Director of the Institute of Statistics. He is
a Fellow of the Institute of Mathematical Statistics, the American
Statistical Association, the Texas Academy of Sciences, and the
American Society for Quality Control. He was President of the
Biometric Society, ENAR, in 1968. He was the 1973 recipient of
the S. S. Wilks Memorial Medal awarded by the American Statistical
Association. He was elected to the International Statistical Institute in 1954. He is author and coauthor of six books and over
100 research papers.

THE IMPACT OF
COMPUTERS ON STATISTICS

H. O. Hartley

Institute of Statistics
Texas A & M University
College Station, Texas

The topic of this paper may be regarded by some as an implication of
something undesirable. Computers are, after all, mechanical tools.
In spite of what we see advertised in glowing colors about these
"electronic brains," they cannot do anything on their own account and
the human brain has to provide all their thinking. The idea, then,
that this mechanical slave should influence the thoughts of scientists
may appear to be altogether dangerous and undersirable. After some
reflection, we may be inclined to admit that this new tool may have
an influence on our methods of computation. However, computational
methods are regarded by many scientists as a sort of second-class area,
a necessary evil to obtain numerical answers, a trivial matter not
worthy of discussion!

Characteristic of this attitude is a casual remark of a statistician:
He was apparently bored when a group of fellow statisticians were discussing convenient computational procedures to obtain statistical estimates. He remarked that the computational procedure most convenient
to *him* was to proceed to the computing laboratory and tell them to get
on with it.

It is true that computers are mechanical tools. As such, they certainly affect the computational aspects of research in statistics. However, will they reach the heart and soul of such research work, will they influence the outlook of statisticians? I think they will.

Let us look at an analogy with which many of you will be very familiar. Soon after Roentgen discovered X-rays, their clinical potentialities were realized. They became a mechanical tool to help physicians in their diagnoses, but did they influence the outlook on medical treatment? No doctor would dispute today that they did. Quite apart from the fact that X-ray therapy is a recognized treatment, the mere fact that X-ray films make it easier to diagnose internal anomalies has influenced the clinical outlook fundamentally. The mere fact then that high-speed computers can carry out computations much faster influences our evaluation of research projects in statistics.

Before turning to the impact of present-day high-speed computers on statistical activities it is appropriate to survey briefly the history of computers or calculating machines.

HISTORICAL COMMENTS

The history of calculating machines can be traced back to its early forerunners in the form of mathematical astronomical instruments and clockworks dating back to classical Babylonian, Egyptian, and Greek eras; (see, e.g., A. Adam, 1973). The first construction of a calculating machine per se was perhaps that by the French mathematician Blaise Pascal (1642), whereas earlier plans, apparently not implemented, can be attributed to the famous German astronomer J. Kepler and his associate H. Schickard (1623). However, the most important developments have occurred in the last century. Specifically, the modern era of the so-called "high-speed computers" commences as late as during World War II. Although it would lead us too far afield to attempt a coverage of history, it may be of some interest to reflect on the era preceding the high-speed computers and dominated by an extensive use of the so-called "adding-listing machines," "desk calculators," and "punched-card equipment."

Adding-listing machines were confined to addition (subtraction) and paper roll printing of items, totals, and subtotals. However, the multiple register versions made provision for transfers of subtotals from register to register.

Most of the desk calculators achieve the four basic operations (+, -, ×, ÷) by keyboard-controlled (repeated) addition or subtraction into a register at electric motor speeds. Practically all models provide for the formation of "sums of products."

The punched card (so well known today that it need not be described here) dominated the statistical oriented computer of the early twentieth century and was truly an American creation. In spite of its forerunners (e.g., the devices of the various machines constructed by the English mathematician Charles Babbage (1792-1871)) the punched card is the invention of the U.S. Census statistician Herman Hollerith who attempted to interest the U.S. Census Bureau to use it for the 1890 U.S. Population Census. Hollerith also invented punched card equipment. He created (among other devices) what is essentially the oldest punched-card machine, namely, the "Sorter-Counter." This machine will separate a pack of punched cards into sub-packs corresponding to the digit punched in a selected "column" on which the "sort is made," while at the same time, counting the cards in each subpack. The punched card equipment of the first half of this century consisted predominantly of multiple register adding machines with punched card input (the so-called tabulators) often having the facility for transfer of totals or subtotals from register to register (rolling).

Most of the machines used in this era were designed for the compilation of statistical tables or for commercial uses (mainly of an accountancy flavor). However, they prompted an intense activity (originated by the late pioneer L. J. Comrie) of adapting these machines to scientific computations including statistical tasks. This resulted in a fairly extensive literature on what today would be called the "software" associated with the above machines. Although the availability of such "software" undoubtedly made a considerable impact on statistical practices of data analysis in the era, practically all

these procedures are rapidly becoming defunct with the gradual dis-
appearance of the hardware for which they were designed, and nowadays
they are only occasionally remembered with some nostalgia!

As an illustrative example of these activities we may describe
briefly the implementation of linear model estimation on punched-card
equipment. If the model is written in the form $y = X\beta + e$, then (under
the customarily accepted assumptions) we wish to compute the least
squares estimate of β in the form $\hat{\beta} = (X'X)^{-1} (X'y)$, and usually the
formation of the matrix $X'X$ represents the largest computational task.
To fix the idea, let us assume first that all elements of the $n \times p$
matrix $X = (x_{ij})$ are positive single-digit integers, then, the procedure (known under the name "progressive digiting") using a Sorter and
Tabulator would be as follows: Assume that the elements of the ith
row of X are punched on p "columns" ($j = 1, \ldots, p$) of the ith punched
card. To obtain the first row of $X'X$, we would sort the n cards on
the column containing x_{i1} into 10 subpacks corresponding to an x_{i1}
digit of $x_{i1} = d = 9, 8, \ldots, 1, 0$. These subpacks would then be
taken from the pockets of the Sorter (nines first, ones last with
zeros left behind) and passed through the tabulator set up for the
addition of all x_{ij} columns ($j = 1, \ldots, p$). If we denote by X_{dj} the
totals of the x_{ij} column for the subpack for which $x_{i1} = d$ then the
tabulator would form the sequence of progressive totals

$$X_{9j},\ X_{9j} + X_{8j},\ \ldots,\ X_{9j} + X_{8j} + \cdots + X_{1j}$$

and if these progressive totals are each transferred (rolled) to a
second register, the latter will contain at the end of the tabulation

$$9X_{9j} + 8X_{8j} + \cdots + 2X_{2j} + X_{1j} = \sum_i x_{i1} x_{ij}$$

If the variables x_{i1} had more than one digit, the above sort would be
repeated on the 10th, 100th, $\ldots$, digits of x_{i1}, and the complete
"digital fields" of all x_{ij} would be added. By sorting on all digits
of all variables x_{ij} the off-diagonal elements of $X'X$ (sums of products) would all be computed twice (for checking) and the diagonal

elements (sums of squares) would be computed once. Of course, the n-vector y would normally be adjoined as one of the x variables to also obtain X'y and y'y.

The literature of the era (e.g., Comrie et al., 1937; Hartley, 1947) is full of detailed operational refinements of "progressive digiting" including their relative computer economic merits in a diversity of data analysis situations.

Similar detailed procedures were carefully evolved and published for such scientific calculations as the solution of simultaneous equations, numerical integration, differentiation, and the solution of systems of partial and ordinary differential equations, time series analysis and serial correlations and many others. The great merit of the activity resided in the fact that the production and maintenance of the machines for their commercial users made them readily (and economically) available for the occasional scientific user as well as for scientific computation on a comparatively large scale.

Leaving, then, this brief historical sketch we are now turning to the influence of the high-speed computers on modern statistical research activities. In my brief report on this influence, I will confine myself to a few selected aspects rather than make an unsuccessful attempt to recite them all. These aspects are now covered in the following sections.

QUALITY CONTROL AND EDITING OF DATA

The ever-growing number of studies and surveys in the social, engineering and life sciences is producing an overwhelming amount of data. A statistical analysis of such information usually consists of the computation of summaries or estimates from such data. For example, we may wish to check on the accuracy of the inventory records of an arsenal by a direct sample survey in the warehouses. This will lead to statistical estimates of the *actual* inventory as opposed to that on record. Again we may wish to estimate incidence rates of certain diseases for certain communities from data collected in the National

Health Interview Survey. If the task of a computer were confined to
the mere doing of "sums and percentages," this would be an easy matter
and one on which a high-powered computer would probably be wasted.
However, as those of you who have had experience with survey data
know only too well, it would be reckless to assume that there are no
errors in the data. It is, therefore, of paramount importance that
at all stages of data collection and processing the quality of the
data be controlled. One such control is the scrutiny of data for
"internal consistency." You have no doubt heard about the famous
"teen-age grandmothers" who turn up occasionally on census question-
naires and are duly eliminated. Here the inconsistency of "relation
to head of household = mother" and "age = 16" is clearly apparent.
There are other inconsistencies that are not as obvious and others
that are conceivably correct such as a "son" whose age exceeds that
of the wife of his father. She may be his stepmother. Until quite
recently such *common sense data* scrutiny and consequential editing
was performed by hosts of clerks. For studies of a more *specialized*
nature, however, inconsistencies can, of course, only be discovered
by personnel with the required expert knowledge. For example, in a
study involving clinical examinations of cancer of the breast the
classification of "stage" may have to be checked against the recorded
"anatomical" or "histological" division codes. Only personnel completely familiar with clinical concepts will, of course, be able to
perform a scrutiny of such records.

With the advent of high-speed computers, more and more of these
functions of data scrutiny are being taken over by these giant machines. For this to be feasible we must convey to the computer in
minute detail the complete logical sequence of the involved check
procedure including all the "know how" as to "what the expert would
be looking for." Moreover, for it to deal with cases which may be
errors or may be correct data of an unusual nature the computer must
be able to refer to statistical information so that it can gauge a
suspect discrepancy against a statistical tolerance.

After all such information has been stored in the computer's
"memory" the data from the particular study, survey, or census are
passed through the computer for automatic scrutiny. As soon as the
computer encounters an inconsistency it is instructed to either (1)
record (on tape or punched card) the details of the suspected inconsistency in the data and list it for human inspection and reconciliation; or (2) immediately "correct" any inconsistent item (or compute
a missing item) with the help of statistical estimation procedures
and using the data which it has already accepted. Most organizations,
such as the Census Bureau, using automatic data scrutiny and editing,
employ a judicious combination of (1) and (2). The procedure (1) is
usually preferred in situations when human eliminations of reconciled
inconsistencies are administratively feasible as it could be with
smaller and/or rather specialized studies. The method (2) called
"imputation" of suspect data is adopted when the merging of a "correction tape" with the original data tape becomes practically infeasible
as is the case with certain Census operations. With an adequate control on the frequency with which imputations are made, such a method
has, in fact, been in successful use since the 1960 Population Census.
Today, the Census Bureau uses this and similar methods of data editing
as an integral part of a tight quality control on data from which its
releases are tabulated. It can be said that these activities constitute one of its main uses of high-speed computers. In other situations,
it is regarded necessary to *always* follow up a suspected error in the
data by human inspection. For example, if a statistical scrutiny of
inventory records encounters a "number of parts in warehouse" = 1360
which on comparison with the previous inventory is about ten times too
large, and has a unit figure 0 one would be hesitant to instruct the
machine to automatically divide that number by 10 assuming that a column shift has occurred in punching. Here one would prefer to record
the suspect number and instruct personnel to chase the trouble or to
satisfy themselves about the correctness of the unusually large record.

You may rightly say that you do not regard such uses of computers
as a breakthrough in research. After all the computer is used for

functions that could (with great effort) be performed by other means.
Moreover, it has to borrow its intelligence from its human program-
mers.   However, we must not underestimate the tremendous speed and
accuracy with which the computer scrutinizes data.   By freeing trained
personnel for more challenging tasks, it enormously enchances the potentialities of a research team engaged in studies which involve extensive data analysis.   Moreover, it permits the analysis of data
which in an uncontrolled and unedited state would have been too unreliable for the drawing of inferences.

Now to the outlook in this area.   With the advent of bigger and
faster computers our systems of automatic quality control of data
will become more and more ambitious.   Although this will result in
more searching error scrutiny, it is clearly impossible to provide a
control system that will detect any error, however unusual.   Much ingenuity is therefore needed by using the knowledge of the experts to
guide the computer logic to search for the errors and error patterns
most likely to be found in any body of data.

As a side glance into the future, the ability of computers to in-
spect vast bodies of information with lightning speed has, as many of
you will know, led to recent attempts at computer diagnosis of symptoms of patients.   I fear, however, that we are still a long way from
feasibility here.   The main difficulty to my mind is that a good clinician will proceed sequentially--The next question he is going to ask
and the next test he is going to make will depend on what he has found
up to that point.   If such a procedure is simulated on a computer this
would normally require a large number of computer runs.   These, in
turn, would necessitate that clinical examinations and tests on a patient would have to be scheduled in a large number of short sessions.

For some time to come, therefore, I would expect that the use of
computers in this area will probably be confined to large (postclinical)
statistical analysis of the associations between symptoms, tests and
confirmed diagnosis.   The findings of such studies should, of course,
be of great help to the clinician in diagnosing pathologies.

ANALYSIS OF EXPERIMENTAL DATA

The techniques most frequently used in this activity are analysis of variance and regression analysis. Most excellent computer systems are now available for performing these computations. Undoubtedly, the availability of computers has increased the capabilities of research teams of having their data analyzed where previously desk computers could only cover the analysis of a fraction of their data. Moreover, computers have more or less eliminated the so-called "short-cut" analysis (such as an analysis of variance based on range in place of mean squares). Such short-cut methods have justified their lower statistical efficiency by their rapid execution on desk computers or by pencil and paper methods. Unfortunately, along with these advantages there are associated serious pitfalls two of which I wish to discuss very briefly below.

Use of Inappropriate "Canned Programs"

As statisticians, we find all too frequently that an experimenter takes his data directly to a computer center programmer (usually called an analyst) for a so-called statistical analysis. The programmer pulls a canned statistical program out of his file and there result extensive machine outputs, all of which are irrelevant to the purpose of the experiment. This deplorable situation can only be avoided through having competent statistical advice, preferably in the design stage and certainly in the analysis stage. Often the statistical analysis appropriate to the purpose of the experiment is not covered by a canned program and it is appreciated that with time schedule pressures it may be necessary to use a canned program which gives at least a relevant basic analysis. For example, it may be decided to use a basic factorial analysis of variance and subsequently, pool certain components on a desk computer to produce the "appropriate ANOVA." Or again it may be decided to use a general regression program to analyze unbalanced factorial data, although certain factors are known to be random and not fixed.

This brings up the question of how many and what kind of programs should be "canned." Such questions are so intricately linked with the nature of the research arising at the respective institutions that general guidelines are difficult. However, there is one general question that may well be raised--Should there be a "general analysis of variance" system (such as, for example, AARDVAK) making provision for a great variety of designs that may be encountered, or should there be a large number of special purpose programs "custom made" for a particular design? I suppose the best answer is *both* should be available. The custom made programs should be used when "they fit the bill," the general purpose program which must obviously take more computer time when they *do not* fit the bill. In a sense, therefore, the general purpose program is an answer to how should we analyze the unusual experiment? However, we must remember that even general purpose analysis of variance systems are restricted in their scope--Many unbalanced data situations are not covered in such general programs, certainly not a program for unbalanced mixed model data of any kind.

Loss of Contact Between Experimenter and His Data Analysis

Here it is argued that "in the good old days" when experimenters "did their sums of squares," they were "learning a lot about their data," and the computer destroys this "intimate contact." Now we must clearly distinguish between "performing computations on desk computers" as opposed to an "intelligent scrutiny of the data," preferably an "inspection of error residuals." The former is, to my mind, pointless, the latter highly desirable. Indeed, I would venture to say that all analysis of variance and regression programs should provide options for both tabulation of all individual error residuals (for inspection) as well as statistical outlier tests on all residuals which flag unusually large residuals in case the experimenter overlooks them. It is very strange that the possibilities of "faulty records" is clearly recognized in the area of censuses and surveys and all too often overlooked in the analysis of experimental data. But the intelligent inspection of residuals should not only provide a monitor for faulty records. It should also be used by the experimenter to "learn something about his data." Systematic patterns of large errors residuals often provide useful pointers for the modification of models. For example, the form of the residuals in linear regression may indicate neglected quadratic or higher order terms. A factorial analysis of variance of a response y in which the main effects of two quantitative factor inputs x_1 (e.g., the temperature of exposure) and x_2 (e.g., the time of exposure) are insignificant but their interaction is significant often suggests that the relevant input is a function of the two inputs x_1 and x_2. In the above example, the product $x_1 \times x_2$, representing the amount of heat administered, may well be the relevant input and an inspection of a table of residuals will often demonstrate such features.

The next item is more closely related to the effect of computers on research in Mathematical Statistics.

SOLUTION OF STATISTICAL DISTRIBUTION PROBLEMS BY MONTE CARLO METHODS

Monte Carlo methods may be briefly described as follows: Given a mathematical formula that cannot be easily evaluated by analytic reduction and the standard procedures of numerical analysis, it is often possible to find a stochastic process generating statistical variables whose frequency distributions can be shown to be simply related to the mathematical formula. The Monte Carlo method then actually generates a large number of the variables, determines their empirical frequency distributions and employs them in a numerical evaluation of the formula.

An excellent and comprehensive account of these methods is given in a book edited by Meyer (1954) and a research memorandum by Kahn (1954), as well as in numerous articles a small fraction of which is given in the References.

In view of the fast growing literature on these techniques this section must be confined mainly to a very special area of their application, namely, the numerical solution of statistical distribution problems. Moreover, our definitions of statistical distributions do not aim at any generality in terms of measure theory but are, for

purposes of simplicity, confined to distribution density functions
which are all integrable in the classical Rieman sense. The concepts
are explained in terms of statistics depending on independent univariate samples.

The Role of Monte Carlo Methods in Solving Statistical Distribution Problems

In the special case when Monte Carlo methods are used for the solution
of statistical distribution problems the "mathematical formula" to be
evaluated is the frequency distribution of what is known as "a statistic"

$$h = h(x_1, x_2, \ldots, x_n) \tag{1}$$

that is a mathematical function (say, a piecewise continuous function)
of a random sample of n independent variate values x_i drawn from a
"parental" distribution with ordinate frequency $f(x)$ and cumulative
distribution

$$F(x) = \int_{-\infty}^{x} f(v) \, dv \tag{2}$$

In this particular case, the "mathematical formula" to be evaluated
is the n-dimensional integral

$$G(H) = P_r\{h \leq H\} = \int \ldots \ldots \int_{i=1}^{n} f(x_i) \, \pi \, dx_i \tag{3}$$

where the range of the n-dimensional integration in Eq. [3] is defined
by

$$h(x_1, x_2, \ldots, x_n) \leq H \tag{4}$$

An "analytic solution" of the distribution problem [3] would consist
in a simplification of the formula [3] to make it "amenable" to numerical evaluation, a concept not clearly defined as it depends on
the tabular and mechanical aids available for evaluation. A solution
of Eq. [3] by Monte Carlo methods would consist of generating a large
number of samples, $x_1, x_2, \ldots, x_n$, of computing Eq. [1] for each sample

and using the proportion of statistics $h \leq H$ as an approximation to
Eq. [3]. With statistical distribution problems the "stochastic process" mentioned above is therefore trivially available by the definition of the problem. In fact, it is the process of generating variables x_i from the parental distribution. To illustrate the above
concepts by a simple example for which an analytic solution for Eq.
[3] is well known, consider a random sample of independent values from
the Gaussian $N(0,1)$ so that

$$f(x) = (2\pi)^{-1/2} \exp\{- \frac{1}{2} x^2\} \tag{5}$$

and consider the χ^2 statistic

$$h(x_1, \ldots, x_n) = \chi^2 = \sum_{i=1}^{n} x_i^2 \tag{6}$$

then

$$P_v\{h \leq H\} = \Gamma^{-1}(\frac{1}{2} n) \int_0^H \exp\{- \frac{1}{2} h\} (\frac{1}{2} h)^{(n/2)-1} d(\frac{1}{2} h) \tag{7}$$

which will be recognized as the incomplete gamma function extensively
tabulated for statisticians under the name of the probability integral
of χ^2. Whereas in the above example an analytic reduction of Eq. [3]
to a simple form [7] (which can be expanded in a Poisson series for
even n) enabled its numerical evaluation, there are numerous instances
when no exact analytic reduction is possible but the approximations
of numerical analysis such as the Euler-MacLaurin formula of numerical
integration can be used effectively.

Monte Carlo Procedures for Evaluating Statistical Distributions

It is clear from the description of Monte Carlo procedures given in
the preceding section that the principal steps of computing estimates
of frequency distributions for statistics $h(x_1, \ldots, x_n)$ are as follows:

1. The generation of random samples $x_1, \ldots, x_n$ drawn from the
   parent population with ordinate frequency $f(x)$.

2. The computation of the statistic h for each sample and computation of a frequency distribution [3] for varying H by
counting the proportion of h values with h $\leq$ H.

The standard procedure in step 1 is first to generate sets of
random numbers or digits and interpret these as the decimal digits of
a uniform variate u_i. The most frequently used method of generating
the u_i is a method well known under the name of the Power Residue
Method.

In order to compute from the uniform variates u_i random variates
x_i following a given distribution f(x), it is customary to employ the
inverse $F^{(-1)}$ to the probability integral F(x) given by (2), and compute the random variates x_i from

$$x_i = F^{(-1)} (u_i) \tag{8}$$

using either a table of $F^{(-1)}(u)$ or a computer routine for this purpose. No *general* guidelines can be given for the computation of
$h(x_1, \ldots, x_n)$, but effective methods of reducing the computational
labor are available in special cases.

Methods of Reducing "Sample Sizes"

As is well known, a very large number N of random values of the statistic $h(x_1, \ldots, x_n)$ is required in order that the empirical frequencies of the N values of h provide even moderately accurate estimates
of its cumulative probability distribution. An idea as to the magnitude of N can be obtained by applying the well known Kolmogorov-Smirnov
criterion of goodness-of-fit. This criterion measures the maximum
discrepancy D_N between the true cumulative distribution Pr{h $\leq$ H} and
its empirical approximation that is the proportion of h values below
H. It can be shown (see, e.g., Massey, 1951) that approximately

$$Pr\{D_N \leq 1.63/\sqrt{N}\} \doteq 0.99 \tag{9}$$

This formula shows that the error in our Monte Carlo estimates decreases
with $1/\sqrt{N}$. To give an example, suppose it is desired to compute a

Monte Carlo distribution which, with 99% confidence, has three accurate decimals, then

$$1.63/\sqrt{N} = 5 \times 10^{-4} \quad \text{or} \quad N = 1.06 \times 10^{7}$$

Numbers of samples of this magnitude may be prohibitive even on computers. It is not surprising, therefore, that considerable efforts were made by the "Monte-Carlists" to modify their methods in order to reduce the number N of sample sequences required to obtain estimates of adequate precision. An excellent account of these methods is given by Meyer (1953) and by Kahn (1954) as well as in numerous journal articles dealing with such methods. Of these we may mention here:

Importance or Corrective Sampling	(Kahn and Marshall, 1953)
Multistage Sampling	(Marshall, 1954)
Conditional Sampling	(Trotter and Tukey, 1954)
Antithetic Variables	(Hammersley and Morton, 1955)
Control Variables	(Fieller and Hartley, 1954)
Stratified Sampling	(Ringer and Smith, 1970)

For details of these methods, we refer to the papers in question (see list of references). Let it suffice here to say that with all these methods sampling can be reduced considerably at no loss of precision and that the relative merits of these methods depend on the circumstances of the sampling problem and on the gadgetary of the high-speed computer which is available. It is of interest that most of the above methods are closely related to devices which sample surveyors use when sampling life populations, i.e., stratification, regression estimates, optimum allocations, and the like. Designers of sample surveys have always been concerned with reducing the variance of estimates at constant cost or sample size. We may in the future look forward to further blending of efforts between the sample surveyor and the mathematical "Monte-Carlist."

The Effect of Monte Carlo Methods on Research in Distribution Theory

The fact that any distributional problem, whatever its analytic difficulties, can be solved by Monte Carlo Methods has undoubtedly a
profound effect on the orientation of mathematical researches in this
area. Indeed, it has been argued that we do no longer require the
mathematically ingenious methods of characteristic functions, combinatorial techniques and the like to obtain numerical answers for $G(H)$.
Such a viewpoint is, however, fallacious, for apart from the large
number of samples N that are usually required in Monte Carlo, practically all distribution problems depend on a number of parameters--
The sample size n is one such parameter, and usually all parameters
involved in the parental cdf $F(x)$ (given by Eq. [2]) will also affect
$G(H)$, and there may be others such as those describing groupings and
truncations. For example, with a moderate number of 20 levels for
each of two parameters in $F(x)$ and 20 different sample sizes $20^3 = 8,000$
Monte Carlo runs would be required, each based on a large number of N
samples. In searching for methods to reduce the computational load
one of the most attractive procedures is to derive an analytic approximation $G^*(H)$ to $G(H)$ including its dependence on all parameters θ
involved and then perform Monte Carlo computations $G_N(H)$ for "key"
combinations of θ, H and obtain a correction to $G^*(H)$ by fitting an
analytically suggested model to the computed differences $G_N(H) - G^*(H)$.

THE JUDICIOUS COMBINATION OF MATHEMATICAL ANALYSIS
AND NUMERICAL COMPUTER ANALYSIS

The example of Monte Carlo computations discussed in the preceding
section illustrates an important point, namely that a solitary reliance
on a numerical computer approach is just as fruitless as the insistence
on a mathematical dogma stating that numerical answers prove nothing.
The *combination* of the two approaches is vital, usually in the form
of obtaining isolated exact numerical corrections to a good analytic
approximation. How futile a pure computer approach can be if it is
not guided by analytic results is well illustrated by the following
example in the construction of experimental designs--Shortly before

the three "Euler spoilers" (R. C. Bose, S. S. Shrinkande and E. T.
Parker) proved with the help of advanced group theory that Euler's
conjecture about $(2n + 2) \times (2n + 2)$ Graeco-Latin squares was wrong,
an attempt was made to construct a 10×10 Graeco-Latin square on a
computer: A program was set up to generate 10×10 Latin squares
attempting to pair them up to satisfy the Graeco-Latin square condition. For several 100 hours the unfortunate computer tried to "marry"
a Latin square to a Greek square (!) and failed to do so. The Euler
Spoilers on hearing about this proved by quite straightforward means
that the way the computer had been set up, if it had been let to run
for half a century it might have had a 50:50 chance to construct a
Graeco-Latin square!

THE IMPACT OF COMPUTERS ON MATHEMATICAL MODEL BUILDING

Perhaps the most significant impact of computers is in this area:
classical analysis used to insist on the use of mathematical models
for which compact mathematical solutions could be found. This generated a tendency on the part of analysts to restrict their studies
to such tractable models, even at the expense of making them unrealistic. With the computer's capabilities we need not be afraid of
formulating more complex models, thereby freeing the scientists from
the fetters of analytic tractability. There are many instances which
could be quoted to illustrate this point, and we must confine ourselves
to a brief discussion of two such examples both involving the generalization of linear to nonlinear models.

Generalizations of the Linear Hypothesis Model

The well-known "linear hypothesis model" is concerned with an n vector
y of "responses" related to an $n \times p$ matrix X of known input variables
via an unknown p-element parameter vector θ through the linear model

$$y = X\theta + e \qquad\qquad [11]$$

where the n-vector e and its elements e_i are usually assumed to satisfy
$E(e_i) = 0$ and $E(ee') = \sigma^2 \Sigma$ with the $n \times n$ covariance matrix Σ known

and the scalar σ^2 unknown. Usually it is assumed that e is multivariate normal. The estimation theory for θ based on Eq. [11] is
aesthetically eminently satisfactory because of its analytic tractability. Unfortunately, as is well known, Eq. [11] often represents
a gross oversimplification of the physical, chemical, biological, etc.,
mechanism generating the data. One unsatisfactory feature (but by no
means the only one) is the assumption that $E(y)$ is a linear function
of θ. Thus, for example, if y_i is the concentration of a chemical at
a process running time x_{i1} and x_{i2} is its initial concentration at
time $x_{i1} = 0$, we may prefer an expected relation of y_i with x_{i1} and
x_{i2} of the form

$$EY_i = (x_{i2} - \theta_2)e^{-\theta_1 x_{i1}} + \theta_2 \tag{12}$$

where θ_2 is the steady-state concentration and θ_1 the coefficient of
the linear rate equation

$$E \frac{dy_i}{dx_{i1}} = \theta_1(\theta_2 - EY_i) \tag{13}$$

Clearly, Eq. [12] is nonlinear in θ_1. As a generalization of Eq. [13],
we should mention the statistical theory of chemical reactions governing the time dependence of *several* concentrations via the so-called
rate equations (a system of differential equations). A completely
analogous situation arises in the deterministic theory of "compartmental analysis" and here the stochastic analog also leads to expected concentrations which, similarly to Eq. [12], are exponential mixtures of time.

Frequent encounters of nonlinear relations for $E(y)$ have then
spawned a considerable literature on nonlinear (regression) models
of the form

$$y_i = f(x_i, \theta) + e_i \tag{14}$$

where now f is a k + p parameter function and x_i denotes a k element
row vector of inputs. Even if the e_i are assumed to be multivariate

normal (which they are clearly not in compartmental analysis), the estimation theory of θ is almost entirely computer based. Point estimation via maximum likelihood estimators $\hat{\theta}$ leading to the Aitken least square set up of the form

$$\min_{\theta} [y - f(x,\theta)]' \, \Sigma^{-1} \, [y - f(x,\theta)] \tag{15}$$

which is tractable by iterative computer methods only. (For the case $\Sigma = I$ see, e.g., Box, (1958); Hartley, (1961); Marquardt, (1963).) Only approximate covariance matrices are computable and as a consequence inference procedures are based on approximate test criteria. No BLUE optimality properties are generally available for small sample sizes n and we rely on Monte Carlo evaluations of the properties of $\hat{\theta}$. Perhaps the only exact small sample theory is provided by the exact confidence regions for θ that can be constructed by methods developed by Williams (1962), Halperin (1962), and Hartley (1964). However, the numerical or graphical delineation of the boundaries of such exact confidence regions also requires computers.

Although without computers statisticians were often forced to use oversimplified unrealistic models, we are now at least in the position to give a somewhat imperfect estimation and inference theory for a more realistic model.

Nonlinear Model Functions Defined by Algorithms

To illustrate this concept by an example let us return to the nonlinear regression model [14] where $f(x,\theta)$ was assumed to be a "mathematical function" of the arguments x and θ such as, for example, Eq. [12]. Normally we would expect this to mean that $f(x,\theta)$ can be implicitly defined via elementary and/or higher mathematical functions or in some "closed form." Unfortunately, there are many situations where this is not the case. For example, in chemical or enzyme reaction work the "rate equations" may not be linear as in Eq. [13], but may be a system of nonlinear differential equations of the form

$$\frac{dy}{dt} = F(y,t; \theta) \tag{16}$$

where $F(y,t; \theta)$ is often given in tabular form. It will normally not
be possible to solve the system [16] for the n-vector y in closed
analytic form but a suitable algorithm for the numerical solution of
a system of first order differential equations (such as, for example,
the Runge Kutta method) may be used to obtain the elements $Ey_s(t)$ of
the vector $Ey(t)$ in the form

$$Ey_s(t) = f_s(t,\theta) \tag{17}$$

and if the index i, as before, denotes the ith chemical reaction ter-
minated at time t_i there will result a system of nonlinear regression
equations of the form

$$y_{si} = f_s(t_i,\theta) + e_{si} \tag{18}$$

with the e_{si} usually intercorrelated for a given i. If such a system
is used for the estimation of θ by, say, the iterative methods of nonlinear regression estimation, the repeated computation of the $f_s(t_i; \theta)$
for trial values of θ requires the repeated application of the algorithm for the solution of systems of differential equations. Because
practically all of the iterative methods also require the repeated
computation of the partial derivatives $\partial f_s(t_i; \theta)/\partial\theta$, the use of the
so-called "parameter perturbation techniques" in the solution of dif-
ferential equations should profitably be used to obtain these. Such
techniques are also of service in the final stages of the iterations
when θ converges to a solution vector $\hat{\theta}$.

With the advancement of computer codes providing speedy implementation of such algorithms, more models involving numerically defined
functionals become amenable to at least numerical solution and Monte
Carlo computation.

We should, as a second example, mention the solutions of linear
and nonlinear programming problems which usually cannot be defined
analytically but only as terminal values of the efficient algorithms
developed by mathematical programming. Here, the equivalent to the

perturbation technique for differential equations is provided by the methods of "parametric programming." However, as is well known, discontinuities in the derivatives arise when the "base system changes." When such techniques are used for the solution of problems in stochastic programming, we are indeed facing problem formulations that would have been impractical without the use of high-speed computers.

CONCLUSION

With the computer's capabilities we need not be afraid of formulating more complex models, thereby freeing the scientists from the fetters of analytic tractability. Future research will therefore be able to search more freely for realistic models incorporating the full spectrum of information that is at the disposal of the scientist. Indeed, he will use the computer as a powerful tool in trying alternate model theories, all of a complex but realistic form to advance his theories on empirical phenomena.

REFERENCES

Adam, A. (1973). *Vom Himmlischen Uhrwerk zur Statistischen Fabrik*, Munk, Vienna. (Circulated at the 1973 ISI meeting in Vienna.)

Bose, R. C., Shrinkhande, S. S., and Parker, E. T. (1960). Further results on the construction of mutually orthogonal latin squares and the falsity of Euler's conjecture. *Can. J. Math. 12*, 189-203.

Box, G. E. P. (1958). Use of statistical methods in the elucidation of basic mechanisms. *Bull. Inst. Int. Statist. 36*, 215-225.

Comrie, L. J., Hey, G. B., and Hudson, H. G. (1937). Application of Hollerith equipment to an agricultural investigation. *J. Roy. Statist. Soc. Suppl. 4*, 210-224.

Fieller, E. C., and Hartley, H. O. (1954). Sampling with control variables. *Biometrika 41*, 494-501.

Hammersley, J. M., and Morton, K. W. (1955). A new Monte Carlo technique; antithetic variates. *Proc. Cambridge Phil. Soc. 52*, 449-475.

Halperin, M. (1962). Confidence Interval Estimation in Nonlinear Regression, Program 360, Applied Math. Dept., SRRC-RR-62-68 Sperry Rand Research Center.

Hartley, H. O. (1947). The application of some commercial calculating machines to certain statistical calculations. *J. Roy. Statist. Soc. Suppl. 8*, 154-183.

Hartley, H. O. (1961). The modified Gauss-Newton method for the fitting of non-linear regression functions by least squares. *Technometrics 3*, 269-280.

Hartley, H. O. (1964). Exact confidence regions for the parameters in non-linear regression laws. *Biometrika 51*, 347-353.

Hastings, C., Hayward, J. T., and Wong, J. P., Jr. (1955). *Approximations for Digital Computers*, Princeton University Press, Princeton.

IBM Reference Manual C20-8011 (1959). Random Number Generation and Testing.

Kahn, H., and Marshall, A. W. (1953). Methods of reducing sample size in Monte Carlo computations. *J. Oper. Res. Soc. Amer. 1*, 263-278.

Kahn, H. (1954). Applications of Monte Carlo, Rand Corporation Research Memorandum AUCU-3259, prepared under Contract with USAEC No. AT(11-1)-135.

Marshall, A. W. (1954). Application of multi-stage sampling procedures to Monte Carlo problems. In *Symposium on Monte Carlo Methods*, H. A. Meyer (Ed.), pp. 123-144, Wiley, New York.

Massey, F. J. (1951). The Kolmogorov-Smirnov test for goodness of fit. *J. Amer. Statist. Assoc. 46*, 68-78.

Meyer, H. A. [Ed.] (1954). *Symposium on Monte Carlo Methods*, Wiley, New York.

Trotter, H. F., and Tukey, J. W. (1954). Conditional Monte Carlo for normal samples. In *Symposium on Monte Carlo Methods*, H. A. Meyer (Ed.), pp. 64-88, Wiley, New York.

Williams, E. J. (1962). Exact fiducial limits in nonlinear estimation. *J. Roy. Statist. Soc. Ser. B 24*, 125-139.

variable selection, 116, 435
variance (See analysis of
    variance, components of
    variance and generalized
    variance)
variance estimation, 96
vector space, 36
vide infra, 352
vide supra, 354, 361, 389
Virginia, 132
Virginia Polytechnic Insti-
    tute, 132, 137
Volterra, A., 166, 192, 393,
    399
von Borkiewicz, L., 304, 312
von Mayer, G., 303
von Mises, R., 31, 153, 156,
    190, 250, 251, 260, 305,
    314
von Mises collectives, 203
von Neumann, J., 271, 340,
    341, 342, 344, 346

W

Wadley, F. M., 136
Wadsworth, G. P., 272, 276
Wagner, P., 10, 25
Waksberg, J., 93, 100, 102
Wald, A., 75, 137, 139,
    140, 145, 163, 191,
    216, 223, 249, 251,
    263, 277, 336, 337,
    340, 341, 342, 343,
    344, 346, 349, 352,
    362, 365, 371, 372,
    373, 375, 407
Wald lectures, 374
Walker, H., 137
Wallace, D. L., 138
Wallace, H. A., 134, 145
Wallis, W. A., 139, 140,
    144, 239, 242, 259
Walsh, J. E., 112, 113, 242,
    245, 263
Wargentin, P., 300
War Production Board, 137
Washington University, 138,
    144
Watson, H. W., 382, 399
weak convergence, 252

weather forecasting, 271
Webb, S. R., 65, 71
Weida, F., 134, 135
Weiss, A. H., 279
Weiss, G. H., 126, 364,
    365, 388, 399
Weiss, L., 341, 346, 349,
    352, 365
Welch, B. L., 20, 25
Welch, P. D., 279
Wesler, O., 189
Wesleyan University, 138
Westergaard, H., 300, 304,
    314
Western Electric Company,
    142
Weyers Cave, Virginia, 132
Weyl, Joe, 405, 414
Whitney, D. R., 239, 260
Wicksell, S. D., 174, 192
Wiener, N., 204, 267, 270,
    271, 273, 276, 279
Wilcoxon, F., 142, 217, 239,
    241, 242, 263
Wilcoxon statistic, 246,
    251, 253
Wilcoxon test, 240, 245,
    248 (See also Mann-
    Whitney)
Wilk, M. B., 54, 280
Wilks, S. S., 44, 49, 134,
    137, 140, 169, 175, 178,
    216, 223, 245, 349, 352,
    406
Wilks, S. S., Memorial
    Medal, 2
Willcox, W. F., 138, 171
Williams, E. J., 121, 439,
    442
Wilson, K. B., 51, 52
Wilson, K. J., 64, 69
Winsor, C. P., 135, 139
Winston, C., 125
Wirtanen, C. D., 326, 330
Wishart, J., 420
Wittgenstein, 43
Wolfe, D. A., 237, 258
Wolfowitz, J., 51, 53, 69,
    70, 139, 189, 213, 222,
    249, 250, 251, 259, 263,
    341, 346, 349, 352, 362,